Integrated Water Resources Management

Integrated Water Resources Management

Editor: Raven Spoon

R CALLISTO REFERENCE

www.callistoreference.com

Callisto Reference,
118-35 Queens Blvd., Suite 400,
Forest Hills, NY 11375, USA

Visit us on the World Wide Web at:
www.callistoreference.com

ISBN: 978-1-63239-876-5 (Hardback)

The publisher's policy is to use permanent paper from mills that operate a sustainable forestry policy. Furthermore, the publisher ensures that the text paper and cover boards used have met acceptable environmental accreditation standards.

Trademark Notice: Registered trademark of products or corporate names are used only for explanation and identification without intent to infringe.

Printed in the United States of America.

Cataloging-in-Publication Data

Integrated water resources management / edited by Raven Spoon.
    p. cm.
Includes bibliographical references and index.
ISBN 978-1-63239-876-5
1. Integrated water development. 2. Water-supply--Management. 3. Water resources development.
4. Water quality management. I. Spoon, Raven.
TC409 .I58 2017
333.91--dc23

Table of Contents

Preface

Water resources management is the engineering of landscapes and built-environments that regulate and control water resources. It is a very important branch of civil engineering. This book on integrated water management discusses the safe-storage and transportation of water from natural channels while keeping in mind environmental impacts. The contents in this book seek to address both the conceptual aspects of water resource planning and management as well as its practical approaches. This book is a valuable compilation of topics, ranging from the basic to the most complex advancements in the field of water resource management. It presents the complex subject of water management in the most comprehensible and easy to understand language. This text is appropriate for students seeking detailed information in this area as well as for experts.

Various studies have approached the subject by analyzing it with a single perspective, but the present book provides diverse methodologies and techniques to address this field. This book contains theories and applications needed for understanding the subject from different perspectives. The aim is to keep the readers informed about the progress in the field; therefore, the contributions were carefully examined to compile novel researches by specialists from across the globe.

Indeed, the job of the editor is the most crucial and challenging in compiling all chapters into a single book. In the end, I would extend my sincere thanks to the chapter authors for their profound work. I am also thankful for the support provided by my family and colleagues during the compilation of this book.

Editor

Analyzing rainfall events and soil characteristics for water resources management in a canal irrigated area

K. G. Mandal*, J. Padhi, A. Kumar, D. K. Sahoo, P. Majhi, S. Ghosh, R. K. Mohanty, M. Roychaudhuri

Directorate of Water Management (ICAR), Bhubaneswar, Odisha, India

Email address:

mandal98kg@yahoo.co.in (K. G. Mandal)

Abstract: Rainfall analysis is essential for water resources management and crop planning. An attempt has been made to analyse the rainfall of Daspalla region in Odisha, eastern India for prediction of monsoon and post-monsoon rainfall by using 6 different probability distribution functions, forecasting the probable date of onset and withdrawal of monsoon and finally crop planning for the region. Soil parameters were assessed for head, mid and tail reaches of the command area. Results revealed that, for prediction of monsoon and post-monsoon rainfall, Log Pearson Type-III and Gumbel distribution are found as the best fit probability distribution functions. The earliest and delayed most week of onset of rainy season was 20th SMW (14-20th May) and 25th SMW (18-24th June), respectively. Similarly, the earliest and delayed most week of cessation of rainy season was 39th SMW (24-30th September) and 47th SMW (19-25th November) respectively. The soils have the clay contents ranging from 29.6 to 48.8%. The bulk density ranged from 1.44 to 1.72 Mg.m-3 irrespective of different sites. The saturated hydraulic conductivity decreased significantly with soil depth due to greater clay contents in lower layers; whereas water retention at field capacity and PWP increased significantly with increase in soil depths. Soil organic carbon varied from 0.34 to 0.95%; it was the highest in the surface (0-15 cm) layer and then decreased down to the soil profile. The results of rainfall analyses and soil properties would help in management of rainfall and canal water in an effective way.

Keywords: Rainfall, Probability Distribution, Soil Analyses, Water Resources Management

1. Introduction

India ranks first among the countries that practice rainfed agriculture both in terms of extent (86 Mha) and value of production [1]. To meet the future food demands and growing competition for water among various sectors, a more efficient use of water in rainfed agriculture will be essential. Rain water management and its optimum utilization is a prime issue of present day research for sustainability of rainfed agriculture. To ensure water availability and dependability to all the farmers by proper utilization and conservation of rain water in storage tank during rainy season and by recycling the same for life saving irrigation during post rainy season. It is necessary that some storage tanks to harvest runoff and subsurface water are made in command area. This will reduce the risk of dry periods in cropping seasons and increase the production. Kothari et al. [2] opined that on the basis of water harvesting, water can be utilized for saving the crops during severe moisture stress

and also to raise the post-monsoon crops thereby cropping intensity and net returns from the cultivated lands can be increased. In order to address the issue, detail knowledge of rainfall distribution can help in deciding the time of different agricultural operations and designing of water harvesting structures to meet out irrigation requirement [3, 4]. For sustainable crop planning, rainfall was characterized in term of its variability and probability distribution by many researchers [5, 6] for different places of India.

Odisha, an eastern Indian province, is mainly an agrarian state where about 70% of the population is engaged in agricultural activities and 50% of the state's economy comes from agricultural sector [7]. In order to stabilize the crop production at certain level, it is essential to plan agriculture on a scientific basis in terms of making best use of rainfall pattern of an area. This necessitates studying the sequences of dry and wet spells of an area so that necessary step can be taken up to prepare crop plan in rainfed regions. Scientific prediction of wet and dry spell analysis for proper crop planning may prove useful to farmers for improving productivity and cropping intensity and in turn their economic returns. Dry and wet spells could be used for analyzing rainfall data to obtain specific information needed for crop planning and also for carrying out agricultural operations.

The knowledge on soil properties viz. particle size distribution, bulk density, hydraulic properties, water retention characteristics is also essential for land use planning, water resources management [8,9,10] in a canal irrigated commands. Assessment of soil water regime is an important step in making water management decisions [11].

Therefore, attempts were made to study the rainfall characteristics of canal irrigated area in the Daspalla block of Nayagarh district in Odisha, an eastern Indian state to manage the water resources for efficient crop planning. The soil characteristics are made to study the site-specific management of soils and crops, and also for development of

water storage structures in the area.

2. Materials and Methods

2.1. The Study Area

The study was carried out in Kuanria Medium Irrigation Project (KIP) at Daspalla, Nayagarh district of Odisha (Fig. 1). It is located at $20°21'$ N latitude and $84°51'$ E longitude at an elevation of 122 m above mean sea level. The geographical area of Daspalla block is 571.57 km2. This study site comes under Agro-Eco Sub-Region 12.2 and Agro-Climatic Zone 7 according to NBSS&LUP (ICAR) and Planning Commission, Govt. of India classification, respectively.

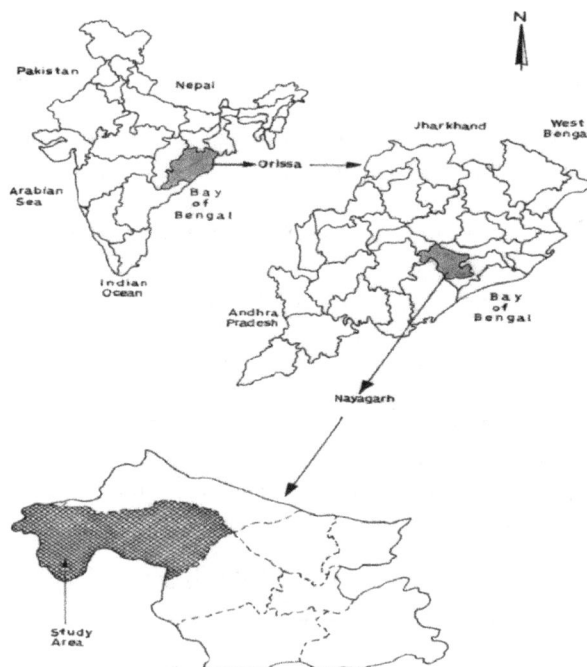

Figure 1. The study area i.e., Daspalla region of Nayagarh district in Odisha.

2.2. Rainfall Data and Analyses Methods

Rainfall data for years from 1995 to 2010 have been collected from meteorological observatory of Kuanria dam, Daspalla, Nayagarh. Rainfall data are categorized into four seasons such as pre-monsoon (March-May), monsoon (June-September), post-monsoon (October-December) and winter (January-February) season. The monthly effective rainfall was calculated using the equations 1 and 2 following USDA Soil Conservation Service method. The same method has been used for calculation of effective rainfall for rainfed districts of India by Sharma et al. [12].

$$Pe = Pt /125 (125-0.2 Pt) \text{ (when Pt < 250 mm)} \quad (1)$$

$$Pe = 125 + 0.1 Pt \text{ (when Pt > 250 mm)} \quad (2)$$

where, Pe = monthly effective rainfall (mm) and Pt = total monthly rainfall (mm).

In this study, rainfall were predicted by six probability distribution functions (PDF) i.e. Normal, 2-Parameter Log Normal, 3-Parameter Log Normal, Pearson Type III, Log Pearson Type III and Gumbel distribution by using DIS-TRIB 2.13 component of SMADA 6.43 (Storm Water Management and Design Aid). Different probability distribution functions [13] are given as follows.

Normal distribution

$$p_x(x) = \frac{1}{\sigma\sqrt{2\pi}} e^{\left[-\frac{(x-\mu)^2}{2\sigma^2}\right]} \quad (3)$$

where, μ = mean of the population of x and σ = variance of the population of x.

Two-Parameter Log Normal distribution

$$p_x(y) = \frac{1}{\sigma_y \sqrt{2\pi}} e\left[-\frac{(y-\mu_y)^2}{2\sigma_y^2}\right] \qquad (4)$$

where, $y = \ln(x)$, μy = mean of the population of y and σy = variance of the population of y.

Three-Parameter Log Normal distribution

$$p_x(y) = \frac{1}{\sigma_y \sqrt{2\pi}} e\left[-\frac{(y-\mu_y)^2}{2\sigma_y^2}\right] \qquad (5)$$

where, $y = \ln(x-a)$, μy = mean of the population of y and σy = variance of the population of y.

Pearson Type III distribution

$$p_x(x) = p_o (1 + x/\alpha)^{\alpha/\delta} e^{-x/\delta} \qquad (6)$$

where, δ = difference between mean and mode ($\delta = \mu - Xm$), Xm = mode of population x, α = scale parameter of distribution, and po = value of px(x) at mode.

Log Pearson Type III distribution

$$p_x(y) = p_{yo} (1 + y/\alpha)^{\alpha/\delta y} e^{-y/\delta y} \qquad (7)$$

where, δy = difference between mean and mode ($\delta y = \mu y - Ym$), Ym = mode of population y, α = scale parameter of distribution and pyo = value of $p_x(y)$ at mode.

Gumbel distribution (also referred to as Fisher-Tippett Type I, Double Exponential, Gumbel Type I and Gumbel extremal distribution) is characterized by the probability density function,

$$p_x(x) = \frac{\alpha}{\beta-\gamma}\left(\frac{x-\gamma}{\beta-\gamma}\right)^{\alpha-1} e\left[\frac{x-\gamma}{\beta-\gamma}\right]^{\alpha} \qquad (8)$$

where, α = scale parameter of distribution, and β = location parameter of the distribution.

All six PDFs were compared by Chi-Square test for goodness of fit as given in the following equation.

$$\chi 2 = \sum \frac{(O-E)^2}{E} \qquad (9)$$

where, O is the observed value obtained by Weibul's method and E is the estimated value by probability distribution functions.

2.3. Computation Method for Onset and Withdrawal of Rainy Season

The onset and withdrawal of monsoon largely determine the success of rainfed agriculture. A prior knowledge of possible onset and withdrawal of effective monsoon is valuable in crop planning. Such knowledge in advance helps in deciding cropping pattern and choice of suitable crop varieties and also to plan comprehensive strategies for proper and efficient rainwater management for improving crop production per unit of available water [14]. Therefore, onset and withdrawal of rainy season was computed from weekly rainfall data by forward and backward accumulation methods as per the procedure suggested by Dash and Senapati [15].

Each year is divided into 52 standard meteorological weeks (SMW). The first SMW of any year starts from 1-7th January and 52nd SMW is from 24-31st December. Weekly rainfall was summed by forward accumulation (20+21+ --- +52 weeks) method until 75 mm of rainfall was accumulated. This 75 mm of rainfall has been considered as the onset time for sowing of rainfed crops [16]. The withdrawal of rainy season was determined by backward accumulation of rainfall (48+47+46+...+30 weeks) data. Twenty millimetres of rainfall accumulation was chosen for the withdrawal of the rainy season, which is sufficient for ploughing of fields after harvesting the crops [17, 18]. The percent probability (P) of each rank was calculated by arranging them in ascending order and by selecting highest rank allotted for particular week. The following Weibull's formula has been used for calculating percent probability.

$$P = (m/N+1) \times 100 \qquad (10)$$

where, m is the rank number and N is the number of years of data used.

2.4. Methods of Soil Sampling and Analyses

Soil samples were collected from head, mid and tail reaches of the canal commands from different sites having cropping systems like rice-fallow, rice-sugarcane and rice-green gram etc. The exact location of different plots with respect to their latitude and longitude were measured with a GPS meter (model, Garmin eTrex Vista, Germany). Soil samples were collected during dry periods of the year 2010-11 and 2011-12. Samples were collected with the help of auger and down to the profile depth up to 90 cm from 5 different locations (5 x 5 m grid per sample) within one representative in a zig-zag pattern and also from four depth increments (i.e. 0-15, 15-30, 30-60 and 60-90 cm) from the soil profile to study the soil properties. Soil particle size distribution was determined by the hydrometer method [19] and soil texture class was determined by following the procedure of USDA classification. Soil pH was measured with the help of a digital pH meter (pHTestr30, Malaysia). Field capacity and permanent wilting point were determined by a pressure plate apparatus (Eijkelkamp, Model 505). The available water capacity (AWC, cm3.cm-3) of soils, expressed as volume of water per unit volume of soil, was estimated as the difference between field capacity (FC) and permanent wilting point (PWP). Saturated hydraulic conductivity (Ks) was measured by constant head method. Five replicates of bulk density (BD, expressed as Mg m-3) samples down to the profile depth up to 90 cm was also collected using soil cores and core samplers (Eijkelkemp Agrisearch Equipment) from four different layers and carried to laboratory. Organic carbon was determined by wet digestion method [20].

The analysis of variance technique was carried out on the data for each parameter as applicable to statistical design. The significance tests were carried out using appropriate standard methods.

3. Results and Discussion

3.1. Annual Rainfall and Distribution over Seasons

Total annual rainfall in Daspalla region ranged between 993.5 to 1901.8 mm with an average of 1509.2 mm and the coefficient of variation (CV) was 14.8%. If rainfall received in a year was equal to or more than the average rainfall plus one standard deviation for 16 years of rainfall (i.e. 1509.2+223.8=1733 mm). On four occasions (1995, 2001, 2003 and 2008), this region had received rainfall of more than 1733 mm; these years were considered as excess rainfall years. Only 25% of total years of analyses under this study had received rainfall of more than 1733 mm for this

region. It is also observed that 44% of the total years of rainfall were below average (1509.2 mm) which were considered as the deficit rainfall years.

Monthly average and effective rainfall of Daspalla region for 16 years are presented in Fig. 2. It is revealed that, mean rainfall of July was 351.4 mm, which was the highest and its contribution was 23.3% to the average annual rainfall (i.e. 1509.2 mm). August rainfall was slightly lower than July rainfall (i.e. 20.6% of annual average rainfall). Average rainfall was lowest in the month of December. Total annual effective rainfall (ER) is 858.2 mm which is 56.9 percent of the total annual rainfall. Therefore, 651 mm of rainfall water is lost in the form of surface runoff, deep percolation and evaporation.

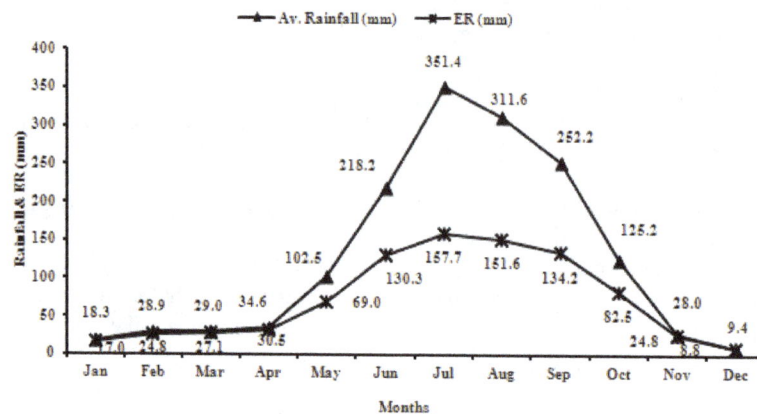

Figure 2. *Long-term average monthly rainfall and effective rainfall for study area, Daspalla; ER is the effective rainfall.*

The distribution of rainfall for different seasons from 1995-2010 is shown in Table 1. The normal southwest monsoon, which delivers about 75.7% of annual rainfall, extends from June to September. This is also the main season (rainy season) for cultivation of rainfed crops. The monsoon rainfall (1133.3 mm) is spreaded over a few rainy days with fewer rain events of high intensity. It causes surface runoff and temporary water stagnation in agricultural fields. Winter season contributes only 3.1% of the total annual rainfall; 10.8 and 10.4% of the total annual rainfall occurred during pre- and post-monsoon season, respectively.

Table 1. *Rainfall distribution in Daspalla region over different seasons (data of 16 years for the period from 1995 to 2010).*

Seasons	Average rainfall (mm)	Percentage of total rainfall
Pre-monsoon (Mar-May)	166.1	10.8
Monsoon (Jun-Sep)	1133.3	75.7
Post-monsoon (Oct-Dec)	162.6	10.4
Winter (Jan-Feb)	47.2	3.1

3.2. Prediction of Rainfall Using Probability Distribution

Functions

Annual rainfall for the region was predicted by using DISTRIB 2.13 component of SMADA 6.43 for 6 different probability distribution functions. Six predicted annual rainfall values were obtained for 6 PDFs by running DISTRIB 2.13. After that Chi-square value for each PDF was estimated by using equation 9. Chi-square values varied from 29.2 to 58.8 for 6 PDFs. Least Chi-square values were observed in Log Pearson Type III distribution for prediction of annual rainfall in Daspalla. Therefore, for this region, Log Pearson Type III considered as best fit PDF for prediction of annual rainfall. In this region, about 86% of the total annual rainfall occurs during monsoon and post-monsoon season and agriculture is totally dependent on the performance of south-west monsoon. Therefore, prediction of monsoon and post-monsoon rainfall is more important than annual rainfall for raising crops successfully with high and stable yields. For this reason, monsoon and post monsoon rainfall were predicted for 6 PDFs by running DISTRIB 2.13 and results are presented in Fig. 3 & 4.

It is revealed that, during monsoon season, the observed monsoon rainfall was 1049.6 mm at 70% probability level and all the probability distribution functions predicted almost comparable rainfall. But in total, least Chi-square value was observed in Log Pearson Type III distribution. With regard to the post-monsoon rainfall, lowest Chi-square value was obtained in case of Gumbel distribution (Fig. 4). In this study, for prediction of monsoon and post-monsoon rainfall, Log Pearson Type III and Gumbel distribution are found as best fit PDFs. Observed rainfall at different probability levels for monsoon and post-monsoon months were determined by using Weibul's formula and presented in Table 2.

Table 2. Observed rainfall at different probability levels for monsoon and post- monsoon months.

Probability (%)	June	July	August	September	October	November	December
10	415.2	724.0	523.3	450.0	322.0	105.5	47.6
20	333.3	469.8	413.1	381.5	274.1	47.7	29.7
30	217.0	365.8	356.5	354.0	123.1	29.7	2.8
40	196.6	337.5	321.1	301.3	116.2	20.9	0.3
50	194.8	282.8	299.7	227.4	87.5	14.8	-
60	182.8	270.0	275.2	177.8	77.4	11.8	-
70	158.0	244.4	274.2	164.9	57.9	6.2	-
80	132.4	217.3	180.6	141.6	28.0	1.0	-
90	105.9	181.9	164.5	75.4	16.9	-	-

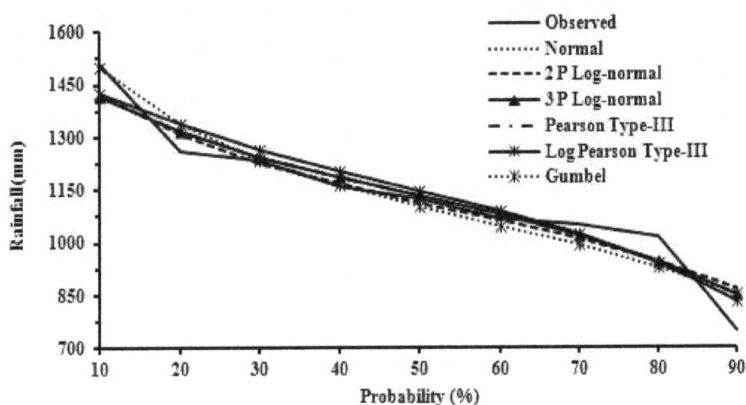

Figure 3. Observed and predicted rainfall (mm) at different probability levels for the monsoon season.

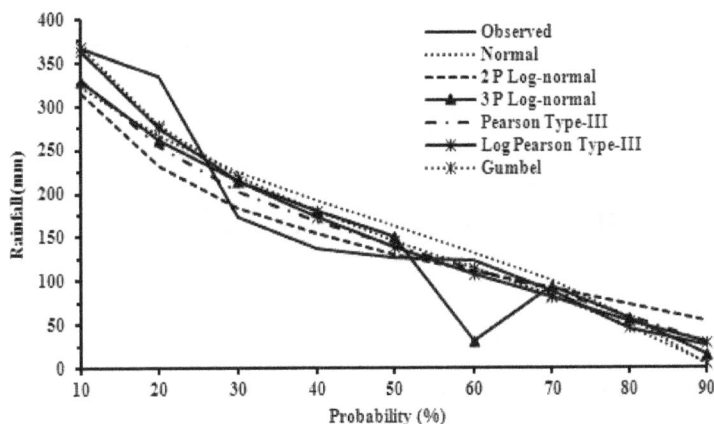

Figure 4. Observed and predicted rainfall (mm) at different probability levels for post- monsoon season.

3.3. Analyses of Rainfall for onset and Withdrawal of Monsoon Season

The data on onset, withdrawal and duration of the rainy season (difference between onset and withdrawal time) and its variability in Daspalla region are presented in Table 3. Weekly rainfall data of 16 years (1995-2010) indicated that the monsoon starts effectively from 23rd SMW (4-10th June) and remains active up to 43rd SMW (22-28th October). Therefore, mean length of rainy season was found to be 21 weeks (147 days). The earliest and delayed week of onset of rainy season was 20th SMW (14-20th May) and 25th SMW (18-24th June), respectively. Similarly, the earliest and delayed week of cessation of rainy season was 39th SMW (24-30th September) and 47th SMW (19-25th November), respectively. The longest and shortest length of rainy season was coincided with 26th and 17th weeks, respectively. The probabilities of onset and withdrawal of rainy season was calculated by using Weibull's formula and results are presented in Table 4. There is a 94% chance that the onset and withdrawal will occur during 25th and 47th SMW, respectively.

Table 3. Characterization of rainy season at the study site, Daspalla (1995-2010).

Particulars	Week No.	Date
Mean week of onset of rainy season	23	4-10 June
Earliest week of onset of rainy season	20	14-20 May
Delayed week of onset of rainy season	25	18-24 June
Mean week of withdrawal of rainy season	43	22-28 October
Earliest week of withdrawal of rainy season	39	24-30 September
Delayed week of withdrawal of rainy season	47	19-25 November

Table 4. Soil properties of head reach soils in the right distributaries.

Soil depth cm	Sand %	Silt %	Clay %	Textural class	Bulk density Mg m^{-3}	EC dS/m	pH	K_s cm/hr	SOC %
0-15	66.6	5.7	27.7	scl	1.50	0.02	6.93	0.140	0.43
15-30	61.6	8.2	30.2	scl	1.48	0.01	7.64	0.128	0.24
30-45	56.6	8.2	35.2	sc	1.50	0.03	7.88	0.121	0.23
45-60	56.6	8.2	35.2	sc	1.48	0.02	7.91	0.103	0.21
60-90	61.6	3.2	35.2	sc	1.48	0.01	7.44	0.104	0.19
90-120	61.6	3.2	35.2	sc	1.50	0.06	7.71	0.100	0.14

3.4. Studies on Soil Characteristics of the Irrigated Command Area

3.4.1. Head-reaches

The soil properties in the head-reaches of right distributaries in the command area are determined and presented (Table 4). It is observed that, texture in the 0-30 cm soil layer was sandy clay loam while in other layers it was sandy clay in the head reach of the Kuanria command area. There was little variation of bulk density among different soil layers. EC values were quite less for all the soil layers (i.e. 0.01 to 0.06 dS m-1). Soil was slightly acidic to alkaline in nature. Soil organic carbon (SOC) content gradually decreased from top to bottom soil layers. Saturated hydraulic conductivity was highest for 0-15 cm soil layer and lowest for 90-120 cm depth of soil.

3.4.2. Mid-reaches

It is observed that, texture in the 0-15 cm soil layer was sandy clay loam; 15-45 cm soil layer clay loam while in other layers it was clay. Bulk density of the soil varied between 1.48 to 1.53 Mg m-3 where as EC varied from 0.23 to 0.62 dS m-1 in all the soil layers. In general, soil was slightly alkaline in nature and soil organic carbon (SOC) content varied between 0.27 and 0.67%. Saturated hydraulic conductivity varied from 0.011 to 0.026 cm hr-1 for all the soil layers (Table 5).

Table 5. Soil properties of mid-reach command area under left distributaries.

Soil depth cm	Sand %	Silt %	Clay %	Textural class	Bulk density Mg m^{-3}	EC dS/m	pH	K_s cm/hr	SOC %
0-15	48.4	16.8	34.8	scl	1.48	0.28	7.63	0.026	0.67
15-30	43.4	16.8	39.8	cl	1.48	0.23	7.96	0.018	0.49
30-45	43.4	16.8	39.8	cl	1.50	0.28	8.22	0.019	0.27
45-60	38.4	16.8	44.8	c	1.52	0.23	8.30	0.012	0.39
60-90	38.4	16.8	44.8	c	1.52	0.55	8.45	0.013	0.29
90-120	38.4	16.8	44.8	c	1.53	0.62	8.60	0.011	0.36

3.4.3. Tail-reaches

The soils of tail end command areas were analyzed for determination of their physical and chemical properties, and are presented in Table 6. There was a definite trend that sand content was greater in every soil layer than silt and clay contents, but there was slight difference in different layers of the soil profile. It is revealed that, soil was moderately alkaline in nature. The soil organic carbon (SOC) content decreased as soil depth increased for the site under WUA 5. Bulk density of soil varied between 1.41 to 1.47 Mg m-3 and EC varied between 0.01-0.11 dS m-1. Saturated hydraulic conductivity was highest in 0-15 cm soil layer and decreases towards lower depth of soil.

Table 6. Soil properties of tail-reach command area in left distributaries.

Soil depth cm	Sand %	Silt %	Clay %	Tex-tural class	Bulk density Mg m^{-3}	EC dS/m	pH	K$_s$ cm/hr	SOC %
0-15	44.2	22.3	33.5	cl	1.41	0.11	8.15	0.023	0.58
15-30	44.9	22.8	32.3	cl	1.42	0.05	8.55	0.021	0.46
30-45	44.8	18.7	36.5	cl	1.44	0.05	8.50	0.022	0.39
45-60	43.5	9.7	46.8	c	1.45	0.01	8.54	0.017	0.39
60-90	41.8	8.9	49.3	c	1.46	0.01	8.53	0.015	0.33
90-120	44.3	8.7	47.0	c	1.47	0.02	8.45	0.014	0.29

The soils in the tail reach of right distributaries are predominantly clay up to 45 cm soil depth and sandy clay in the 45-120 cm soil depths due to higher proportion of sand and clay contents (Table 6). Soil pH in this site was moderately alkaline. Bulk density of the soil increased gradually towards the lower depth of soil profile as amount of clay content was less in 45-120 soil depth in comparison to 0-45 cm soil depth. EC varied from 0.02-0.07 dS m-1 where as organic carbon (SOC) content varied between 0.23-0.39 percent. Saturated hydraulic conductivity was highest within 0-15 cm soil depth and lowest in 90-120 cm soil depth.

Saturated hydraulic conductivity values of tail reach soils of left distributaries were lower in comparison to tail reach soils of right distributaries. As the soil properties were different for head, mid and tail reach soils of left and right distributaries under Kuanria command area, different management of irrigation is required for optimal use of canal, rainfall and ground water. Through proper management, improvement of water productivity can be achieved.

4. Conclusion

The option to go for harvesting the rainfall is useful. Harvesting a small fraction of the excess rainfall and utilizing the same for supplemental irrigation to mitigate the impacts of devastating dry spells offer a good opportunity in the fragile rainfed regions. Based on experiences from watershed management research and large-scale development efforts, practical harvesting of runoff is possible only when the harvestable amount is more than 50 mm or greater than 10% of the seasonal rainfall. In this study area, as runoff amount was more than 10% of the seasonal rainfall, therefore practical harvesting of runoff is possible. Therefore, it is essential to construct water harvesting structures for conservation of rainfall and runoff water to ensure water availability and dependability to all farmers. Instead of growing rice, short duration and low water-requiring crops like groundnut, maize, sorghum, green gram, soybean, sunflower, field bean, cowpea and other low water required

crops which have high return value can be grown during monsoon season. Pigeon pea is a very good crop for growing in this area under upland situation and on the bunds separated by rice fields. Another advantage of growing short duration cereals, pulses and oilseeds in the first fort-night of June is that these can be harvested by the end of September (39th SMW) and short duration post-monsoon crops can be sown during 40–43rd SMW (1–28th October). Since, post-monsoon rainfall is more uncertain and erratic than southwest monsoon, growing of high value post-monsoon crops without supplementary irrigation would be very risky. The significant contribution of weekly rainfall (>46 mm) during 36–40th SMW and high consecutive wet week probability during 36–40th SMW, there is a potential for harvesting excess runoff water for supplemental irrigations. Similarly, greater probabilities of consecutive dry weeks after 44th SMW, hints for need of supplementary irrigations and moisture conservation practices to be taken up. Even in the event of mid season dry weeks, mulching and other moisture conservation practices would help in reducing soil evaporation and conserve moisture in the soil.

The present study characterized the soil properties like particle size fractions, bulk density, saturated hydraulic conductivity, soil organic carbon and also pH for the cultivable area under Kuanria command area in Odisha, an eastern Indian state. This information will be useful for soil and water management decisions to be undertaken for the area. The measured soil properties and information would be utilized as input variables for many models for studying the changes in climate, soil hydraulic properties, soil environmental issues, water balance and solute transport etc. The organic carbon information would help for future planning on the cropping systems concerning carbon sequestration and policy making process for soil organic carbon restoration and other soil properties especially for the Kuanria command. Soil water retention is a major soil hydraulic property that governs soil functioning in ecosystems and greatly affects soil management. The information on Ks would be useful for making any decision for construction of water storage structure or open wells where water storage would be possible and better management of water would lead to better cropping practices.

Acknowledgements

Authors acknowledge the help and cooperation of Er. S.C. Sahoo, Kuanria Irrigation Sub-division Officer, Daspalla, Odisha, India. Authors are thankful to the INCID, Ministry of Water Resources, Govt. of India for providing financial support in carrying out the research work.

References

[1] B.R. Sharma, K.V. Rao, K.P.R. Vittal, Y.S. Ramakrishna and U. Amarasinghe, "Estimating the potential of rainfed agriculture in India: prospects of water productivity improvements", Agric. Water Manage., 2010 97(1), pp. 23-30.

[2] A.K. Kothari, M.L. Jat and J.K. Balyan, "Water balanced based crop planning for Bhilwara district of Rajasthan". Indian J. Soil Cons., 2007, 35(3), pp. 178-183.

[3] C. Prakash and D.H. Rao, " Frequency analysis of rainfall data for crop planning-Kota". Indian J Soil Cons., 1996, 14(2), pp. 23-26.

[4] H.C. Sharma, H.S. Chauhan and Sewa Ram, "Probability analysis of rainfall for crop planning". J Agric. Engg 1979, 16(3), pp. 87-94.

[5] S. Mohanty, R.A. Marathe and S. Singh, "Probability models for prediction of annual maximum daily rainfall for Nagpur", J. Soil Water Cons., 2000, 44 (1&2), pp. 38-40.

[6] R.S. Rana and D.R. Thakur, "Rainfall analysis for crop planning in Kulu valley, Himachal Pradesh", Indian J. Soil Cons., 1998, 26(2), pp.144-146.

[7] D. Panigrahi, P.K. Mohanty, M. Acharya and P.C. Senapati, "Optimal utilization of natural resources for agricultural sustainability in rainfed hill plateaus of Odisha", Agric Water Manage., 2010, 97, pp. 1006-1016.

[8] A. K. Singh, "Use of pedotransfer functions in crop growth simulations", J. Water Manag. 2000, 8, pp. 18–21.

[9] R. Kaur, S. Kumar, H.P. Gurung, J.S. Rawat, A.K. Singh, S. Prasad and G. Rawat,. "Evaluation of pedotransfer functions for predicting field capacity and wilting point soil moisture contents from routinely surveyed soil texture and organic carbon data", J. Indian Soc. Soil Sci., 2001, 50 (2), pp. 205–208.

[10] U.S. Saikia, and A.K. Singh, "Development and validation of pedotransfer functions for water retention, saturated hydraulic conductivity and aggregate stability of soils of Banha watershed", J. Indian Soc. Soil Sci. 2003, 51, pp. 484–488.

[11] F. Ungaro, C. Calzolari, and E. Busoni, "Development of pedotransfer functions using a group method of data handling for the soil of the Pianura Padano–Veneta region of North Italy water retention properties", Geoderma., 2005, 124, pp. 293–317.

[12] D. Sharma and V. Kumar, "Prediction of onset and withdrawal of effective monsoon dates and subsequent dry spells in an arid region of Rajasthan", Indian J. Soil Cons., 2003, 31(3), pp. 223-228.

[13] V.T. Chow, "Hand Book of Applied Hydrology", Mc Graw Hills Book Company, 1964, New York.

[14] H.P. Das, R.S. Abhyankar, R.S. Bhagwal and A.S. Nair, "Fifty years of arid zone research in India", CAZRI, Jodhpur., 1998, pp. 417-422.

[15] M.K. Dash and P.C. Senapati, "Forecasting of dry and wet spell at Bhubaneswar for agricultural planning" Indian J. Soil Cons., 1992, 20(1&2), pp. 75-82.

[16] B. Panigrahi and S.N. Panda, "Dry spell probability by Markov chain model and its application to crop planning in Kharagpur", Indian J. Soil Cons., 2002, 30(1), pp. 95-100.

[17] P.N. Babu and P. Lakshminarayana, "Rainfall analysis of a dry land water shed-Polkepad: A case study", J. Indian Water Res Soc., 1997, 17, pp. 34-38.

[18] G.V. Srinivasareddy, S.R. Bhaskar, R.C. Purohit and A.K. Chittora, "Markov chain model probability of dry, wet weeks and statistical analysis of weekly rainfall for agricultural planning at Bangalore", Karnataka J. Agric Sci., 2008, 21(1), pp. 12-16.

[19] G. J. A. Bouyoucous, "Recalibration of the hydrometer for making mechanical analysis of soil", Agron. J., 1951, 43, pp. 434-438.

[20] A. Walkley and I. A Black, "An examination of the method for determining soil organic matter and proposal modification of the chromic acid titration method", Soil Sci., 1934, 37, pp. 29–38.

Flood frequency Modeling using Gumbel's and Powell's method for Dudhkumar river

Md. Abdullah Asad[*1], Mohammad Ahmeduzzaman[1], Shantanu Kar[1], Md. Ashrafuzzaman Khan[1], Md. Nobinur Rahman[2], Samiul Islam[3]

[1]Dept. of Civil Engineering, Stamford University Bangladesh, Dhaka 1217, Bangladesh
[2]Dept. of Civil Engineering, Rajshahi University of Engineering & Technology, Rajshahi 6204, Bangladesh
[3]Office Engineer, BETS Consulting Services Ltd., Dhaka, Bangladesh

Email address:
abdullah.asad03@gmail.com(M. A. Asad), maz060086@gmail.com(M. Ahmeduzzaman)

Abstract: The results of a study on an international river Dudhkumar (shared by Bhutan, India and Bangladesh) analyzing flood frequency of 14 years using Gumbel and Powell distribution have been presented in this paper. Flash flood occurrence over recent years had washed away fields making vulnerable life safety. It was assumed that, Dudhkmar flood flows obey the Gumbel and Powell distribution. The scale and shape parameters of the distribution were estimated using method of moments. . A Chi-square test results (p =1.000) between observed and predicted flood flows which is considered to be not statistically significant by the conventional criteria. Due to goodness of fit of the Gumbel and Powell distribution, it was assumed to be appropriate for modeling frequency of Dudhkumar River floods. However, the magnitudes of the 100, 200 and 1000 year floods were significantly differed in the two mentioned methods.

Keywords: Dudhkumar River, Flood Frequency Gumbel and Powell Distribution, Recurrence Interval

1. Introduction

Dudhkumar River the Raidak or the Sankosh river of WEST BENGAL (India) enters Bangladesh near Pateshwary and is renamed as the Dudhkumar. The river receives the Gadadhar and the Gangadhar as tributaries at Pateshwary and travels along a 52 km long meander course and joins the BRAHMAPUTRA at KURIGRAM SADAR upazila. Most of the main course of the Dudhkumar lies in India. The river is free from tidal influence, but often overflows. The average slope of the river is about 10 cm/km. This high slope makes Dudhkumar a flashy type river during rainy season when onrush of surface runoff due to monsoon rain cause flooding to the flood plain of the river causing bank erosion and destruction of houses and settlement of the people living on both banks. Destruction caused by this flashy river has increased in recent years. This paper aims at estimating return period associated with flood peaks of varying magnitudes from recorded floods using statistical methods.

2. Literature Review

Flood frequency analysis (FFA) is the estimation of how often a specified event will occur. Before the estimation can be done, analyzing the stream flows data are important in order to obtain the probability distribution of flood (Ahmad et. al., 2010). One of the greatest challenges facing the Hydrology is to gain a better understanding of flood regimes. To do this, flood frequency analysis (FFA) is most commonly used by engineers and hydrologists worldwide and basically consists of estimating flood peak quantities for a set of non-exceedance probabilities. The validity of the results in the application of FFA is theoretically subject to the hypothesis that the series are independent and identically distributed (Stedinger et al., 1993; Khaliq et al., 2006). Nevertheless, to determine flood flows at different recurrence intervals for a site or group of sites is a common challenge in hydrology. Although studies have employed several statistical distributions to quantify the likelihood and intensity of floods, none had gained worldwide acceptance and is specific to any country (Law and Tasker, 2003). Ferdows et. al. analyzed (1964-2000) discharge data for four specific river location in Bangladesh. Design flow and stage computation for Teesta River was done by Rahman et. al. using frequency analysis and MIKE 11modeling. However, this study involves the flood frequency analysis of Dudhkumar river for its flashy nature in the recent years (1996-2010) using two methods Gumbel's and Powell's respectively. To the best of author's knowledge, no previous study was done to model dudhkumar's flashy nature. It was assumed that, Dudhkumar River flood flows fit Gumbel and Powell distribution model. Location of Dudhkumar River in Bangladesh map is shown in Fig. 1.

3. Data Collection and Interpretation

Discharge data (in m^3/s) for 14 water years of record for SW 81 gauging station on Dudhkumar River were collected from Bangladesh Water Development Board (BWDB). The flow recording station SW 81 was equipped with an automatic recorder. Flow data were expressed in terms of exceedence probabilities and recurrence intervals. Denoting Qi as the annual maximum flood in year i, the quantile Qi (F) is the value expected Qi to exceed with probability F, that is, $P(Qi \geq Qi$ $(F)) = F$ during the year of interest. Thus, there is a $F\%$ chance that $Q \geq Q$ (F). Conversely, there is a $(1-F)$ % chance that $X < Q(F)$. The return period of a flood, $1/(F)$ is the reciprocal of the probability of exceedence in one year (Haan, 1977; Shaw, 1983).

4. Gumbel Distribution

This is one of the most widely employed distributions to describe the flood data. As per this distribution, following equations are used to calculate recurrence interval and corresponding flood magnitude.

$$Q_T = Q_{mean} + K\sigma \qquad (1)$$

$$K = (y-yn)/\sigma \qquad (2)$$

$$y = -0.834 - 2.303\log\log(T/T-1) = -0.834 - 2.303X_T, \qquad (3)$$

$$Y_T = \text{Log Log } (T/T-1) \qquad (4)$$

Where,
y= reduced variate.
T= recurrence interval
Q_T = magnitude of the flood with recurrence interval of T
Q_{mean} = mean of the maximum instantaneous flow
k= frequency factor
σ = standard deviation
Y_T = reduced variate corresponding to a recurrence interval.

5. Powell Distribution

It is also an effective method to describe flood data. As per this method, the magnitude of the flood with recurrence interval of T and frequency factor is given by

Figure 1. *Dudhkumar river location*

$$Q_T = Q_{mean} + K\sigma \qquad (5)$$

$$K = \sqrt{6}/\pi \, [\lambda + \ln \ln(T/T-1)] \qquad (6)$$

Where,
T= Recurrence Interval, σ= standard deviation, K= frequency factor
λ = Euler's constant = 0.57722

6. Goodness of Fit

It was assumed that, the discharge data fit the Gumbel and Powell distribution. A chi-square test was carried out to find the goodness of fit between the measured and predicted flood. After computing the goodness of fit for the distribution, flood magnitudes were calculated for exceedence probabilities of 0.040, 0.080, 0.120, 0.160, 0.200, 0.240, 0.280, 0.320, 0.360, 0.400, and 0.440, 0.480, 0.520, 0.560, 0.600, 0.640, 0.680, 0.720, 0.760, 0.800, 0.840, 0.880, 0.920, 0.960.

7. Results

The maximum instantaneous flow of 2753.68 m^3/s was

recorded at SW 81 in 2005 whereas the lowest flood flow of 800.35 m^3/s occurred in 1997.The mean instantaneous flow is 1506.47504 m^3/s. The values of the flood data for 14 years are presented in table 1.

Table 1. Observed and predicted floods in corresponding years

Years	Observed discharge (m^3/s)	Predicted discharge (m^3/s)
1996	1201.61	1181.61
1997	800.35	780.35
1998	821.56	771.56
1999	1120.90	1060.9
2000	1834.02	1754.02
2001	1210.61	1160.52
2002	1020.73	970.73
2003	1603.03	1573.03
2004	1357.29	1347.29
2005	2753.68	2805
2006	1118.03	1200
2008	2519.31	2700
2009	1193.84	1300
2010	2535.63	2535

Observed and predicted flood flows show no significant (p=1.000) differences. Using Gumbel and Powell method the magnitude of flood corresponding to the recurrence interval are calculated.

Sample size= 14

Q_{mean} =   1506.47504

Standard Deviation σ = 654

Y_n=   0.50928

σ_n= 0.86562

A software based chi-square test results chi-square (X^2= 0.685) having two tailed P value of 1.0 with 13 degrees of freedom. The flood magnitudes corresponding to the exceedence probabilities were calculated. The flood magnitude was decreasing with an increase in exceedence probability values. Results have shown that, Dudhkumar River flood flows were variable in 14 mentioned years of study. A chi-square test shows a satisfactory fit between observed and estimated data. Besides, increasing exceedence probability and increasing flood magnitudes indicates the flashy nature of Dudhkumar.1000 year flood will be most violent however extreme care including river dredging, bank protection, channelization and overall river training works should be carried out to have sustainable solution in future.

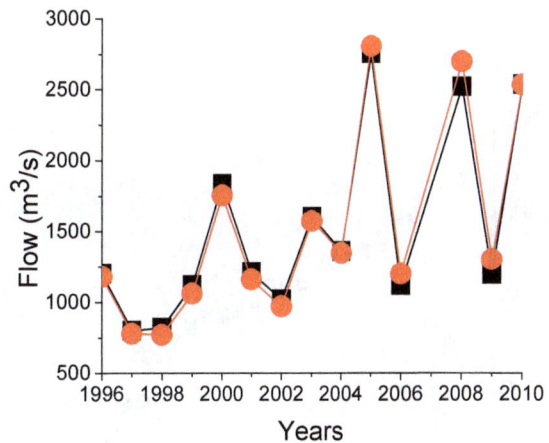

Figure 2. Observed discharge is shown by black squared symbol and predicted discharge is shown by red circular symbol.

Table 2. Floods corresponding to recurrence interval

Recurrence interval (T)	Magnitude of flood (m^3/s) (Gumbel)	Magnitude of flood (m^3/s) (Powell)
2	1399	1399
2.33	1559	1507
5	2255	1977
10	2822	2360
20	3366	2727
25	3539	2843
50	4071	3202
100	4598	3558
200	5124	3913
1000	6342	4734

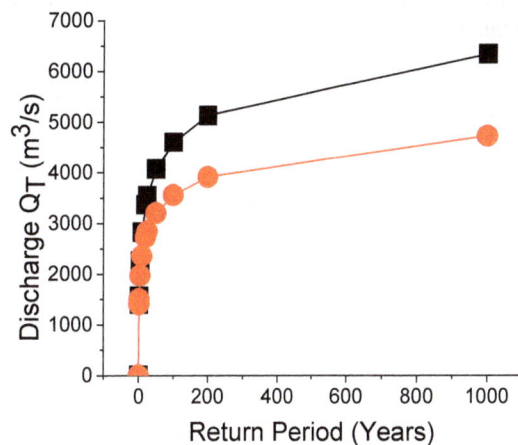

Figure 3. Discharge data versus return period, black squared symbol represents Gumbel's flood frequency and red circular symbol shows Powell's flood frequency.

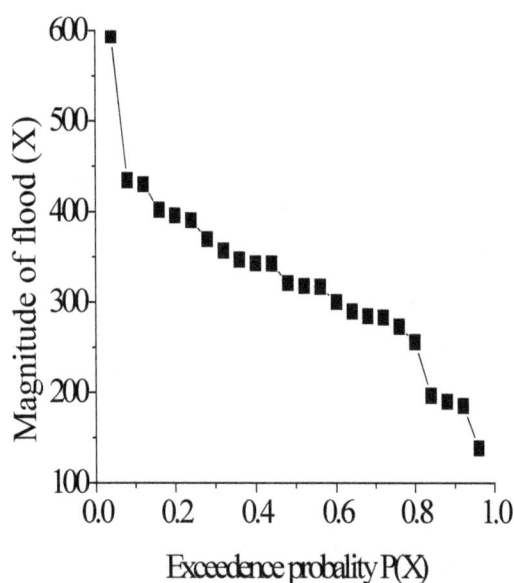

Figure 4. *Magnitude of flood versus exceedence probability, black squared symbol represents flood magnitude corresponding to exceedence probabilities.*

All these works could be implemented after being judged by the flood modeling methods like Gumbel and Powell.

8. Conclusion

Flood Frequency Analysis (FFA) has been done for Dudhkumar River for 14 years discharge data.The results tells that, Gumbel and Powel distribution clearly describes the flood magnitude while a chi square test derives no significant differences (P=1) between the predicted and observed floods. Probability distribution function also fit with the flood data. Due to goodness of fit and probable fitted value of the flood, the distribution should be used in calculating design flood magnitude. Hence the distribution models can be used to predict the occurrence of flood event for Dudhkumar River.

References

[1] U. N. Ahmad, A. Shabri, and Z. A. Zakaria, "Flood frequency analysis of annual maximum stream flows using L-Moments and TL-Moments." Applied Mathematical Sciences, vol. 5, pp. 243– 253, 2011.

[2] J. R. Stedinger and R. M. Vogel, "Frequency analysis of extreme events." Handbook of Hydrology, chapter 18, McGraw-Hill, New York, 1993.

[3] D. S. Reis and J. R. Stedinger, "Bayesian MCMC flood frequency analysis with historical information." Journal of Hydrology, vol. 313, pp. 97–116, 2005.

[4] M. Khaliq, T. Ouarda, J. Ondo, P. Gachon, and B. Bobée, "Frequency analysis of a sequence of dependent and/or non-stationary hydro-meteorological observations: A review." Journal of Hydrology, vol. 329(3-4), pp. 534–552, 2006.

[5] G. S. Law, and G. D Tasker, "Flood-Frequency prediction methods for unregulated streams of Tennessee." Water Resources Investigations Report 03-4176, Nashville, Tennessee, 2003.

[6] M. Ferdows, M. OTA, R. Jahan , M. Bhuiyan and M. Hossain, "Determination of probability distribution for data on rainfall and flood levels in Bangladesh." , Journal - The Institution of Engineers, Malaysia, vol. 66, pp. 61–72, March 2005

[7] M. Rahman, D. Arya, N. Goel and A. Dhamy, "Design flow and stage computations in the Teesta River, Bangladesh, using frequency analysis and MIKE 11 modeling." Journal of Hydrologic. Engineering.", vol. 16(2), pp. 176–186, 2011.

[8] C. T. Haan, "Statistical Methods in Hydrology." Iowa State University Press, Ames, Iowa. I '. Haefner, Journal of Water, 1997.

[9] S. B. Shaw and S. J.Riha, "Assessing Possible Changes in Flood Frequency Due to Climate Change in Mid-sized Watersheds." School of Civil and Environmental Engineering, Hollister Hall, Cornell University, Ithaca, NY 14853-3501, USA.

[10] N. Mujere, "Flood frequency analysis using the Gumbel distribution." International Journal on Computer Science and Engineering (IJCSE), vol. 3, pp. 2774–2778, July 2011.

3

Interactive website with systems analysis environment for prefeasibility studies of small scale water and power production units integrating renewable energy

Djamal Boudieb, Kamal Mohammedi, Abdelkader Bouziane, Youcef Smaili

MESOteam, LEMI, M. Bougara University Boumerdes, Algeria

Email address:
djamelboudieb@yahoo.fr(D. Boudieb)

Abstract: This paper focuses on *RESYSproDESAL* systems analysis environment (SAE) for the prediction of technical, economic and ecological performance of water and power point systems including desalination (e.g. membrane and thermal processes), renewable energy sources for power (e.g. wind energy and photovoltaics) and conventional power supply (e.g. Diesel GenSet).This tool was developed within EU FP6 projects in cooperation between EU-MENA countries. The SAE is applied to a small scale container system for 10 m³/day seawater reverse osmosis desalination powered from Diesel and photovoltaics. Starting from a reference design case three alternative configurations and size are developed and analysed for comparison. The results show a considerable potential for economic improvement of the plant concept, bringing the project closer to affordability for the target population: Optimized Diesel and battery sizes reduce levelised water cost by about 15 %. Up-sizing the whole system from 10 to 50 m³/d and power recovery reduce specific power consumption by about 45 % and integration of water production with village power supply may meet user needs better and increase reliability of back-up.

Keywords: Desalination, Renewable Energy, Co-Generation, Performance, Systems Analysis

1. Introduction

Most Middle East and Northern Africa (MENA) countries are facing growing problems of water supply. Impressive efforts are dedicated to the implementation of large scale equipment with well proven cost-effective technologies for central sea water desalination at coastal sites or brackish water desalination near inland cities. However there are many technically neglected places remote from the countries' centres of water and power production. Typically such settlements of few hundred people with no clean underground water depend from long distance transport of water by truck with high risks due to limited reliability of driver, vehicle and fuel supply as well as hygienic deficiencies of equipment. The true cost of such methods of supply is often not evaluated by the responsible authorities. If grid power connection is not at reach the village may have a simultaneous problem of water and power supply. The inhabitants of such places deserve safer and cost-effective solutions for satisfaction of their needs

for an acceptable standard of living. Water and power production should be implemented on site employing appropriate technologies and making best use of local resources of energy, material and labour. Therefore European experts in small and medium scale desalination and hybrid power generation from conventional and renewable energy sources are developing engineering methods for Integrated Water and Power Point (IWPP) systems, characterised by flexible design, fast implementation, energy efficiency and low emissions. A consortium composed by ZSW (Germany), NERC (Jordan), MESOteam/UMBB (Algeria) and SimTech (Austria)has been involved in the development of the systems analysis environment (SAE) *RESYSproDESAL* for the prediction of technical, economic and ecological performance of water and power point systems including desalination and water treatment (e.g. membrane and thermal processes), renewable energy sources for power (e.g. wind energy conversion and photovoltaic power generation) and conventional power generator (e.g. Diesel GenSet) Mohammedi, 2006.

The process simulation within this tool is done with the commercial software *IPSEpro* from company SimTech in Austria. The standard library of this programmewas extended with special models for desalination equipment and components for conversion of renewable energy.

The challenge of such systems analysis and engineering stems from the necessarily integrative character for the solutions: Usually only simultaneous water and power production and hybrid utilisation of conventional and renewable energy sources make reliable and cost-effective solutions feasible. The integrative character of the engineering approach is recognised from. Fig. 2, showing a typical case of integrated water and power production by a sea water reverse osmosis (SWRO) desalination plant powered from hybrid fossil/renewable energy conversion from Diesel engine and photovoltaics (PV).

There are two subsystems in this example:

- The water treatment system including sea water intake (or well), pre-treatment tank and dosing of chemicals, high pressure pump, pressure vessels for reverse osmosis, post-treatment and storage of permeate, disposal of concentrate.
- The power supply system including PV generator, Diesel GenSet, electric energy storage (accumulator), inverters DC/AC and reverse, power

busbar with load management control system. The busbar may be equipped with connections for another power source, e.g. Wind energy Converter (WEC), and for export of power into a local (village) power grid.

2. Modelling, Simulation and Analysis Environment

The simulation and optimisation of the the the 48 m³/day brackish water Reverse Osmosis small scale desalination unit powered from PV and Diesel generatoris done with *RESYSproDESAL* tool under *IPSEpro* System Analysis EnvironmentRheinländer,2003. *RESYSproDESAL* tool for simulation can help design and optimize water-renewable production systems. The standard library of this software was extended using MDK toolkit with Resysprospecfic models for desalination equipment and components for conversion of renewable energy.The challenge of such systems analysis and engineering stems from the necessarily integrative character for the solutions: Simultaneous water and power production and hybrid utilisation of conventional and renewable energy sources make reliable and cost-effective solutions feasible.

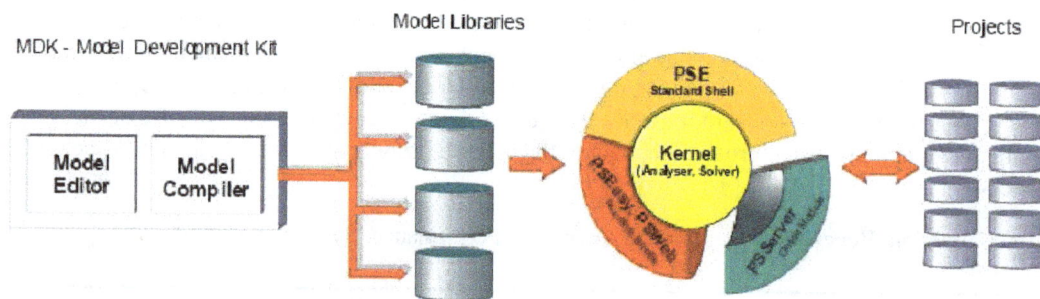

Figure 1: IPSEpro architecture

3. Case Studies in Algeria

3.1. Case 1: Azeffoun

Located on the mediteranean coast, 150 km east of Algiers and 20 km east of Azzefoun small village, Asseklu site is in the heart of the ZET (Zone d'ExpansionTouristique) nearby the Guraya National Park. With an annual average of wind velocity at 10 m height of 8.4 m/s, the site should be suitable for wind energy conversion.

The main indicative data of Azzefoun (Asseklu) site case study are summarized below:

Coordinates: 36.9 N,4.7 E, time zone 1
Climate:Mediterranean
Conventional Energy:none
Elevation: 30 m
Pumping head including height of the tank: 10m
Raw water: sea water

WaterSalinty: 40000 ppm TDS
Irradiation:4500 Wh/day min., 5400 Wh/day max.

Population are semi scattered. Houses may have around 60 m distance from each others. Estimated number of population is around 200. Daily demand for potable water is around 40 m³/day.

3.2. SWRO Desalination

The operation of the system shall aim at a daily production of 10 m³ of potable water. A performance simulation of the integrated RO+PV+Dieselwas done with the RESYSproDESALtool for systems analysis. The process simulation in this tool is done with the commercial software IPSEpro extended with special models for desalination equipment and components for conversion of renewable energyKershman, 2002. The process scheme for the integrated RO+PV+Diesel system is shown in Fig. 1. The scheme is designed for more general applications including another renewable power source (e.g. wind energy conversion) and power supply to a village grid. By setting zero power flows for these connections they are excluded from process simulation.

The simulation of the RO process was done assuming the same set of pressure vessels and membranes as reported: 2 streams with 2 vessels each and 2 membranes type SW30-2540 from FILMTEC per vessel. However the (water) recovery rate calculated here is 49 % against 40 % reported, though in both cases a fouling factor of 0.85 was assumed.

Design Results

Energy

Description	value	Unit
total power requested by SWRO at nominal operation conditions	6.8	kW
specific energy consumption for SWRO process (all auxil. included)	3.07	kWh/m³
power generated by PV system (output from inverter for default op.)	12.3	kW
ratio of annual RE supply to demand from process	0.4606	----
annual consumption of fuel for Diesel GenSet	104198	kg/a
annual energy supplied or consumed		
to desalination process and auxiliary loads	59706	kWh/a
from RE sources via busbar to grid	2757	kWh/a
from RE sources to busbar	27499	kWh/a
from village to grid connection node	0	kWh/a
from grid connection node to village	249040	kWh/a
from Diesel to grid connection node	281247	

Cost

Description	value	Unit
Present Worth		
investment (total plant)	186252	€
investment for desalination system	43593	€
fixed O&M costs (total plant)	257998	€
fixed O&M costs for desalination system	184571	€
variale O&M costs (total plant)	42294	€
variable O&M costs for desalination system	38632	€
replacement costs (total plant)	23854	€
replacement costs for desalination system	10692	€
cost of fuel consumed on site	597571	€
water sold	437451	€
electricity sold	667271	€
net (NPW)	-3247	€
levelised electricity cost (LEC)	0.2345	€/kWh
levelised water cost (LWC)	2.182	€/m³
period for payback of investment from discounted net cash flow	20.4	a

Ecology

Description	value	Unit
annual emission of CO2 from Diesel GenSet operation	343615	kg/a
specific CO2 emission from local electricity production	1.113	kg/kWh

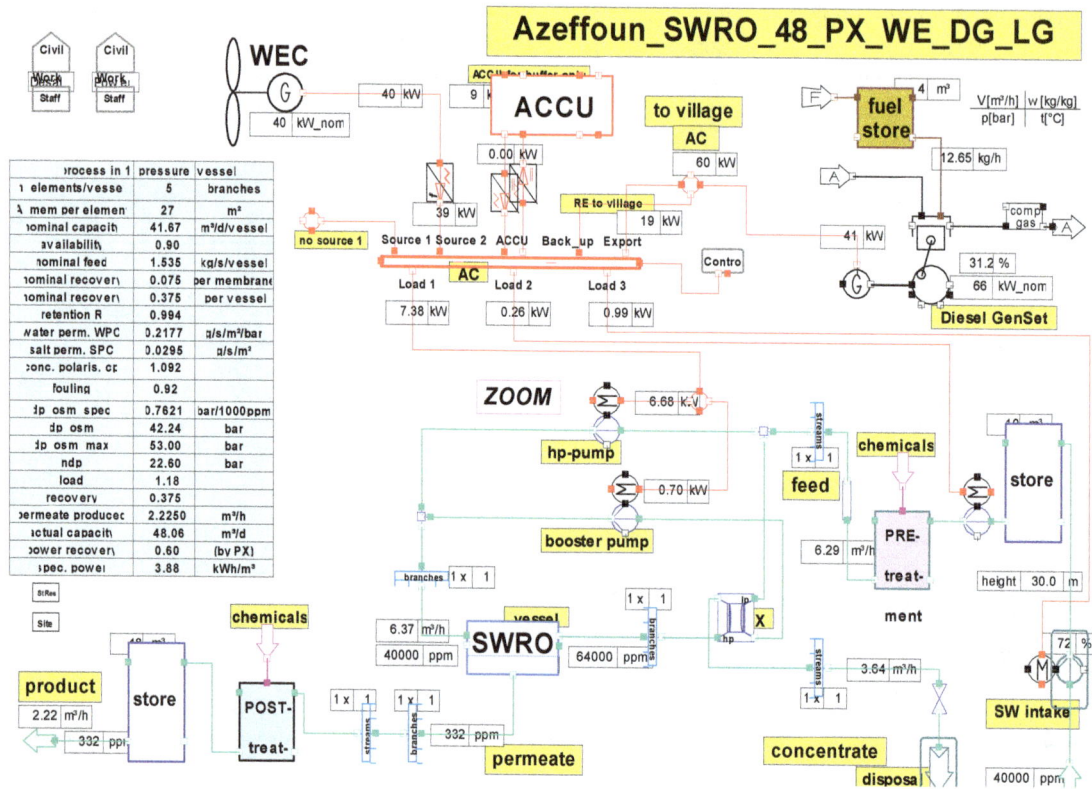

Figure 2 : *Desalination process flowsheet*

3.3. Case2: Hassikhebbi

Coordinates: 29.2 N,5.4 W, time zone 1

Elevation: 90 m

Pumping head including height of the tank: 10m

Water Salinity: 2000 ppm

Populations are semi scattered. Houses may have around 40 m distance from each others, but this is not a problem since there is an already water network existing from the well to the houses. Estimated number of population is around 600. Daily demand for potable water is around 40 m³/day

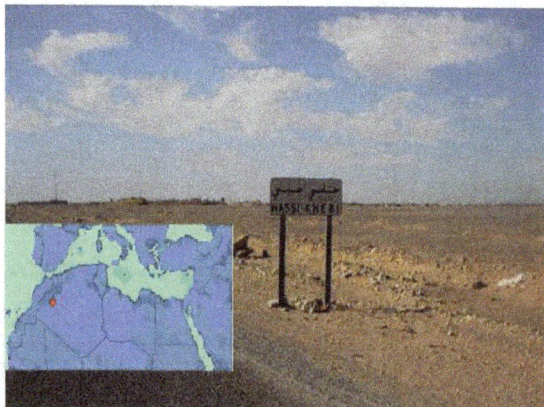

Figure 4. *Hassi Khebbi Location and view*

3.4. BWRO Desalination

The design of the 48 m³/d brackish water RO powered from PV and Diesel generator with connection to local grid assumes demand for simultaneous supply of water and power to a village Boudieb, 2012. The system is composed from 3 subsystems:

- BWRO brackish water desalination
- PV modules power generation
- Diesel generator set

Energy integration of the subsystems is controlled by a power busbar with load management.

The desalination subsystem includes:

- BWRO water treatment with parallel RO vessels (branches) in parallel streams
- High pressure feed pump in every stream
- Chemical pre- and post-treatment
- Raw water and product storage tanks
- Brackish water intake
- Concentrate disposal
- AC electric motors for all pumps
- Headers for splitting and mixing of streams and branches

Design Results

Cost

Description	value	Unit
Present Worth		
investment (total plant)	69255	€
investment for desalination system	33558	€
fixed O&M costs (total plant)	243284	€
fixed O&M costs for desalination system	181867	€
variale O&M costs (total plant)	33407	€
variable O&M costs for desalination system	29974	€
replacement costs (total plant)	10113	€
replacement costs for desalination system	8511	€
cost of fuel consumed on site	529681	€
water sold	310630	€
electricity sold	575864	€
net (NPW)	755	€
levelised electricity cost (LEC)	0.2014	€/kWh
levelised water cost (LWC)	1.543	€/m³
period for payback of investment from discounted net cash flow		

Energy

Description	value	Unit
total power requested by SWRO at nominal operation conditions	2.8	kW
specific energy consumption for SWRO process (all auxil. included)	1.26	kWh/m³
power generated by PV system (output from inverter for default op.)	3.1	kW
ratio of annual RE supply to demand from process	0.3206	----
annual consumption of fuel for Diesel GenSet	92360	kg/a
annual energy supplied or consumed		
to desalination process and auxiliary loads	24513	kWh/a
from RE sources via busbar to grid	0	kWh/a
from RE sources to busbar	7860	kWh/a
from village to grid connection node	0	kWh/a
from grid connection node to village	249040	kWh/a

Ecology

Description	value	Unit
annual emission of CO2 from Diesel GenSet operation	304577	kg/a
specific CO2 emission from local electricity production	1.113	kg/kWh

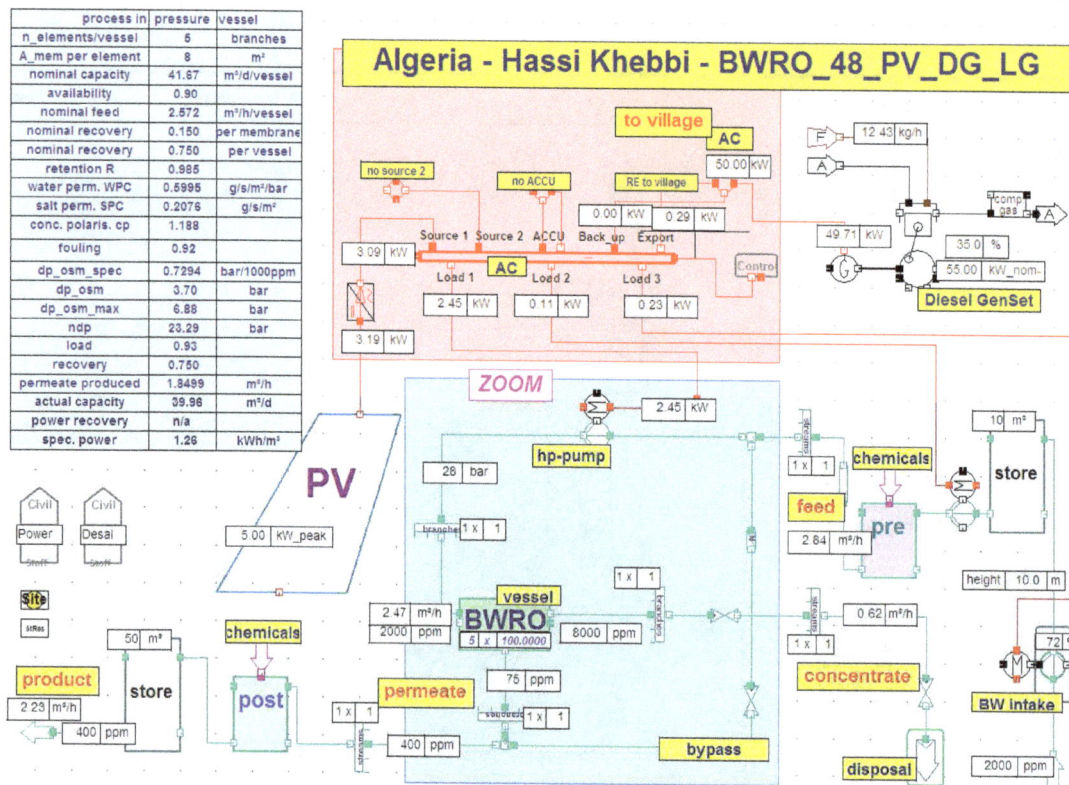

Figure 5. *Hassi Khebbi case study Ipsepro FlowSheet*

4. Conclusion

In this paper, we presented two case studies with SWRO and BWRO desalination respectively. The results of simulations under *IPSEpro* environment with a case study concerned with a 48 m³/day seawater reverse osmosis (SWRO) desalination in Algeria on the east Mediterranean coastwith up to 60 kW additional power supply to the consumers of the water.The desalination subsystem is large enough to include energy recovery by pressure exchanger. The technical performance simulation predicts more than 80% fraction of the wind power contribution to the annual demand. The results show a potential for economic improvement of the system design reducing the expected levelised water (production) cost from the economic analysis under 2 Euros/m³. Naturally, this case study can be extended to other south Mediterranean countries sharing the same conditions.

The Hassi-Khebbi Brackish water Reverse osmosis desalination plant aims a daily production of 48 m³ of potable water. the integrated RO+WE+Diesel simulation was done with the *RESYSproDESAL* systems analysis tool. The process simulation in this tool is done under IPSEpro extended with special models for desalination equipments and components for conversion of renewable energy. The scheme is designed for more general applications including another renewable power source (e.g. wind energy

conversion) and power supply to a village grid. By setting zero power flows for these connections they are excluded from process simulation. With a solar fraction of less than 1/3 of the annual electricity supplied to the BWRO process the originally projected hybrid system more likely is a BWRO Diesel-System with PV for back-up than the opposite.

References

[1] Boudieb, D, Mohammedi, K, Sadi, A, Smaili,Y," Analysis and optimization of a small scale bwro desalination plant integrating renewable energy", International Conference on Nuclear & Renewable Energy Resources, Istanbul, Turkey, 20-23 May 2012

[2] Kershman, S.A., Rheinländer, J., Gabler, H.: Seawater reverse osmosis powered from renewable energy sources – hybrid wind/photovoltaic/grid power supply for small-scale desalination in Libya,. Desalination 153 (2002) pp. 17-23.

[3] Mohammedi,K., Rheinländer, J., Sadi, A.,: Performance *Performance Analysis of De-central Water and Power Production by BWRO Integrating PV Solar Energy*.Eurosun2006, Glasgow UK, June 2006.

[4] Rheinländer, J., Perz, E., Goebel, O., "Performance simulation of integrated water and power systems – software tools IPSEpro and RESYSpro for technical, economic and ecological analysis", European Conference on Desalination and Environment: Fresh Water for All, Malta, 4-8 May 2003.

4

Response of river flow regime to various climate change scenarios in Ganges-Brahmaputra-Meghna basin

Rajib Kamal[1,*], M. A. Matin[1], Sharmina Nasreen[2]

[1]Department of Water Resources Engineering, Bangladesh University of Engineering and Technology, Dhaka, Bangladesh
[2]Bangladesh Water Development Board, Dhaka, Bangladesh

Email address:
rajibkamal@wre.buet.ac.bd(R. Kamal)

Abstract: The potential climatic variability over Ganges-Brahmaputra-Meghna (GBM) basin like alterations in precipitation and temperature are expected to have a significant impact on the natural flow regime of its rivers. The Lower Meghna River, being a major drainage outlet of the basin, is likely to be affected by such variability and hence its response to climate can be studied through the use of plausible scenarios of climate change. In this study, an artificial neural network (ANN) model, based on future climate projections of HadCM3 GCM, was constructed to examine the potential changes in the river flow regime assuming that climate tend to change as per the SRES scenarios A1B, A2 and B1. The results showed a trend of increasing monsoon flows for these scenarios during the periods of 2020s, 2050s and 2080s with a projected shift in the seasonal distribution of flows. Examining the monthly projected flows for different scenarios and comparing with the observed condition, it was found that the peak flow may increase 4.5 – 39.1% in monsoon and the dry period low flows may drop by 4.1 – 26.9% indicating high seasonality as a result of climate change. Due to seasonal variation of precipitation and temperature, i.e., excess precipitation in monsoon and lack of precipitation along with higher temperature in the dry season, the flood peaks are likely to shift towards earlier months and the rate of change of flows during the rising and recession of flooding would be much higher compared to current state of the river. These results also indicate the exacerbation of flooding potential in the central part of Bangladesh due to the largest increase of peak flows during monsoon.

Keywords: Climate Change, GBM Basin, Lower Meghna River, Flow Regime

1. Introduction

Rivers take up an important place when investigating the climate as a whole. For the last century, natural flow regimes of rivers have been modified continuously by different anthropogenic factors such as dam construction, water withdrawals for irrigation, electricity production, manufacturing, domestic purposes and others[1]. Many rivers have also been artificially modified by channelization, embanking, straightening, widening or deepening with further impacts on flow characteristics. Now-a-days climate change constitutes another factor for flow regime alteration and will interact with other anthropogenic flow modifications. The variation of river flow depends on several physical and hydrological processes of a basin and hence is likely to be affected by the magnitude and direction of climate change. The anticipated change in climate is likely to lead to an intensification of the global hydrological cycle and to have a major impact on regional water resources system. The Fourth Assessment Report (AR4) of Intergovernmental Panel on Climate Change (IPCC) mentions with high likelihood that the observed and projected increases in temperature and precipitation variability are the main causes for the reported and projected impacts of climate change on water resources, resulting in a significant impact on a river basin and associated river systems[2]. Hence, it is important to identify the response of river flow regime to climate change because such changes have significant environmental and socio-economic implications for planning and sustainable water resources management of a basin.

The five components of flow – magnitude, frequency, duration, timing and rate of change affect the natural regime and dynamics of rivers directly and indirectly. Among them the quantity and timing of river flow are critical components to water availability, water quality and the ecological integrity of a river basin[3]. River flows respond to changes in basin runoff which is more sensitive to changes in precipitation and evaporation than other climate variables. For a catchment with a low runoff ratio, the effect of a 10% reduction in precipitation may range from a 50% reduction in river discharge with no direct CO_2 effect, to a 70% increase in discharge with a maximum direct CO_2 effect[4]. Assuming no changes in land use, long term changes in basin runoff are largely driven by changes in precipitation and those climatic factors controlling evaporation, in particular, temperature. Variation in temperature also affects the melting of glacier-snows in the upper headwater region. Changed basin water balance due to variation of precipitation and glacier melts may alter the discharge hydrographs of rivers, and such alterations may cause significant changes in the flow regime of a river. Since the principal climatic factors that control the streamflow are precipitation and temperature, the response of river regime to climate can be studied through the use of plausible scenarios of climate change.

In the present study, the river flow response to changes in climate was evaluated by considering the Lower Meghna River as a case study. The river, being a major drainage outlet of the Ganges-Brahmaputra-Meghna (GBM) basin, is one of the largest rivers in the world in terms of river flow. The river carries the combined flows of Ganges, Brahmaputra and Upper Meghna which receives over 90 percent of the water from the GBM basin[5]. The GBM basin is likely to be sensitive to potential climate change impacts where the hydrological regime is strongly influenced by the variables like precipitation, temperature

or evaporation. A modification of the prevalent climate can considerably affect this regime and induce important impacts on the associated river systems like the Lower Meghna River. To analyze such response, a data-driven artificial neural network (ANN) model was developed using climate-flow data. The objective of this study was to assess the potential impacts of future climate scenarios on river at basin scale and to examine potential changes in flow regime assuming that the climate tend to change as per the plausible scenarios mentioned in the Special Report on Emission Scenarios (SRES) given by IPCC.

2. The Ganges-Brahmaputra-Meghna River System

The Ganges-Brahmaputra-Meghna basin is located between 22 degree 3.5 minutes and 31 degree 50 minutes north latitudes and 73 degree 10.5 minutes and 97 degree 53 minutes east longitudes[5]. The basin is characterized by diversified geographic features covering five countries – China, India, Nepal, Bhutan and Bangladesh of the South Asian region. Topographically it is extended in three characteristic areas: the Hindukush Himalaya, the Ganges Delta and the Peninsular Basin of central India (Fig. 1). The mountainous areas of Himalaya and some hilly regions of central and eastern India are the major sources of the rivers in this region. The three main rivers - the Ganges, the Brahmaputra and the Meghna cover 907,000 sq. km, 583,000 sq. km and 65,000 sq. km areas of the basin respectively[6].

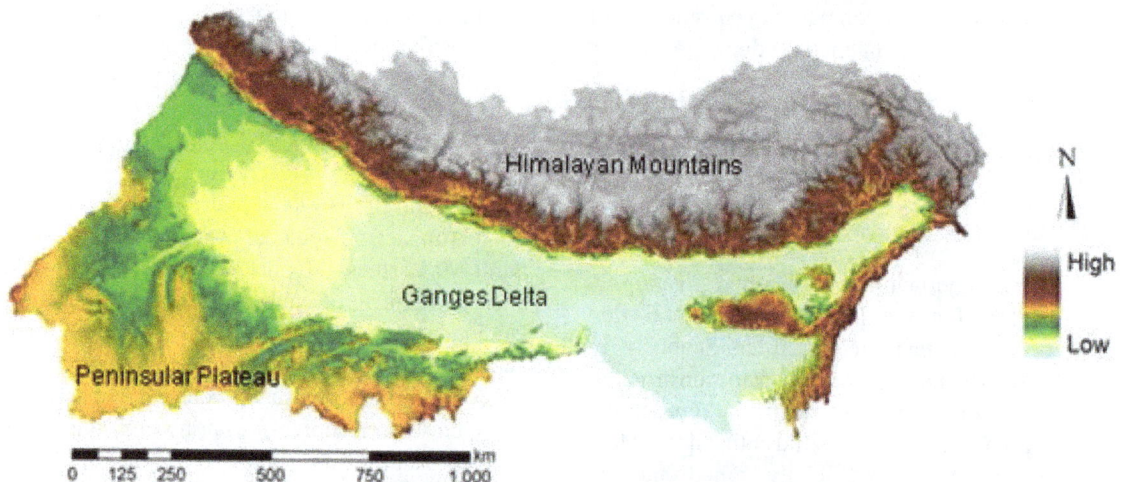

Figure 1. Topography of the Ganges-Brahmaputra-Meghna basin

The headwaters of the Ganges and Brahmaputra originate in the Tibetan China area of the Himalayan Mountain range. The Ganges originates in the Uttaranchal Himalayas after the confluence of six rivers, flows south-west into India after the confluence of six rivers, turns southeast joining its major tributaries and enters Bangladesh through the western border (Fig. 2).

Originating from the great glacier mass of Chema-Yung-Dung in the Kailas range of southern Tibet, the Brahmaputra traverses south and west into India and then directs south into Bangladesh. The Meghna rises in the Manipur Hills of northeast India, flows west and is formed inside Bangladesh above Bhairab Bazar by the combination of the Surma and Kushiyara rivers. It meets the combined flow of Ganges and Brahmaputra (known as the Padma) near Chandpur and is drained into the Bay of Bengal through the Lower Meghna River. The combined discharge of these three major rivers is among the highest in the world. While the discharge within the Lower Meghna (the part downstream of the confluence near Chandpur) typically varies between 8,000 m^3/s in February-March and 100,000 m^3/s in July to September, the peak discharge exceeds 120,000 m^3/s in the year of severe flood[7]

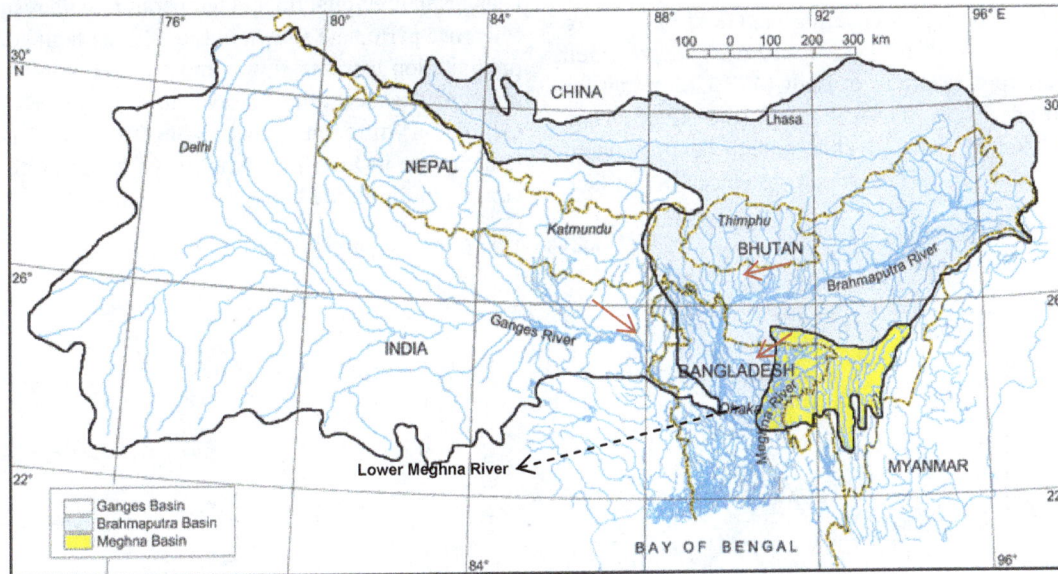

Figure 2. The GBM basin with principle channels of Ganges, Brahmaputra and Meghna

3. Climate Change Scenarios

Among the major river systems of the world, the impact of climate change on the Ganges-Brahmaputra-Meghna basin is expected to be particularly strong[8]. The water supply of GBM rivers is dependent on the rains brought by the southwesterly monsoon winds as well as on the flow from melting Himalayan snows. The Ganges basin is characterized by low precipitation in the northwest of its upper region and high precipitation in the areas along the coast. Annual rainfall ranges from 760 mm at the western end to more than 2000 mm at the eastern end. The Brahmaputra basin, excluding the Tibetan portion, forms high precipitation zone and dry rain shadow areas as an integral part of the southeast Asian monsoon regime with a mean annual rainfall of 2600 mm. The world's highest precipitation area is situated in the Meghna basin[5, 9]. When intensive rainfall occurs simultaneously over these basins, the combined runoff causes high floods in the downstream rivers. Regional variation of temperature is also significant in these basins. Mean temperatures in the Himalayas ranges are 10–15 °C. The temperature of other parts of the basin is much milder, with highs ranging from 23 °C to 26 °C and lows averaging from 8 °C to 14 °C[5]. Variation in temperature also affects the basin water storage through melting of Himalayan snows and evaporation. Therefore any climatic variability like alterations in precipitation and temperature are hypothesized to be responsible for the streamflow variation within the GBM rivers, both in terms of magnitude of flow and in timing of onset, peak and recession of flooding.

To assess the response of flow regime of Lower Meghna River to changes in GBM basin climate, it is necessary to construct future climate projections as per IPCC guidelines. Currently General Circulation Models (GCMs) are the most credible tools available for simulating the response of the global climate system to increasing greenhouse gas concentrations and to provide estimates of climate variables such as temperature, precipitation etc. For the present study, the transient HadCM3 GCM, developed by Hadley Center for Climate Prediction and Research, were used for the projected emission scenarios of SRES A2, A1B and B1. These scenarios represent a plausible range of conditions where A2 corresponds to relatively unconstrained growth in global emissions and B1 corresponds to reduction of global emissions over the next century[10]. A1B assumes a balanced energy adoption between fossil fuels and other energy sources to drive the expanding economy and has been used in many impact studies for Bangladesh.

4. Methodology

In this study, the response of Lower Meghna River flow regime was evaluated following a linear approach - feeding climatic inputs into a system model, comparing system performance with observed conditions and estimating impacts with alternate climate inputs. Using artificial neural network (ANN) technique, a model was developed to approximate the relation between the observed historical precipitation, temperature and river flow data and to estimate the future river flows driven by anticipated changes in climate according to the SRES scenarios.

4.1. Data Collection and Processing

The observed temperature and precipitation data over the GBM basin were obtained from the global database developed by Climate Research Unit (CRU). The CRU provides gridded dataset of monthly precipitation and temperature[11]. These data covering the GBM basin were collected for the period of 1975 to 1998. The projected temperature and precipitations of HadCM3 GCM model for different periods were retrieved from the IPCC Data Distribution Center (DDC). This model has a spatial resolution of 2.5° x 3.75° which produces a surface

representation of about 417 km x 278 km. To estimate the future streamflows of the river, 30-year averaged model projections were used for three time slices namely 2020s for 2010-2039, 2050s for 2040-2069 and 2080s for 2070-2099 with reference period of 1961-1990. The basin averaged changes in precipitation and temperature with respect to this reference period are shown in Fig. 3. This figure depicts that precipitation increase is maximum in the month of July for most of the scenarios. However, scenario B1 in 2050s exhibits a different pattern due to lesser increase of temperature in July. The variation of temperature change is consistent for all the scenarios.

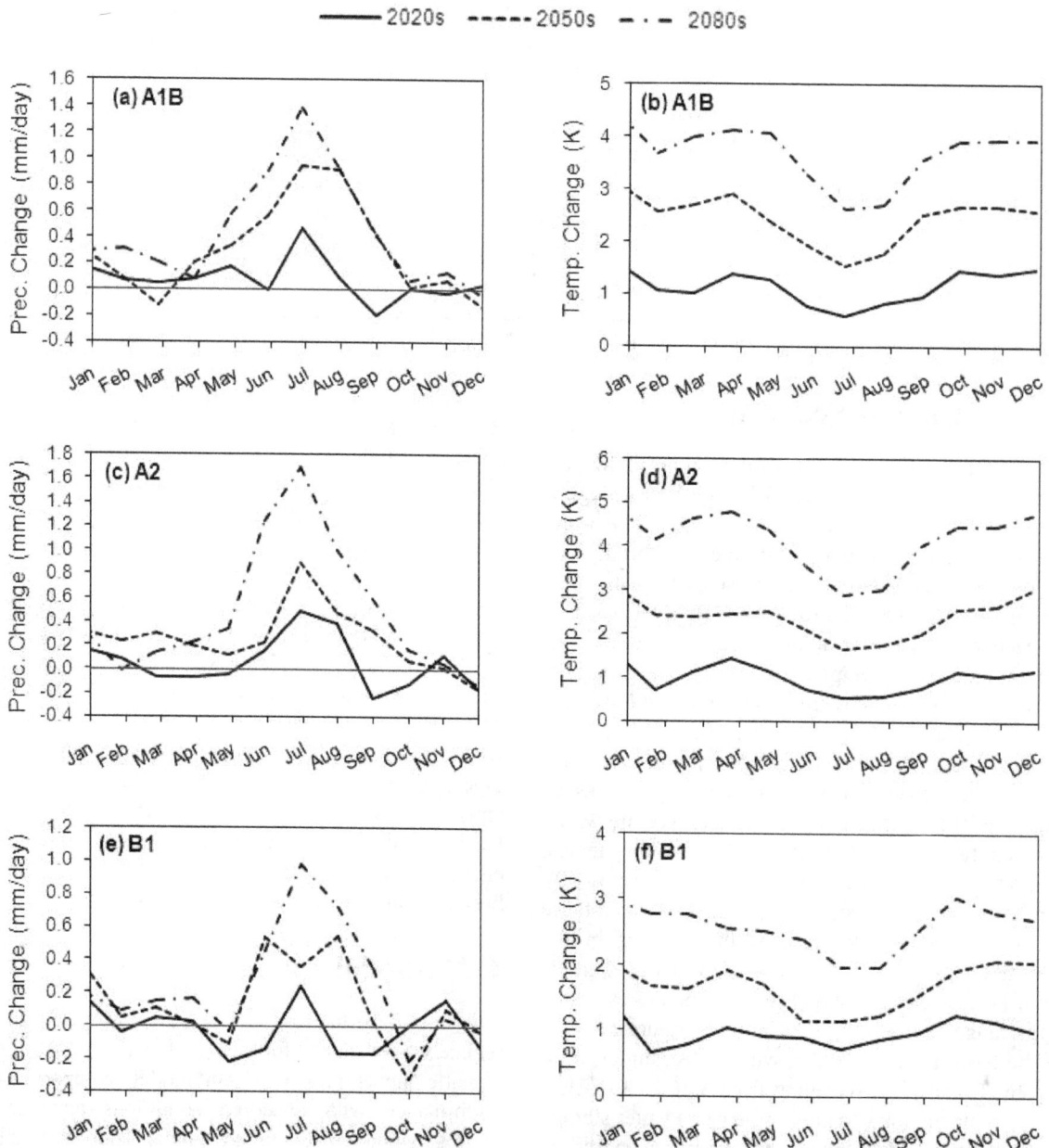

Figure 3. Basin averaged changes in monthly precipitation and temperature for SRES scenarios A1B, A2 and B1

Various hydrographic and river flow data was collected from Bangladesh Water Development Board (BWDB) for the period of 1975 to 2008. The river discharges at Chandpur

were derived form 1-D model HEC-RAS by incorporating discharge of Padma and Upper Meghna River and were employed as observed streamflow records to validate the simulated outputs obtained from the ANN.

4.2. Setting of Basin Grids

To identify the implicit connection between the river flow regime and the climate variability of GBM basin and to understand the nonlinear complex interactions among them, the study area was divided into total of 32 grids of resolution 2.5° x 3.75° (latitude by longitude). In order to maintain consistency among the grids and the resolution of the GCM projections, the observed precipitation and temperature data was also converted in the same resolution. United States Geological Survey produced drainage maps of GBM basin of 0.5 degrees grids along with major river networks of the region[12]. These grids along with superimposed flow network was modified and simplified into 3.75×2.5 degree resolution in order to match with the climate inputs over GBM basin. Fig. 4 shows both the flow network and grids of GBM basin with the polygon of Bangladesh. Flow network was constructed in such a way that it carries water (runoff) from one cell to the next discharging cell based on the drainage directions.

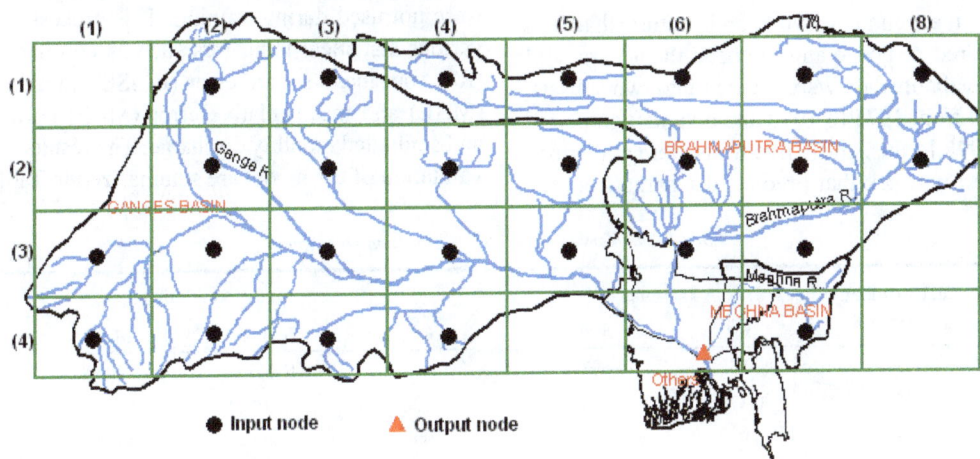

Figure 4. *Simplified GBM grids with flow network and input-output nodes of ANN model*

In a firm ANN architecture, key variables must be introduced and unnecessary variables must be avoided in order to prevent confusion in training (calibration) and validation process. The input node for the present study was the precipitation and temperature over GBM basin. Based on the flow direction on the grids shown in Fig. 4, a total of 26 out of 32 grids were selected as input nodes. 6 grids (grid number 11, 12, 54, 64, 83 and 84) do not contribute to the flow and hence were excluded from the analysis. Assuming all the precipitation over GBM contributes to the basin runoff and the resultant combined flow drains out through the Ganges, Brahmaputra and Meghna River system, the Lower Meghna River can be considered as the outlet of the basin. Therefore the discharge of the river at Chandpur was taken as the output node of the neural network model. Both the input and output nodes, i.e. temperature, precipitation and discharge, were the input variables of the model.

4.3. The Artificial Neural Network Model

A data-driven artificial neural network model was developed to evaluate streamflow responses of Lower Meghna River to climate variations of GBM basin. Although the application of physical or conceptual models could have been a better representation of the hydrological features of the basin, ANNs can be considered an alternate to such models because of its ability to simulate nonlinear complex system without detailed watershed data. Therefore in recent years, many researchers have demonstrated the successful application of the ANN in basin wide streamflow forecasting[13-15]. For the present study, the basin was schematized as a system, whose inputs were precipitation and temperature (which influence the evaporation and snowmelt) and its output was the river flow. Then the neural network model can be written as –

$$y = f(u_j) \text{ with } u_j = \sum w_i x_i - \theta_j \tag{1}$$

Here, x_i = inputs to flow, w_i = weight of x_i and θ_j = critical value. The output of node j, y_j, can be obtained by computing the activation function that determines the response of a node to the total input signal it receives. The most commonly used activation function is the sigmoid function[16]. Historical analyses of the streamflow (Q), precipitation (P) and temperature (T) variations in the GBM river basin indicated that its streamflow lagged the precipitation by one month[6]. This leads to the relation of Q at time (t), treated as a function of P and T at time (t) and (t-1) as follows –

$$Q(t) = f(P_t, P_{t-1}, T_t, T_{t-1}) \tag{2}$$

In this study, feedforward neural network technique was employed as this is closely related to statistical models that are a data-driven approach and more suited for forecasting applications[17]. To determine an appropriate set of weights in (1), the model was trained using the error backpropagation algorithm and momentum was used for speeding convergence to a minimum error. The calculation was done using MATLAB software. A total of 26 nodes representing monthly temperature and precipitation and a single node representing monthly discharge were used in the input and output layer of the model respectively. The number of hidden layers and their nodes depend on the performance of the model and was determined by trial and error basis for the present study. Due to the nature of the sigmoid function, all input data of the model were normalized to an internal representation between 0 and 1 in order to receive equal attention in the training of the model.

4.4. Model Training and Validation

To determine the neural network architecture that best matches the desired response and to check the accuracy of the model, the available data were divided into two subsets: a training period from 1975 to 1990 and a validation period from 1991 to 1998.In this study, five trials were performed to obtain the best network that predicts discharge from the

temperature-precipitation inputs. For each trial, different networks were used by changing number of hidden layers, number of processing elements or nodes in each hidden layer, number of iterations etc. During training, initial weights were randomly generated and the goal for error criterion was 0.0001. These were adjusted based on the mean squared error (MSE) between the ANN outputs and the observed discharges and continued until a weight space was found, which results in the minimum MSE and best overall prediction of discharge. For all the trials, the calibrated networks were verified for the period of 1991 to 1998 by imposing the monthly temperature-precipitation data that were not used during training. This process was guided by various statistical indicators such as correlation coefficient (R^2), root mean square error (RMSE), mean absolute error (MAE) and mean relative error (MRE) between observed and simulated monthly discharge. The results of training and validation of the model are summarized in Table 1.

Table 1. Summary of the model training and validation

Trial No.	No. of Hidden Layers	Training		Validation			
		R^2	RMSE	R^2	RMSE	MAE	MRE
1	3	0.857	6937.2	0.723	10405.8	5493.4	0.349
2	5	0.878	4754.3	0.720	8385.2	5451.1	0.359
3	5	0.917	4355.5	0.749	8091.9	5427.4	0.316
4	7	0.895	4031.3	0.742	7310.7	5033.0	0.213
5	10	0.928	3148.4	0.806	7548.5	4338.2	0.144

From the table it appears that the optimal network topology resulted in trial 5 of one output, ten hidden-layer neurons with R^2 of 0.928 and RMSE of 3148.4 m^3/s in the training period. For the validation period from 1991 to 1998, the simulated streamflows were close to the observation with a R^2 value of 0.806 and absolute error of the monthly

mean discharge of 4338.2 m^3/s, merely 13% of the monthly mean value. However, the model systematically underestimates the observed streamflow by an average of 1330 m^3/s (4 %). Fig. 5 shows the measured and simulated monthly discharge record for the Lower Meghna River from the period of 1975 to 1998.

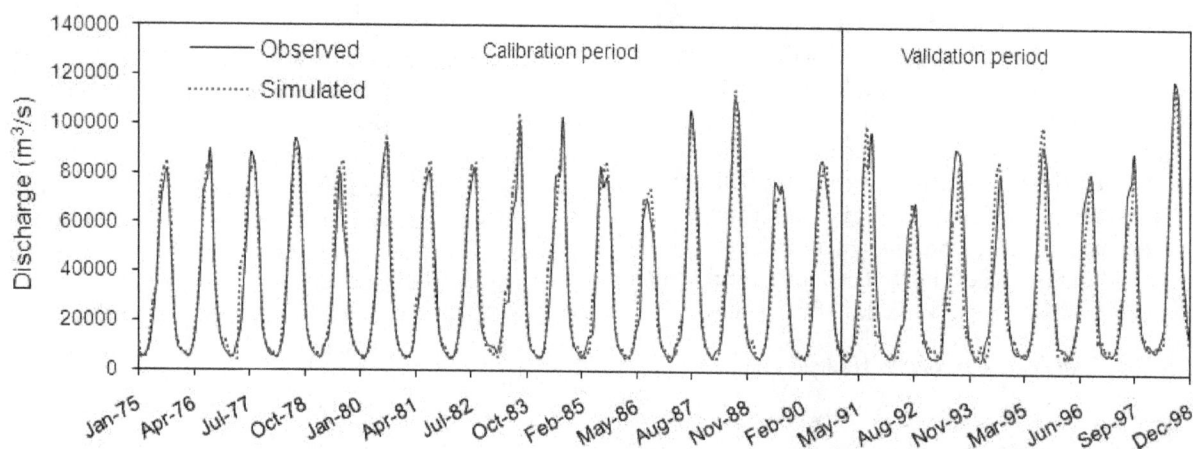

Figure 5. Comparison between observed and simulated discharge during training and validation

From the figure it is evident that the model is able to capture the rising and falling limbs of hydrograph, i.e., it predicts flow for both wet and dry season with considerable accuracy and deviations are hardly visible when looking at the seasonal pattern. These results suggest that the

developed artificial neural network model is reliable to describe the hydrological processes of the basin and to address their effect on the flow variations as well as overall flow regime of the river.

5. Results and Discussion

The developed neural network model was employed to describe the hydro-climate of GBM basin and to examine the influence of precipitation and temperature change on the river flow regime. Fig. 6 shows the ANN model results of projected flows of Lower Meghna River under HadCM3 SRES scenarios A1B, A2 and B1 for the periods of 2020s, 2050s and 2080s. To compare the streamflow variation, the 30-year average of 1975-2004 was considered as the observed base condition which represents the current state of the river.

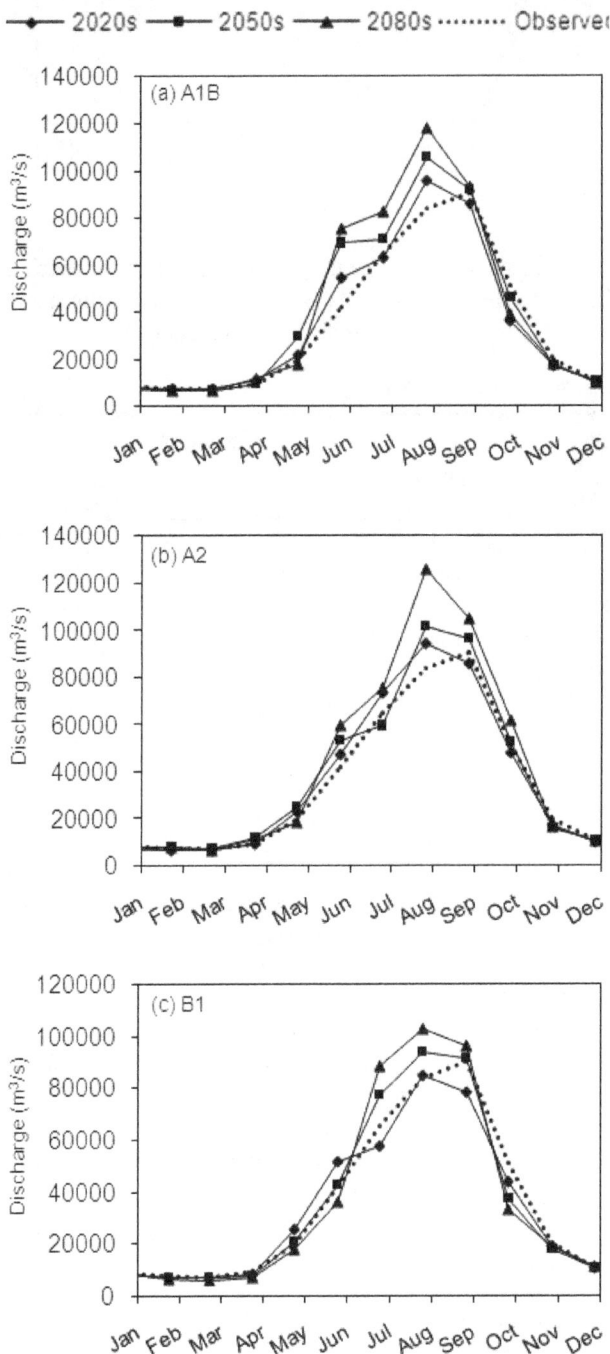

Figure 6. Projected streamflows for different climate change scenarios

The model projected streamflows of the river show that the discharge in monsoon increases progressively for the periods of 2020s, 2050s and 2080s for scenarios A1B and A2. For scenario A1B (Fig. 6a), maximum discharge was found as 96009 m³/s, 105586 m³/s and 118271 m³/s for the projected periods respectively which are 6.1, 16.7 and 30.7% higher than the observed condition. The rising limbs of the hydrographs tend to shift leftward indicating the occurrence of early floodwaters due to increase of precipitation in the premonsoon (upto 13% in May and 22% in June at 2080s) as seen from Fig. 3a. As a result of higher increase of summer-winter temperature than monsoon (Fig. 3b), the effect of evaporation becomes more pronounced than glacier snowmelt and hence the dry period low flows can decrease by 7.1, 6.3 and 16.8% respectively.

Similar trend is also observed for scenario A2 (Fig. 6b) with maximum streamflows of 94539 m³/s, 101730 m³/s and 125835 m³/s which are 4.5, 12.4 and 39.1% higher compared to observed condition. In this case, the peak flow in 2080s are maximum compared to other scenarios due to maximum increase in precipitation (upto 26% in July at 2080s) during pre-monsoon (Fig. 3c). The discharge hydrograph becomes wider and the falling limbs tend to shift towards right. This prolonged floodwater is caused mainly by the increase in precipitation in September-October (about 10% in 2080s). The flow may reduce 12.5, 4.1 and 19.3% respectively during the dry months of the projected periods.

On the other hand, peak streamflow in monsoon for scenario B1 (Fig. 6c) decreases 6.5% in the 2020s but increases 3.5% in 2050s and 13.7% in 2080s. For these periods, the maximum discharge can be found as 84611 m³/s, 93672 m³/s and 102830 m³/s respectively. Due to lack of precipitation and moderate temperature change in summer-winter dry periods (Fig. 3e and 3f), flow in the dry periods decreases 11.4%, 18% and 26.9% for the stated periods. Hence the streamflows are more concentrated within the shorter span of monsoon resulting in longer periods of low flows in the river due to climate change.

The projected streamflows also indicate that the variations of flows are much higher in monsoon period than the summer-winter dry season when compared with the observed condition of the river. For all the three scenarios the streamflows increases considerably during the period of July -August-September. Fig. 7 shows the comparative variation of high flows for scenarios A1B, A2 and B1 for different time slices. In 2020s, the high flows decrease due to lack of monsoon precipitation for scenario B1 but increases for A1B and A2. Afterwards, the high flows increase progressively in 2050s and 2080s for all the scenarios. In comparison, the peak flows in 2020s and 2050s are maximum for scenario A1B, while in 2080s the largest peak is found for scenario A2. This is due to the fact that as the temperature rises with time, the contribution from snowmelt base flow reduces and the streamflows become more responsive to precipitation increase which is maximum for scenario A2 in 2080s. The rise of peak flows may result in severe floods of high intensity in the Lower Meghna River.

Due to the largest increase of flood peaks in August-September, a large area, particularly the central part of Bangladesh, may be inundated. Such effect will deteriorate the situation of a flood prone country like Bangladesh.

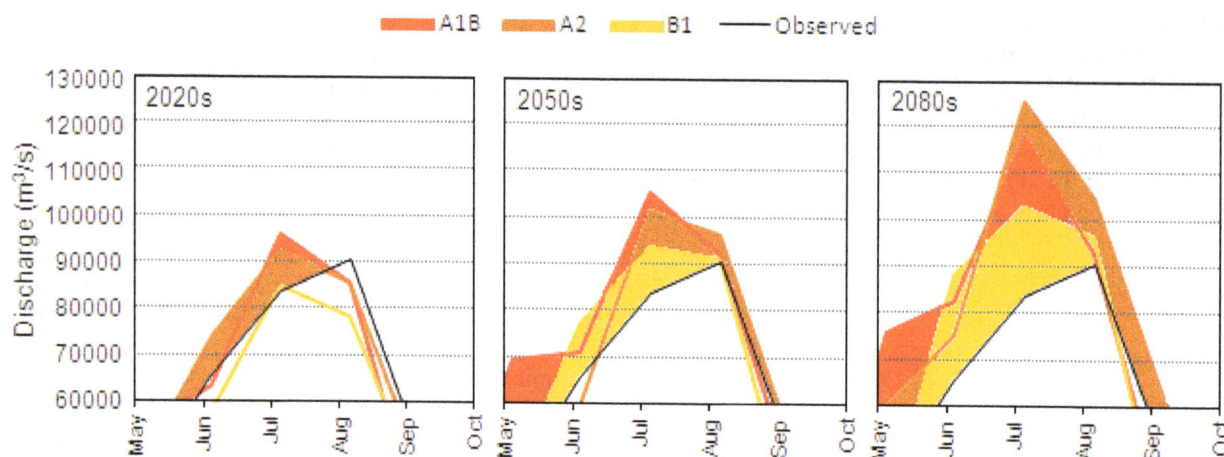

Figure 7. Increase of high flows for different scenarios compared to observed condition

The variation of flow during the dry and wet periods, specially the high flows in monsoon, affects the various components of river flow regime and alters the flood characteristics. These changes in flow characteristics such as timing of peak flow and its onset and recession under different climate scenarios are shown in Table 2.

Table 2. Hydrograph characteristics for the projected river flows

Scenario	Period	Timing of Peak Flow	Rising Limb		Falling Limb	
			Start of onset	Gradient (m^3/s/month)	End of recession	Gradient (m^3/s/month)
Observed	-	Sep	May	21555	Nov	35171
A1B	2020s	Aug	Apr	20506	Nov	34154
	2050s	Aug	Apr	23234	Nov	37319
	2080s	Aug	Apr	27650	Nov	37651
A2	2020s	Aug	May	23169	Nov	34227
	2050s	Aug	May	23601	Nov	40156
	2080s	Aug	May	33749	Nov	44179
B1	2020s	Aug	May	18392	Oct	29713
	2050s	Aug	May	28293	Oct	43648
	2080s	Aug	May	35364	Oct	53018

From the table it is seen that the timing of flood peaks are likely to shift more often towards earlier month when compared to the observed condition. These changes in high flows can be explained by rising temperatures which cause earlier snowmelt in addition to the increased precipitation. As the temperature rises, more of the precipitation falls as rain and less water is stored as snow. Hence, the peaks tend to arrive in advance for the projected periods. For the same reason the variation of high flows for scenarios A1B and A2 are more rapid than scenario B1 due to greater increase in temperature. The steeper gradients of the rising and falling limbs for various scenarios indicate that the rate of change of flow is likely to be greater in contrast to current state of the river. For B1, the rate of change is maximum as floodwater is more concentrated within the shorter period of time compared to other scenarios.

The results of this study suggest that the flow regime of Lower Meghna River has a stronger response to the increase of precipitation than to the increase of temperature, specially in the monsoon period. These results are consistent with the fact that the variation of precipitation increase is much greater than the temperature change during monsoon and the total volume increase in basin runoff is largely contributed by the precipitation other than the effect of evaporation and melting of glacier-snows. However, the impact of temperature and consequent evaporation becomes more pronounced in the dry periods.

6. Conclusion

The anticipated change in Ganges-Brahmaputra-Meghna (GBM) basin climate is likely to have a significant impact on its rivers, particularly on the lower riparian rivers like the Lower Meghna. The projected streamflows of the river, based on HadCM3 GCM, showed that the changes in basin climate tend to affect the timing and magnitude of peak flows as well as the high flow and low flow events at monthly or seasonal time scale resulting in dramatically altered flow regime. Observing the monthly projected flows of Lower Meghna River for different SRES scenarios and comparing with the observed condition, it was found that the peak flow may increase 4.5 – 39.1% in monsoon and the low flows may decrease 4.1 – 26.9% in dry season indicating high seasonality as a result of climate change. Seasonal variation of precipitation and temperature, i.e., excess precipitation in monsoon and lack of precipitation along with higher temperature in the dry season also affects the timing of peak, onset and recession of flooding. The flood peaks are likely to shift towards earlier months and the rate of change of flows during the rising and recession of floodwaters are likely to be much higher compared to current state of the river.

The results of the study also indicate the exacerbation of flooding potential, particularly during monsoon due to the largest increase of peak flow in August-September. In monsoon, the confluences between the Brahmaputra and the Ganges (known as the Padma) and the Meghna become two huge water pools. Due to climate change, the projected increase in Lower Meghna River flow depicts a critical situation where the synchronization of peak flows of the major rivers will induce severe floods of high intensity, particularly in the central part of Bangladesh. Changes in various components of flow regime such as the timing of onset, peak and recession will also affect the pattern, intensity and duration of such floods. The situation may become worse if it is associated with global warming accelerated sea level rise[18].

Since similar trends of flows appear for all the three scenarios A1B, A2 and B1, the results of the present study are quite indicative about the alterations of flow regime and high flow-low flow events of the river due to climate change. However, substantial uncertainty lies in the magnitude of projected streamflows which is primarily associated with GCM structure, magnitude of global warming, emission scenarios and response of regional climate to the global climate. Proper selection of scenarios, sufficient understanding of the basin processes and careful training and validation of the model can reduce these uncertainties considerably. Although this is a single GCM, single impact model study, this has important implications for the indicative and qualitative assessments of the response of flow regime of GBM rivers to climate change. The impact such responses of rivers, particularly the floods or droughts, can be managed through regional cooperation among the concerned countries, long term basin wide planning and sustainable management of water resources of the basin. The type of analyses reported here can provide the planners and managers with the information they need to make evidence-based decisions about meeting demands for water resources, managing flood risks and protecting socio-economic and environmental balance of the basin in future.

Acknowledgements

The author is thankful to IPCC Data Distribution Center (DDC; http//www.ipcc-data.org), Climate Research Unit (CRU; http//www.cru.uea.ac.uk) and Bangladesh Water Development Board (BWDB) for providing necessary data to carry out this research work.

References

[1] B. Malmqvist and S. Rundle, "Threats to the Running Water Ecosystem of the World", Environmental Conservation, vol. 29, pp. 134–153, 2002.

[2] IPCC, "Climate Change 2007: The Physical Science Basis", Contribution of Working Group I to the Fourth Assessment Report of the Intergovernmental Panel on Climate Change[Solomon, S., D. Qin, M. Manning, Z. Chen, M. Marquis, K.B. Averyt, M. Tignor and H.L. Miller (eds.)]. Cambridge University Press, Cambridge, United Kingdom and New York, NY, USA, 2007.

[3] C. A. Gibson, J. L. Meyer, N. L. Poff, L. E. Hay and A. Georgakakos, "Flow Regime Alterations Under Changing Climate in Two River Basins: Implications for Freshwater Ecosystems", River Research and Applications, vol. 21, pp. 849–864, 2005.

[4] F. H. Verhoog, "Impact of Climate Change on the Morphology of River Basins", Proceedings of the Vancouver Symposium, IAHS, Publ. no. 168, 1987.

[5] CCC, "Impact Assessment of Climate Change and Sea Level Rise on Monsoon Flooding", Report Prepared by Climate Change Cell, DoE, MoEF; Component 4b, CDMP, MoFDM, Dhaka, 2009.

[6] M. R. Chowdhury and N. Ward, "Hydro-Meteorological Variability in the Greater Ganges–Brahmaputra–Meghna Basins", International Journal of Climatology, vol. 24, pp. 1495–1508, 2004.

[7] Ministry of Water Resources (MoWR), "Hydro-Morphological Dynamics of the Meghna Estuary", Report Prepared by DHV Consultants, Bangladesh Water Development Board, Bangladesh, 2001.

[8] A. K. Gain, W. W. Immerzeel, F. C. Sperna-Weiland and M. F. P. Bierkens, "Impact of Climate Change on the Stream Flow of Lower Brahmaputra: Trends in High and Low Flows Based on Discharge Weighted Ensemble Modeling", Hydrology and Earth System Sciences Discussion, vol. 8, pp. 365–390, 2011.

[9] M. Q. Mirza, R. A. Warrick, N. J. Ericksen, and G. J. Kenny, "Trends And Persistence in Precipitation in The Ganges, Brahmaputra and Meghna River Basins" Hydrological Sciences, vol. 43, no 6, pp. 845-858, 1998.

[10] N. Nakicenovic et al., "Emissions Scenarios", Special Report of Working Group III of the Intergovernmental Panel on Climate Change, Cambridge University Press, Cambridge, UK and New York, NY, USA, 2000.

[11] T. Mitchell and P. Jones, "An Improved Method of

Constructing a Database of Monthly Climate Observations and Associated High-Resolution Grids", International Journal of Climatology, vol. 25, pp. 693-712, 2005.

[12] United States Geological Survey (USGS), "HYDRO1k Elevation Derivative Database – Asia", Distributed by the Land Processes Distributed Active Archive Center (LP DAAC) located at the USGS EROS Data Centre, from http://LPDAAC.usgs.gov, 2001.

[13] O. Kişi, "River Flow Modeling Using Artificial Neural Networks", Journal of Hydrologic Engineering, vol. 9, no. 1, pp. 60–63, 2004.

[14] J. S. Wu, J. Han, S. Annambhotla and S. Bryant, "Artificial Neural Networks for Forecasting Watershed Runoff and Stream Flows." Journal of Hydrologic Engineering, vol. 10, no. 3, pp. 216–222, 2005.

[15] X. Chen,J. Wu and Q. Hu, "Simulation of Climate Change Impacts on Streamflow in the Bosten Lake Basin Using an Artificial Neural Network Model", Journal of Hydrologic Engineering, vol. 13, no. 3, pp 180-183, 2008.

[16] ASCE Task Committee, "Artificial Neural Networks in Hydrology I." Journal of Hydrologic Engineering, vol. 5, no. 2, pp. 115–123, 2000.

[17] ASCE Task Committee, "Artificial Neural Networks in Hydrology II." Journal of Hydrologic Engineering, vol. 5, no. 2, pp. 124–137, 2000.

[18] M. A. Matin and R. Kamal, "Impact of Climate Change on River System", Proceedings of the International Symposium on Environmental Degradation and Sustainable Development (ISEDSD-2010), Dhaka, Bangladesh, pp.61-65, 2010.

Quantification of bacteria in domestic water storage tanks in Sharjah

Ala H. Amiri, Ruwaya R. Alkendi, Yasser T. Ahmed

Faculty of Science, Department of Biology, United Arab Emirates University

Email address:

200734525@uaeu.ac.ae(A. H. Amiri), ruwayaa@uaeu.ac.ae(R. R. Alkendi), Yasser.turki@uaeu.ac.ae(Y. T. Ahmed)

Abstract: Maintaining the microbiological quality of water is an important means of preventing water-bornediseases. The aim of the present study was to use the Membrane Filter (MF) method to examine the level of coliforms (*Escherichia coli*)and total bacteria in water storage tanks fitted in different homes in Sharjah. The MF method can determine the presence or absence of bacteria within only 24 hours. A volume of 100ml of water sample is filtered through bacterial retaining membrane; the membrane is then transferred to a selective media and incubated for 24h at 37□C to enable the growth of the bacteria. On the growing plates, Total Bacterial Count (TBC) shows as yellow colonies, Total coliforms (TC)shows as dark red colonies and *E. coli*shows as dark blue colonies. Eleven houses were sampled for along a period of six weeks. Samples were collected from two storage tank levels (ground level and roof top level) located in each house. The results showed that, although none of the samples contained *E. coli*, they did contain other coliforms. The absence of *E.coli* indicate no fecal contamination by animal and/or human, on the other hand, other Coliform bacteria were present in water which are usually introduced by the environment such as *Klebsiella*, *Enterobacter* spp., and *Serratia*. These isolates pose a health risk if they reach the human system. More than half of the samples collected (72.7%) showed a high TBC (>10CFU/100ml), which suggests further investigation is needed to examine the sources of contamination to the storage tanks. The bacterial contaminants found in the storage tanks usually come from the environment which may indicate the presence of other contaminants like chemical contaminants that are also found in the environment and entered via the same route to the storage tanks. The results of this study suggest the adoption of a cleaning system for the water tank at least twice a year to prevent accumulation of contaminants. In addition, the results suggest that chemical contaminants might be present in the water, therefore, chemical analysis is recommended.

Keywords: Fecal Coliform, Total Coliform, E. Coli, Membrane Filter Technique, Sharjah, Total Bacterial Count

1. Introduction

Unsafe drinking water, poor sanitation, and lack of hygiene are major causes of disease in both developed and developing nations. Water-borne diseases cause millions of deaths each year, mainly among children under 5 years-of-age [15];however, many of these diseases are preventable. It is estimated that the global disease burden could be reduced by one-tenth simply byimproving water quality, sanitation, and personal hygiene [20].

A human being requires about 20 liters of freshwater every day to meet basic survival needs (drinking and cooking), and an additional 50 to 150 liters for washing, bathing, laundry, irrigation etc. As the global population and overall living standards increase, the demand for freshwater is approaching its limit (one-third of the global population now live in areas of "water stress"). In addition, increasing pollution from urban, industrial, and agricultural sources is making available resources either unusable or a major health risk. In the developing world, almost 5 million deaths per year are directly attributable to water-borne diseases, but an adequate supply of clean and safe water could prevent many of these deaths[2].

In the United Arab Emirates (UAE), fresh water is supplied in the form of ground water (obtained from wells)or desalinated water; however, the use of both is constrained. Desalinated water is expensive, and is produced using associated gas. The supply of ground water is limited by the total reservoir capacity within the country. These reservoirs are replenished by rainfall, which in the UAE is both scarce and erratic. In general, the requirements of industry and of people living in urban areas are met by the supply of desalinated water[12].Tap water, supplied in the form of desalinated sea water,is generally considered safe to drink, although the government's Food and Environment Laboratory does warn of the risk of contaminated water in buildings that have poorly maintained pipes [8].

Usually, domestic houses are fitted with two large water tanks: one is located at ground level and is filled with treated water, and the other is located on the roof and receives water pumped from the ground level tank. Both tanks supply the house with drinking water and water for cooking, washing clothes, showering, irrigation, fire suppression, and agriculture(plants and livestock). These tanks are often old and are not regularly cleaned or maintained, and therefore water quality is an issue. Although the demand for tap water in the UAE is increasing, the public must be aware of the consequences of poor quality water [21].

The Sharjah Electricity and Water Authority supply all regions of the Emirates with drinking water from both underground and desalinated sources. Modern desalination and production plants produce and treat both seawater and water from underground, which is then distributed to residential, commercial, and industrial consumers via transmission and distribution networks[17].

Worldwide, the microbiological content of water is tested to monitor and control both quality and safety. Such tests are undertaken to ensure that the water used for drinking, food preparation, and bathing is safe. Water contains many potential pathogens; therefore, it is impractical to screen all samples for all possible pathogens. Instead, several "indicator organisms" are used as surrogate markers of risk. Most water-borne diseases are caused by fecal pollution of water sources; therefore, the majority of tests aim to detect coliforms and E. coli[3].

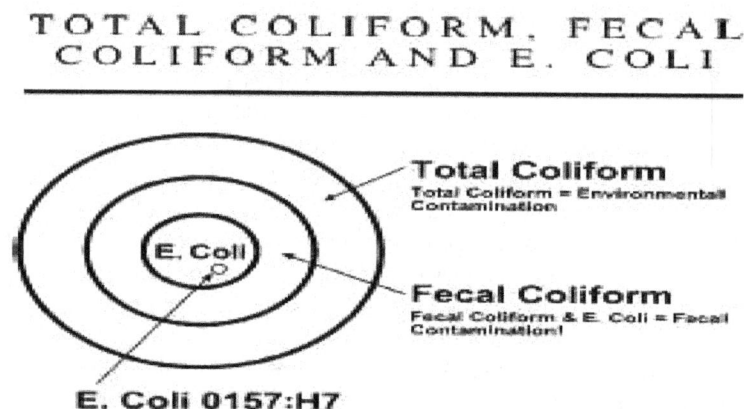

Source: Coliform Bacteria in Drinking Water (2011)

Figure 1. Groups of bacteria that are found in environment and/or intestinal tract of warm blooded animals and human.

There are three groups of coliform bacteria. Each is an indicator of drinking water quality and each carries a different level of risk. Coliform bacteria are common in soil and on vegetation, tend to live longer than pathogenic microbes [4], and are generally harmless; thus, they can be used as indicator organisms. For example, when coliforms are present in a sample of drinking water, it indicates that the source has been contaminated by surface water. The coliform test is considered a reliable indicator of the possible presence of fecal contamination, and the results correlate with the presence of pathogenic organisms[20].

Fecal coliform bacteria are a subgroup of coliform bacteria that are present in the intestines and feces of humans and animals. The presence of fecal coliforms indicates that a drinking water sample has been recently contaminated with feces and that there is an increased risk that it contains pathogens [7].

E. coliare a subgroup of fecal coliforms. Most E. coli reside within the intestines of humans and warm-blooded animals, and are harmless. However, some strains can cause illness. The presence of E. coli in a drinking water sample usually indicates recent fecal contamination, and

therefore an increased risk of infection by pathogenic microbes[7].

Human and animal waste is a primary source of contaminating bacteria, which can enter the water supply via run-off from feedlots, pastures, and other land upon which animal waste is deposited. Natural soil and plant bacteria are also a potential source, along with seepage or discharge from septic tanks and sewage treatment facilities. Bacteria from these sources can also enter wells that are either open at the land surface, or do not have water-tight casings or caps [14]. Wells that are poorly constructed and, or, poorly maintained (particularly shallow wells) are at a high risk of contamination, allowing bacteria and other harmful organisms to enter the water supply[6]. Old wells dug by hand and lined (cased) with rocks or bricks usually have large openings and casings that often are not well-sealed. This makes it easy for insects, rodents or animals to fall into the well, providing a further source of contamination[14].

Infections and illness resulting from recreational water contact are generally mild; therefore, they are difficult to detect using routine surveillance systems. Even in cases of severe illness, it is difficult to identify contaminated water as the cause. Targeted epidemiological studies show that a number of adverse health outcomes, including gastrointestinal and respiratory infections, are associated with recreational water polluted by fecal matter [16]. Such adverse outcomes result in a significant disease burden and substantial economic losses [16].

The microorganisms that cause infection or disease depend upon the pathogen involved. It also depends on the circumstances in which the microorganism is encountered, the conditions of exposure, and the host's susceptibility and immune status. Indeed, in cases concerning viruses and parasitic protozoa, the dose may amount to no more than a few viable infectious units[20].

The quality of drinking water has a direct effect on the well-being of individuals at every social level. A report by the WHO attributed 4.0% of all deaths and 5.7% of the global disease burden to water-related illnesses, which stem from poor water quality, poor hygiene, and poor sanitation[15].These diseases disproportionately affect those in the developing world, particularly young children. In contrast to the developed world, less than half of sub-Saharan Africa has access to safe drinking water [19]. Infants are particularly susceptible to diseases caused by contaminated drinking water because they have not yet developed acquired immunity. Diarrhea, which is the major disease symptom caused by drinking contaminated water, accounts for 2.5 million deaths per year in children under5 years-of-age [11]. The aim of this study is to evaluate the quality of the water stored in storage tanks at Sharjah homes and compare with the quality of other emirates found from previous studies on Abu Dhabi and Al Ain.

2. Methods

Water samples were collected from the ground and roof top water storage tanks fitted in homes located in different regions in Sharjah. Triplicate samples were collected in 500 ml sterile bottles and shipped to the laboratory on ice. Two homes were sampled every week for six weeks. Microbiological analysis was performed using the MF technique, which isan effective and accepted method for testing fluid samples for microbiological contamination. The method involves less sample preparation than many traditional methods, and is one of the few methods that allow microorganisms to be isolated and counted quickly. The results are available within 24 hours.

Briefly, a sterile filter membrane was placed into a funnel assembly attached to a vacuum flask and the sample to be tested was poured in. The sample was filtered through the membrane under vacuum, thereby trapping any microorganisms present in the sample. The membrane filter was then removed from the funnel and placedonto an absorbent pad coated with prepared culture media. A pad coated with one ampoule of m-Endo broth(Millipore) was used for the total coliform cultures, a pad coated with one ampoule of m-ColiBlue24 broth (Millipore)was used to

culture *E. coli*, and a pad coated with one ampoule of trypticase Soy broth (Millipore) was used for the TBC. The pads and filters were then incubated at 35°C for 24 h, and the number of colonies growing on each pad was counted. These steps were repeated for all triplicate samples and the average number of colonies per sample was calculated.

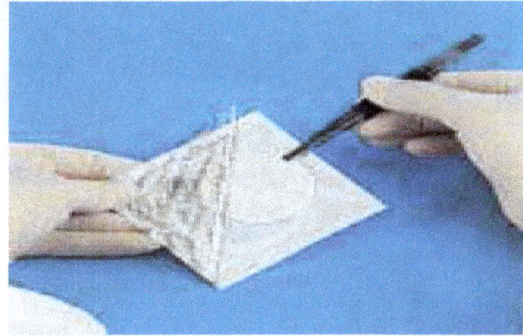

Fig.2. *Remove the membrane from the sterile package.*

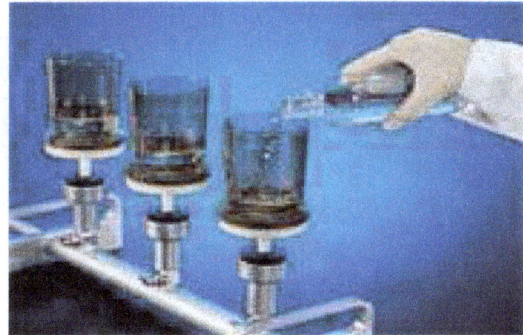

Fig. 3. *Turn on the vacuum and allow the sample to pass completely through the filter.*

Fig.4. *Remove the membrane filter from the funnel.*

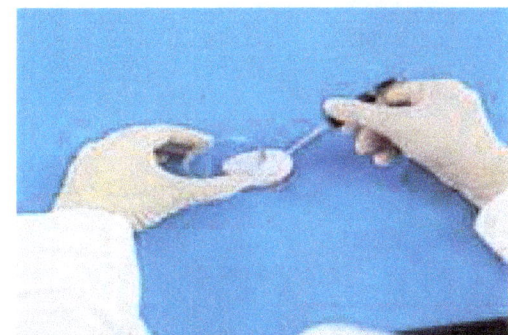

Fig. 5. *Place the membrane filter into the prepared Petri dish.*

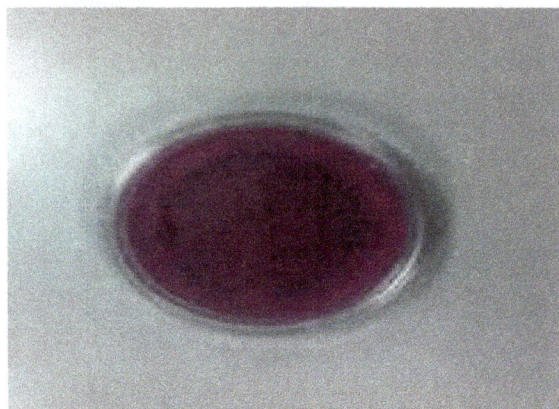

Fig.6. The colonies are in red.

Fig.7. The colonies are in yellow.

3. Microbial Standards

Table 1: Acceptable levels of bacterial contamination in the domestic water supply according to the Regulation and Supervision Bureau.

Parameter	Units of measurement	Maximum Prescribed Value
Total Coliforms	Number/100 ml	0
E.coli or thermotolerent Faecal coliform Bacteria	Number/100 ml	0
Enterococci	Number/100 ml	0
Total Bacterial Count (37□C)	Number/100 ml	10

Source: The Water Quality Regulations. (2009). The Regulation and Supervision Bureau for the Water (Ed.). Abu Dhabi, UAE.

Table 1 show that the Maximum prescribed value for coliform bacteria in drinking water is zero (no coliforms detected in 100 ml of water). It is sometimes difficult to count the number of coliforms and identify the individual species if excessive numbers of other bacteria are present. These samples may be classified as "too numerous to count" or as "confluent growth" [14].

4. Analysis and Discussion

In total, 22 samples from the ground (lower) and roof top (upper) tanks were collected and analyzed over a period of 2 months. As shown in Table 2, each sample was coded by a letter, which refers to the tank position (U, upper; L, lower), followed by the region in which the sample was collected, and a number that refers to the individual sample (1, 2 or 3) within a triplicate.

Table2. Sample details and test results

Sample no.	Date of sampling	Location	Total coliforms [per100ml] ml	*E.coli* [per 100 ml]	TBC [per 100ml]
LA1		Kalba - Al Musalla	TFTC*	BDL□	TFTC
LA2	19/3/2012	Kalba - Al Musalla	TFTC	BDL	TFTC
LA3			TFTC	BDL	TFTC
			-	-	-
UA1			TFTC	BDL	TFTC
UA2	19/3/2012	Kalba - Al Musalla	TFTC	BDL	28
UA3			TFTC	BDL	92
Average			-	-	40
LB1			TFTC	BDL	TFTC
LB2	11/4/2012	Kalba - Al Baraha	TFTC	BDL	TFTC
LB3			TFTC	BDL	TFTC
Average			-	-	-
UB1			157	BDL	TNTC†
UB2	11/4/2012	Kalba -Al Baraha	151	BDL	TNTC
UB3			200	BDL	TNTC
Average			169.3	-	-
LK1			TFTC	BDL	TFTC
LK2	11/4/2012	Kalba -Alkhuwair	TFTC	BDL	TFTC

Sample no.	Date of sampling	Location	Total coliforms [per100ml] ml	*E.coli* [per 100 ml]	TBC [per 100ml]
LK3			TFTC	BDL	TFTC
Average			-	-	-
UK1			178	BDL	203
UK2	11/4/2012	Kalba -Alkhuwair	202	BDL	219
UK3			194	BDL	210
Average			191.3	-	210
LH1			TFTC	BDL	TFTC
LH2	18/4/2012	Kalba -Hutteen	92	BDL	121
LH3			55	BDL	73
Average			49	-	67
UH1			96	BDL	132
UH2	18/4/2012	Kalba -Hutteen	157	BDL	213
UH3			116	BDL	203
Average			123	-	182
LZ1			TFTC	BDL	TFTC
LZ2	23/4/2012	KhawrFakkan-Zubarah	TFTC	BDL	TFTC
LZ3			TFTC	BDL	118
Average			-	-	39.3
UZ1			TFTC	BDL	TFTC
UZ2	23/4/2012 23/4/2012	KhawrFakkan-Zubarah	TFTC	BDL	TFTC
UZ3			41	BDL	77
Average			13.6	-	25.6
LT1			TFTC	BDL	TFTC
LT2	30/4/2012	AlDhaid-Tawi Al Saman	TFTC	BDL	TFTC
LT3			TFTC	BDL	TFTC
Average			-	-	-
UT1			TFTC	BDL	TFTC
UT2	30/4/2012	AlDhaid-Tawi Al Saman	TFTC	BDL	TFTC
UT3			TFTC	BDL	TFTC
Average			-	-	-
LN1			TFTC	BDL	42
LN2	30/4/2012	Sharjah- Am Knorr	TNTC	BDL	TNTC
LN3			TFTC	BDL	TNTC
Average			-	-	14
UN1			TNTC	BDL	TNTC
UN2	30/4/2012	Sharjah- Am Knorr	TNTC	BDL	TNTC
UN3			TFTC	BDL	102
Average			-	-	34
LD1			32	BDL	127
LD2	21/5/2012	Sharjah- Dibba Al Hosn	TFTC	BDL	TFTC
LD3			TFTC	BDL	TFTC
Average			10.6	-	42.3
UD1			TFTC	BDL	TFTC
UD2	21/5/2012	Sharjah- Dibba Al Hosn	TFTC	BDL	TNTC
UD3			TFTC	BDL	149
Average			-	-	49.6
LY1			TFTC	BDL	34

Sample no.	Date of sampling	Location	Total coliforms [per100ml] ml	*E.coli* [per 100 ml]	TBC [per 100ml]
LY2	21/5/2012	Sharjah-Al Yarmouk	115	BDL	TNTC
LY3			TNTC	BDL	TNTC
Average			38.3	-	11.3
UY1			TFTC	BDL	TFTC
UY2	21/5/2012	Sharjah-Al Yarmouk	TFTC	BDL	TFTC
UY3			TFTC	BDL	33
Average			-	-	11
LS1			TFTC	BDL	TFTC
LS2	27/5/2012	Kalba-AlSour	TFTC	BDL	TFTC
LS3			TFTC	BDL	TFTC
Average			-	-	-
US1			TFTC	BDL	70
US2	27/5/2012	Kalba-AlSour	TFTC	BDL	TFTC
US3			TFTC	BDL	TFTC
Average			-	-	23.3
LE1			TFTC	BDL	TFTC
LE2	27/5/2012	Sharjah- Al Kadesia	TNTC	BDL	TNTC
LE3			TNTC	BDL	TNTC
Average			-	-	-
UE1			TFTC	BDL	TNTC
UE2	27/5/2012	Sharjah- Al Kadesia	TNTC	BDL	TNTC
UE3			TNTC	BDL	TNTC
Average			-	-	-

*TFTC: Too Few To Count
☐BDL: Below Detection Limit
†TNTC: Too Numerous To Count

Total Coliforms.The results showed that 5/11 houses (45.4%) met the total coliform standard(0 CFU/100ml). These samples (L/U: A, T, N, S, and K) were collected from Al Musalla (Kalba), TawiAlSaman (Al Dhaid), Am Knorr (Sharjah), Al Sour (Kalba), and Al Kadesia (Sharjah), respectively. Twohouses(18.1%; L: D and Y) collected from Dibba Al Hosn (Sharjah) and Al Yarmouk (Sharjah), respectively, exceededthe standard. Both of these samples were taken from the lower tanks. The corresponding samples taken from the upper tanks did not exceed the standard counts. By contrast, threehouses (27.2%; U: B, K, and Z) collected from the upper tanks in Al Baraha (Kalba), Al Khuwair (Kalba), and Zubarah (KhawrFakkan) exceeded the standard count. Samples taken from the corresponding lower tanks did not exceed the standard counts. One house (9%; L/U: H) out of the 11 collected from Hutteen (Kalba) exceeded the standard count in both tanks.

Total Bacterial count.The results showed that eight out of 11 analyzed samples exceeded the TBC standard count (10 CFU/100ml). Three houses (27.3%; U: A, K, S) showed counts above the TBC standard in the upper level tank only. These samples were collected from Kalba- Al Musalla, Kalba- Al khuwair and Kalba- Al Sour, respectively. On the other hand, five other houses (45.4%; U/L: H, Z, N, D,

and Y) exceeded the TBC standard in both tank levels. These samples were collected from Kalba- Hutten, KhawrFakkan- Zubarah, Sharjah- Am Knorr, Sharjah- Dibba AlHosn, and Sharjah- Al Yarmouk, respectively.

E.coli counts.None of the samples were positive for *E.coli*,suggesting the absence offecal contamination.

More than half of the samples (54.5%) collected from storage tanks fitted to homes in Sharjah emiratewere positive for coliforms;however, none contained*E.coli*, although*E.coli*can reproduce outside its natural environment (the intestine) [7]. Because both the global and local standard counts for coliforms and *E.coli*are 0/100ml,their presencemeant that the water was unsafe for domestic use. This represents a clearrisk to public health, particularly in cases where people do not boil, filter or disinfect the water. It is important to determine the TBC to assess the safety of domestic drinking water. Furthermore, the absence of *E.coli*does not mean that the water is notcontaminated byviruses or protozoa, which are more resistant to disinfection procedures[10].

The TBC is a national water quality standard and must not exceed 10 CFU/100ml. Eight (72.7%) of the water samples tested in this study exceeded this limit (U and/or L: A, K, H, Z, N, D, Y and S). Exceeding the standard is not an indicator of potential pathogenesis, as the count could include normal flora that do not pose a risk to human health. Therefore, it is necessary to identifythe species of bacteria present in water samples to make an accurate risk assessment.

There was a marked difference in theTBCbetween different areas within the same emirate, and between different tankswithin the same house.Bacterial growth is affected by a number of environmental factors, including temperature, pH, salinity, and the availability of nutrients. Thus, bacterial counts may be higher in summer than in the winter.

Several water samples showed TFTC levels below the TBCstandard (10 CFU/100ml). The plate culture can underestimate the number of bacteria in the original sample because the harsh treatment involved may injure the cells, rendering them unable to grow on the plate. On the other hand, if non-coliformmicroorganisms are present in very high numbers, they may inhibit the growth ofcoliforms.

In general, tanks located at roof level showed higher bacterial counts (U: A, K, H, N, D and S). This is in agreement with the results of a similar study by Alkendi and Omer (2011), whichexaminedwater storage tanks in Abu Dhabi and Al Ain [1]. The higher level of contamination in the upper tanks may be explained by thefact that water is drawn from the bottom of the ground level tank and pumped into the upper tank, and therefore the water in the bottom of the lower tank may contain more bacteria, which settle to the bottom along with dust particles.

5. Conclusion

The MFmethod used in the current study to isolate and quantify bacteria has several advantages. The main advantage is that it is fast, yielding results within 24 hours of sampling. The rapid and accurate monitoring of microbes in drinkingwater is essential if we are to safeguard the consumer and improve water treatment and distribution systems.

For fast detection of water bacteria, the IDEXX technique is a breakthrough technology that delivers a fast, clear, visual color change that makes bacterial detection simple, without the need for culturing or colony counting, which can be subjective.

The pathogenicity of isolated bacteria can be assessed by gram staining of isolated bacteria.

Testing could also be improved by taking samples from different water columns within the same tank, which wouldprovide information about how the bacteria are distributed.

In most cases the quality of water deteriorates after it enters the storage tanks. The water in tanks is supposed to be safe to drink; however, we found that many tanks were contaminated with Coliform bacteria. Therefore, we recommend that the government and other relevant organizations in the UAE work to establish a comprehensive system to provide safe drinkingwater,

which should include a robust household water supply, and provide related educational programs regarding hygiene.

In this case, water used for domestic purposes should be boiled, filtered, or disinfectedprior to use.Regular cleaning of tanks, along with appropriate protection, sealing, and maintenance, are highlyrecommended to ensure water quality. We recommend that families use appropriate filters to remove microorganismsfrom drinking water.Finally, a campaign should be instigated to raise public awareness and educate them about ways to protect and manage the quality of their water resources.

References

[1] Alkendi and Omer. (2011).Bacterial Quantification In Homes' Water Tanks. United Arab Emirates University, Faculty of Science, Biology department.

[2] Blacksmith Institute. (2012).*Contamination surface water*. New York, NY 10035 USA

[3] Barrell, Hunter, G Nichols.(2000). Microbiologicalstandards for water and their relationship to health risk.

[4] Bradley Scott.(2009). Total Coliform Bacteria.

[5] Coliform Bacteria in Drinking Water Supplies. (2011). New York State Department of Health, Retrieved 18 January 2012, Retrieved from http://www.health.ny.gov/

[6] Canada, Public Health Agency. (2011). *Methodological Options of Source Attribution*, Retrieved from http://www.phac-aspc.gc.ca

[7] Connecticut Department of Public Health.(2010).*Presence of Total Coliform or FecalColiform/ E. coli Bacteria in the Water Supply.*

[8] Dubai Electricity and Water Authority. (2012).*Electricity and Water.*Dubai Electricity and Water Authority,UAE.

[9] Guidelines for Water Reuse.(2004).*U.S. Environmental Protection Agency.* Washington EPA/625/R-04/108: Retrieved fromwww.epa.gov/

[10] Geneva.(2008).*Guidelines for drinking-water quality,* third edition ,volume 1 recommendation.

[11] Kosek M, Bern C, Guerrant R. (2000). *The global burden of diarrhoeal disease Bull.* World Health Organ. 2003;81(3):197–204.

[12] Moushumi Das chaudhury.(2005).*UAE water consumption one of the highest in the world.*.Khaleej Times

[13] New Hampshire department of environmental service.(2010).*Interpreting the Presence of Coliform Bacteria in Drinking Water.*WDDWGB41.

[14] Oram B.(2012).*Bacteria, Protozoans, Viruses and Nuisance Bacteria.* Retrieved from Water Research Center website: http://www.water-research.net

[15] Prüss-Üstün A, Bos R, Gore F, Bartram J.(2008). *Safer Water, Better Health, Costs,benefitsandsustainability of interventions toprotect and promote health. Geneva,* World drinking-water.

[16] Pruss A, Kay D, Fewtrell L, Bartram J.(2002). *Estimating the burden of disease from water, sanitation, and hygiene at a global level. Environ.* Health Perspect, 110(5):537–542.

[17] Sharjah Electricity and Water Authority:SEWA.(2008),*Water Services*.http://www.sewa.gov.ae/english/services/water.asp

[18] The Water Quality Regulations. (2009). The Regulation and Supervision Bureau for the Water (Ed.). Abu Dhabi, U.A.E.

[19] United Nations.(2005). *Millennium Development Project Report*

[20] WHO (2010).*Optimizing regulatory frameworks for safe and clean drinking water,* World Health Organization.

[21] Wait, I. (2008). *Changing Perceptions: Water Quality and Demand In The United Arab Emirates*. Paper presented at the 13th IWRA World Water Congress.

6

Water sector in Morocco: situation and perspectives

Mohamed Alaoui

Ministry Of Energy, Mines, Water and Environment, Department Of Water; Morocco

Email address:

m.alaoui@water.gov.ma, alaoui.water@gmail.com

Abstract: Due to its geographic location, Morocco is characterized by a strongly contrasted climate, with its rainfall being highly irregular in space and time. The natural renewable water resources is estimated 22.8 billion m³/yearBCM of with the contribution of18 BCM from surface water and 4.8 BCM of groundwater recharge resulting in 730 m³/capita/year which is below the commonly accepted threshold of 1000 m³/capita /year .Despite the success of the past water policies initiated in the sixties, the future water sector management challenge of the water availability and demand must address the most urgent problems and make water a decisive factor in its sustainable development. In this challenging environment, Morocco has succeeded in ensuring the needs in domestic and industrial water and the development of large-scale irrigated agriculture. Despite these achievements, the factors that determine water availability and water needs have changed so much in recent decades that the country must adapt radically in order to prevent a critical situation. Many solutions exist. Their implementation within an innovative and integrated approach to the whole water sector should allow Morocco to address the most urgent problems and make water a decisive factor in its sustainable development. The new water strategy, implemented in 2009, is expected to support the development of the water needs for the development until 2030 through the implementation of integrated policy combining water conservation and resource mobilization in conventional and unconventional water while respecting the environment and the rights of future generations. The strategy will focus on three components; water demand management and water valuation, supply development and management and preservation and protection of the water sources and the environment. Demand management measures is expected to save 2.4 BCM with contribution of 120 MCM to domestic, construction of 50 dams with additional capacity of1.7 BCM, water transfer of 800 MCM, 400 MCM of desalinated water, 300 MCM of wastewater reuse and increased recharge wastewater. Also implementation of all provisions of Law 10-95 on water, preservation of wetland, drought and flood risk management and enhance information system. These measures are expected to achieve sources sustainability and meeting water requirement.

Keywords: Water, Management, Planning, Morocco

1. Introduction

Water availability has been not an obstacle to socio-economic development of Morocco in the past as result of implemented management measures but can pose major challenges for the future demand if nothing is done. Several major challenges and problems are now faced by the water sector relate mainly to the increasing water scarcity of pressured economic development and its impact, exacerbated by climate change, deterioration of water quality due to pollution and, the unsustainable exploitation of groundwater resources, and the waste of consumption trends.

Soil erosion is impacting different regions as, 75% of the 23 million hectares in mountainous areas are affected by erosion, a third of which are critically affected. The erosion impacts have contributed to the silting of dams, resulting in a loss of storage capacity of about 75 Million Cubic Meter (MCM)/year. The dam reservoirs are designed with dead volumes for storing silt during the life of the works and flushes are also carried out during floods to evacuate part of the silt through the bottom outlets. However, despite these measures, the problem is more acute because of the particular emphasis of erosion by increased demand on soil and vegetation cover

The impact of climate change on water resources has been felt over the past three decades through the intensification of extreme weather events, droughts and floods, and is presented as a major challenge that is being felt acutely now and will continue to be felt in the future. Indeed, the kingdom has faced over the past three decades 20 years with 3 dry periods each lasting four consecutive years. This means that the climate situation was exceptionally severe in recent decades and particularly challenging. But Morocco has managed to overcome this critical situation. According to meteorologist's forecasts, Morocco, like the Mediterranean countries, will observe high intensity rainfall generating violent floods threatening lives and infrastructure. In addition to natural hazards, a number of factors exacerbate the vulnerability of flood-prone areas, especially the uncontrolled development of land and urbanization in flood areas

Drought has become a structural phenomenon in Morocco and of the last several sequences of drought, the most severe were recorded for the periods 1944-1945, 1981-1985, 1991-1995 and 1998-2001 and 2006-2007. During these years of drought, the rainfall situation was characterized by a generalized deficit that affected all the country. The deficit reached 50 to 60% in some areas. On the hydrological level, deficits reach more than 70% in some areas. During these periods, significant declines of flow streams were observed, usually during the months of July and August.

To support the development of the country water requirements, Morocco has been committed to controlling the mobilization of water resources in order to ensure water needs without major difficulties. The strategy adopted in the early 1960s enabled the country to provide significant water infrastructure ensuring to meet the drinking water supply for population, industry and the development of large-scale irrigated agriculture. Considerable efforts have been made also in the protection against flooding of vast territories that suffered significant damages in the past. The expertise developed in recent decades has enabled the country to overcome very critical situations of water shortages and manage sources and demands.

2. Morocco Water Resources

Morocco, one of the north Arab countries is characterized by irregular rainfall distribution with 51% of its amount fall on in less than 7% of the national territory, and in time, with alternating sequences of years of strong runoff and years of severe drought.

According to the latest estimates, the natural renewable water resources are nearly 22 billion m³/year(BCM), the equivalent of about 730 m³/capita/year which is below the commonly accepted threshold of 1000 m³/capita/year, indicating the occurrence of shortages and impending crises. Rainfall volume is estimated 140 BCM contributing to 22 BCM of renewable resources of 18 BCM of surface water and 4.8 BCM of groundwater recharge as shown in figure 1 [2].

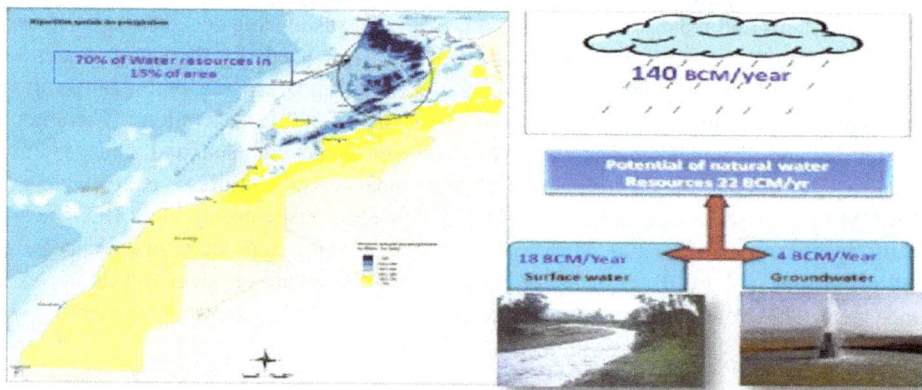

Fig. 1 water potential of Morocco

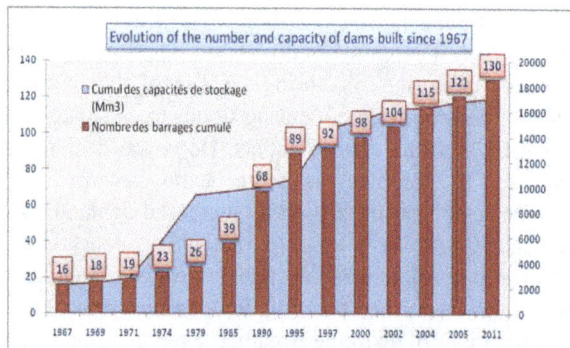

a: Evolution of the number and storage capacity of dams

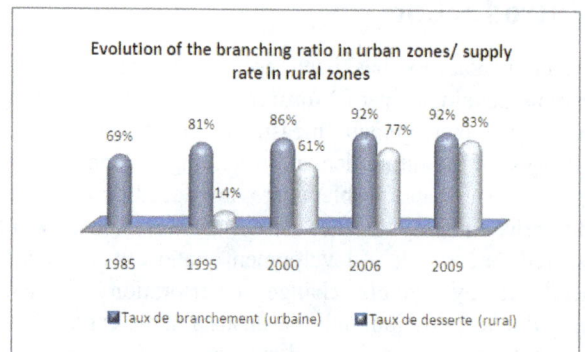

b: Access to drinking water

Fig2: Surface water mobilization (a) and drinking water supply (b)

The water sector in Morocco has been a central concern of economic policy because of its key role in the water security and support of the country's development, particularly irrigated agriculture.

In this context, Morocco has undertaken since the 60s a dynamic policy to give the country a major water infrastructure, improving access to drinking water, meeting the needs of industry and tourism and development of the large-scale irrigation.

Figure 2 shows the achievements in mobilizing surface water [1]and generalization of access to potable water[3].

The current irrigated area is around 1.5 million hectares of which two thirds are equipped by the Government. The distribution by type of schemes and irrigation systems is given below [5]:

Table 1: Distribution of irrigation methods

Type	Gravity irrigation Area (ha)	Spray irrigation Area (ha)	Drip irrigation Area (ha)	Total Area (ha)	Pourcentage
Large Hydraulic	533 900	113 800	34 900	682 600	47%
Small and Medium Hydraulic	327 200	6 900	-	334 100	23%
Private Irrigation	317 600	16 950	106 900	441 450	30%
Total	1 178 700	137 650	141 800	1 458 150	100%

The hydropower energy production is shown in figure 3[4].

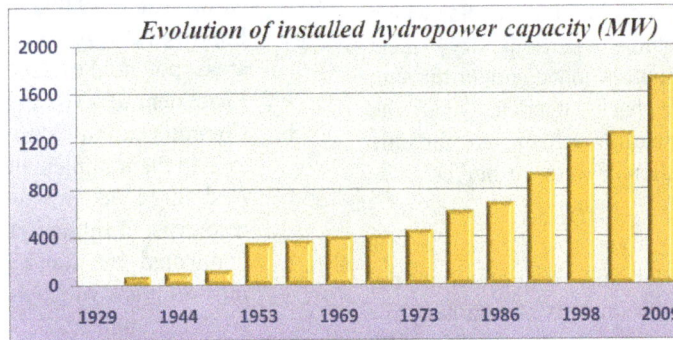

Fig. 3: Evolution of installed hydropower capacity (MW)

Significant progress has also been made in the regulatory and institutional fields through the enactment of the Law 10-95 on water with the creation of river basin agencies with the designated functions of planning, management and preservation of water resources at the level of river basins

In sum, the baseline scenario of the water balance predicts in 2030 a deficit of about 5 billion m3 that will need to be filled by the implementation of action plans in an integrated strategy taking into account the endogenous and exogenous factors affecting water resources in Morocco. Indeed, as shown in figure 3 below,the current water demand, estimated at 13.7 BCM is met from renewable water mobilized (11.7 BCM) and but also by overexploitation of groundwater (2 BCM). Water demand in 2030 would be 16.2 BCM, and if no measures aretaken, the deficit would be 5 BCM[5].

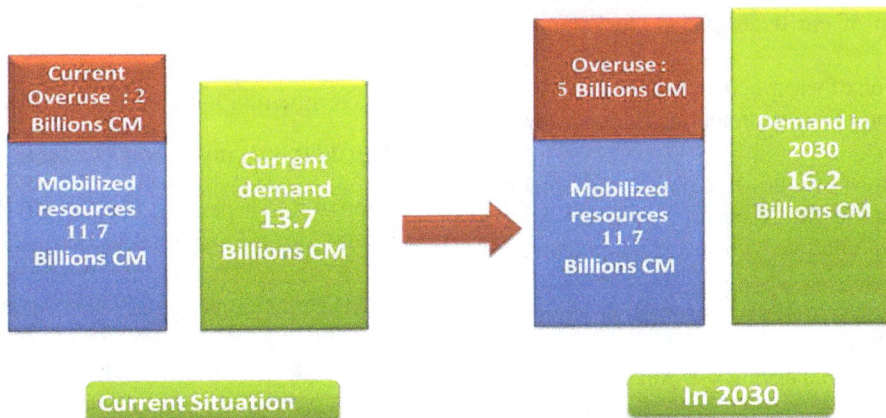

Figure3: Water balance on the horizon 2030_ Baseline scenario

3. Short and Long Term Comprehensive Management Initiatives

3.1. Policy focus

Morocco has been taken key initiatives recognized at the international level to manage its water resources through appropriate efficient models despite a difficult environment. Action taken consisted of:

- Policy of control and mobilization of water resources through the construction of big reservoirs dams and works for water transfer;
- The development of technical skills and applied scientific research;
- A policy of long-term planning started in the early 1980s that allows decision makers to anticipate water shortages by giving government a long-term visibility (20 to 30 years);
- Achieving significant progress in the regulatory and institutional domain, namely the Law 10-95, which consolidated integrated, participatory and decentralized water resources management through the creation of river basin agencies and the introduction of financial mechanisms for the protection and preservation of water resources.

3.2. Water Strategy

To consolidate past successes and to succeed in the aforementioned challenges, a new impetus of a strategy for the reinforcement of the water policy was launched in 2009.

The strategy was elaborated based on three levers, namely:

a.   Much more ambitious objectives to consistently meet the country's water needs, but also to continually and sustainably protect against the effects of climate change

b.   A dramatic shift in the behavior of resource use and management through a coordinated supply and demand management, covering:

- The perpetuation of the protection measures and reconstitution of our underground stocks and lake areas
- Rationalization of water demand
- Generalization of the treatment and the wastewater reuse in cities
- An innovating portfolio of mobilization solutions and of access to the resource, combining all the relevant local solutions with better interconnection between regions
- Proactive protection measures (of the environment, and against floods).

c.   A real long-term management of water through:

- Regularly updated, readily available data (at the national level), of the needs and availabilities in the long run

- Political commitment, and effort from all stakeholders, supported by adequate regulatory framework and governorship
- More ambitious public and private funding.

4. Strategy Analysis

The main lines of the strategy are structured around the following axes[5]:

4.1. Water Demand Management and Water Valuation

In this context of scarcity, ensuring strong water demand management and optimizing water valorization is both urgent and of prime importance; this can be done through technical, legal and financial instruments.

a.   In the agricultural domain for instance, the average water savings potential in irrigation amounts to 2.4 BCM/year

- Conversion to modern irrigation: a potential of 2 BCM/year, if 40 000 ha per year are upgraded ;
- Improvement of the adductions to the irrigated areas: potential of 400 MCM/year ;
- Enactment of a volume-based pricing;
- Communication campaign and training for farmers to help them implement water-saving techniques
- And for a better valorization of mobilized water resources, bridging the gap between the area dominated by dams and the attached associated infrastructure which is yet to be built ; 108000 ha are programmed

b.   For drinking, industrial and touristic water, a potential of a 120 MCM/year savings is possible by:

- Improving the distribution networks, with 80% as a nationwide target;
- Developing norms and incentives to use water efficient devices: pipes, water-closets, etc.
- Reform of the water pricing system: with a pricing that enables a more rational use of drinking water and improved cost recovery;
- Improvement of water use efficiency in industry and tourism units and a push towards larger-scale water reuse;
- Water-efficient standards for equipment and construction for both residential and industrial sites

4.2. Supply Development and Management

Although it is necessary, managing water demand is not the panacea for the quantitative and qualitative challenges bound to water use and one should not exaggerate its virtues. It certainly can defer some investments, but don't allow removing them. Managing demand and water mobilization are a pair of complementary solutions that will have to be optimized. The fundamental question that arises is where to place the cursor for this pair of levers while balancing the need to manage "real" water demand generated by the support of socio-economic development needs. To cope with the increased demand for water, two portfolios of solutions exist: water resources managers must first act on the water demand for water savings and then increase the supply by mobilization of additional water resources to fill the gap. Water-saving measures, although they are inexpensive as mobilization technique, are not sufficient to fill in the gap. It is necessary to ensure the adequacy of water resources and needs by mobilizing additional resources in conventional and unconventional water, although they are more expensive than they were in the past. The challenge is how to manage both solutions together.

a. Morocco has made great efforts in mobilizing water resources, and these efforts will be continued under the new water strategy by the mobilization of new water resources on a large scale through:

✦ The construction of 50 dams by 2030: an additional 1.7 BCM will thus be mobilized.

✦ A North-South transfer aimed at sustaining the socioeconomic development of the Bouregreg, Oum ErRbia and Tensift basins: 1st stage of 400 MCM/year from the oued Sebou, 2nd stage of 400 MCM/year from the Loukkos-Laou ;

b. Small scale water mobilization is also programmed through:

✦ The continuation of the small and mid-size dams program: objective to build 1000 small and mid-sized dams by 2030;

✦ Rainwater capture: pilot projects before a possible larger-scale deployment as has been done in India and Australia;

✦ Extension of the artificial cloud insemination where possible.

The reinforcement of the maintenance of existing infrastructure and the system interconnection will allow for the diversification of water sources and therefore allow for a better security of water supply and important synergies and gains in efficiency.

c. In rural zones, the generalization of access to drinking water will keep to its current pace through the upgrade of existing community systems to ensure that they can continue their operations in the long run and the development of individual systems for the isolated population.

d. Furthermore, the mobilization of unconventional resources is inevitable through:

✦ Desalinization of seawater and demineralization of brackish waters: objective of equipping a production potential of nearly 400 Mm³ of potable water per year.

✦ Reuse of treated wastewater: 300 Mm³/year of treated wastewater for reuse in the watering of golf courses and green space, in addition to reuse in crop irrigation.

✦ For the implementation of these action plans, the new strategy of the water sector in Morocco has advocated contracting and the development of public private partnership, especially for the implementation of major transfer projects, desalination, and the program of small dams.

4.3. The Preservation and Protection of Water Resources, the Environment Sensitive Areas

a. Preservation and Recovery of Groundwater:

Protection of groundwater resources is essential from a strategic point of view. During the long drought that hit the country during the 1980s, groundwater use was the only recourse that allowed farmers to maintain the viability of their farms and thus avoid abandoned fields and the acceleration of the rural exodus. However, these withdrawals reached unsustainable levels, exceeding the volumes of annual recharge and drawing on the stock of non-renewable water.

Concerning preservation of groundwater, the proposed medium and long term strategy is based on the development of sustainable management through:

• Strengthening the control system and penalties in the case of overuse;

• Limiting groundwater pumping (revision cost of water prices, elimination of incentivizing subsidies for overuse, rules prohibiting and restricting pumping, efficient technologies…)

• Strengthening the responsibility of the ABH's (Agences de BassinHydrographique) in groundwater management and generalization of groundwater contracts

• Automatic appeal with the water resources of conventional and nonconventional substitution, preferably with the underground resources in order to ease the pressure on the underground resources;

• Programs to artificially regenerate groundwater (storage of 180 Mm³ per year);

• Re-injection of treated waste water in the coastal groundwater used for irrigation (100 Mm³ by 2030);

• Substitution of the groundwater taken by the ONEP (Office National de l'Eau Potable which is the National Agency for drinking water) to surface water (85 Mm³ per year by 2020).

• Protecting the quality of water resources is a strategic objective of the National Water Strategy. This action is based on a deep knowledge of the quality of water resources and sources of pollution and the proposed program for the prevention of and fight against pollution. Specifically, it consists of:

• Accelerating the pace at which the national plan for sanitation and treatment of wastewaters is adopted: Objective of a sanitation access rate of 90% in 2030;

• National Sanitation Plan for Rural Areas: Objective of a sanitation access rate of 90% in 2030;

• National Plan for the Prevention and Fight Against Industrial Pollution;

• Implementation of the National Plan for the Management of Household and Similar Waste.

b. Safeguarding Watershed, oasis, and Wetlands

Morocco has numerous areas of inestimable ecological value including wetlands, natural lakes, oases and coastal areas. These sensitive areas are threatened by the pressures exerted by various economic activities on natural resources, threatening both their balance and sustainability. To this end, it is essential to pursue a strategy to safeguard these sensitive areas through:

- Protecting watersheds against erosion by managing them upstream of dams;
- Plan for the protection of water sources;
- Plan for the protection of wetlands and natural lakes;
- Preservation of oases and the fight against desertification;
- Coastal protection;
- Limitation and control of pumping in groundwater resources directly affecting the natural lakes;
- Improving the sourcing of lakes by diverting water courses and developing thresholds and small dams upstream.

5. Reduction of the Vulnerability to Natural Risks Related to Water and Adaptation to Climate Change

Given its geographic situation coupled with the effects of climate change, Morocco is confronted with natural risks linked to extreme natural phenomena: flooding and droughts.

i. Improving the protection of persons and property against flooding:
a. Completing the actions set out in the National Plan for Protection against Flooding: target of 20 protected sites per year;
b. Integrating flood risk in planning land use, urban planning and planning of watershed management;
c. Improving knowledge in the field of weather forecasting and urban hydrology;
d. Developing flood warning and emergency plans.
e. Developing financial mechanisms (insurance and disaster funds).
ii. Fight against the effects of drought: management plans by river basin:

Insofar as drought protection is concerned, a concerted national strategy should be initiated by the drought management plans at the level of all river basins, aimed at:

a. Characterization of droughts: identification and proposal of monitoring indicators.
b. Implementation of structural measures: diversification of sources of water supply.
c. Development of contingency plans.
d. Development of financial mechanisms such as insurance and funds for natural disasters.

5.1. Further Regulatory and Institutional Reforms

Significant progress has been made through the implementation of the Water Law 10-95, with the introduction of the principles of user-pays and polluter-pays and the establishment of Hydraulic Basin Agencies. However, these advances need to be supplemented by the completion of the legal framework necessary for the implementation of all provisions of Law 10-95 on water.

5.2. Modernization of Information Systems and Capacity Building and Skills

The objective is to provide the necessary support for the implementation of the actions of the new strategy. The Administration should follow parallel development of the water sector in human and material resources level, and the modernization of the administration and development of information systems, including in particular the implementation of an Information System on Water for professionals and the general public. The action plan also provides for the modernization of networks measures, strengthening applied research and skills development.

6. Highlights of the Diagnosis of Moroccan Water Sector

The analysis of the water sector in Morocco reveals the following structural observations:

- Morocco's climate varies by region; rainfall is characterized by a high variability in time and space. This situation has been exacerbated by climate change that aggravated extreme events, resulting in dry periods increasingly long, intersected by brief and violent rainstorms that can generate catastrophic floods if they are not controlled;
- in this very challenging context, and to support the development of the country, Morocco has long been involved in the process of controlling its water resources through the implementation of major water infrastructure, which has enabled it to ensure its water needs without major difficulties, even in situations of exceptionally severe water shortages. Indeed, the Kingdom has faced these past thirty years 20 years with 3 dry periods that each lasted four consecutive years;
- Morocco, where water and its use are part of the deepest roots of an ancient civilization, is now a great hydraulic nation thanks to these achievements. However, this apparent success has left some effects, moreover repairable as long as the Government makes so much will and determination, as in the past, to surpass them;
- If nothing is done to adapt to this new situation, Morocco may soon find itself in a critical situation in terms of water deficit, overexploitation of groundwater aquifers and environmental degradation.

To avoid compromising its achievements and make water a decisive factor in its sustainable development, Morocco has implemented in 2009 a renewed strategy based on three levers:

+ more ambitious goals;
+ a radical change in behavior vis-a-vis the use and management of water resources;
+ the implementation of a real long-term management of water.

The new water strategy in Morocco is based on an integrated and participatory approach. It should enable the country to support its long-term development, meeting the needs of growth and protecting the country against the unpredictable effects of global warming.

7. Conclusion

The new strategy of the water sector is expected to support Morocco's development economic growth to large projects and protecting the country against unpredictable effects of global warming.

Actions on water demand and the development of supply, including the Green Morocco Plan,is expected to ensure a sustainable availability of water to meet the future demand. It is planned by 2030 to save 2.5 billion m³/year through the management of water demand and generate an additional water resource of 2.5 BCM /year through water mobilization.

Mobilization projects will be benefic to basins of Sebou, Bouregreg, OumErRbia and Tensift that have concentration of the industrial, agricultural and tourism activities in the country. In particular, the inter-basin transfer of water will be a key element for economic and social development to support the development of the region of Marrakech and will significantly increase agricultural added value in Doukkala and the Chaouïa regions.

The planned management measures will be implemented by the combination of water supply and demand management oriented towards reducing water deficits and the availability of sufficient resources while ensuring the preservation of the environment.

References

[1] Direction des Aménagements Hydrauliques (DAH), Ministère de l'Energie, des Mines, de l'Eau et de l'Environnement- Maroc, Grands barrages du Royaume, 2012.

[2] Direction de la Recherche et de la Planification de l'Eau (DRPE), Secrétariat d'Etat chargé de l'Eau et de l'Environnement (SEEE)-Maroc, Evaluation des ressources en eau mobilisables, 2008.

[3] Office National de l'Electricité (ONE)- Maroc, Production de l'Energie Hydro-électrique au Maroc, 2010.

[4] Office National de l'Eau Potable (ONEP)- Maroc Bilan annuel de l'année 2010.

[5] Secrétariat d'Etat chargé de l'Eau et de l'Environnement (SEEE)- Maroc, Etude de la Stratégie Nationale du Secteur de l'Eau, 2008.

7

Using treated wastewater as a potential solution of water scarcity and mitigation measure of climate change in Gaza strip

Eng. Jamal Y. Al-Dadah

M.Sc. in Agriculture & Environmental Science, Head of Planning Department, Palestinian water Authority, Gaza Strip

Email address:

jaldadah@hotmail.com

Abstract: The use of wastewater is one of the most sustainable alternatives to cope with water shortage in Gaza Strip (GS). It would have a number of advantages that include closing the gap between supply and demand, alleviating the pollution of fresh water resources, providing sound solution to water scarcity and potentially cover half of the total agricultural water demand in GS. Wastewater reuse could provide a mitigation solution to climate change through the reduction in green house gases by using less energy for wastewater management compared to that for importing water, pumping deep groundwater, seawater desalination, or exporting wastewater, and enrich the deteriorated soils in GS with more organic matter which lowering the application of chemical fertilizers. This paper investigated the effects of wastewater application on the level of organic matter and soil carbon sequestration which demonstrated by many experiments in Gaza Strip, which induced the possibility of wastewater as a mitigation measure of climate change.

Keywords: Wastewater, Water Scarcity, Mitigation, Gaza Strip, Water Shortage, Integrated Water Management Plan, Climate Change, Wastewater Reuse

1. Introduction

The combination of severe water shortage, contamination of water resources, densely populated area and highly intensive irrigated agriculture characterizedGaza Stripwhich described as one of the most exploited places in the world where the level of demand on water and land resources exceed the capacity of the environment.The water balance records revealed a water deficit of 80 MCM in 2012 (Palestinian Water Authority, 2012).Climate change is expected to aggravate the situation even more. Wastewater effluent is the most readily available and cheapest source of additional water and provides a partial solution to the water scarcity problem. The agriculture sector is the second major consumer of groundwater in the Gaza Strip, where the level of groundwater, the main water resource, is being depleted and its quality is adversely affected. Irrigated agriculture plays a noticeable role in the sustainability of crop production to feed the rapid increasing population in the Gaza Strip.The total abstraction of ground water as estimated is proximately 181 MCM/y, from which 95 MCM/y for domestic use (90

MCM/y from Gaza water wells and about 5 MCM/y from Mekorot, Israeli Water Company), (CSO-G, 2011), while the total water supplied for agriculture use was about 86 MCM/y. This over extraction from the aquifer has resulted in drawdown of the groundwater with resulting intrusion of seawater and up-coning the underlying saline water. As a result of all current and expected problems, there is an urgent need to adopt solutions to achieve conservation of water quantity, improve water quality, and achieve sustainability. Selected solutions may be one or more of the following: i) Water use conservation, ii) Desalination of sea water, iii) Storm water collection, iv) treated waste water (TWW) for agricultural uses. Palestinian Water Authority (PWA) reported a range of conclusions as to the required future interventions in the water sector. The first conclusion pertains to the existing situation in relation to water supply in Gaza, with strong rejection the continuation of the "status quo" as an acceptable option. This reflects the fact that the groundwater which is the only source of fresh water in Gaza at the present time is being massively over-pumped and the aquifer is showing clear signs of imminent failure or collapse, with rapidly advancing degradation of the water resources in terms of quality and quantity as shown in Figure 1. Severe contamination mainly from disposing raw or partially wastewater to the sea or adjacent water coursesis also evident, and almost none of the groundwater meets internationally accepted guidelines for use as a domestic supply. The population of approximately 1.7 million Palestinians in Gaza, (PWA, 2012) is therefore exposed to very high levels of risk and high levels of waterborne disease continue to be prevalent amongst the Gaza population.

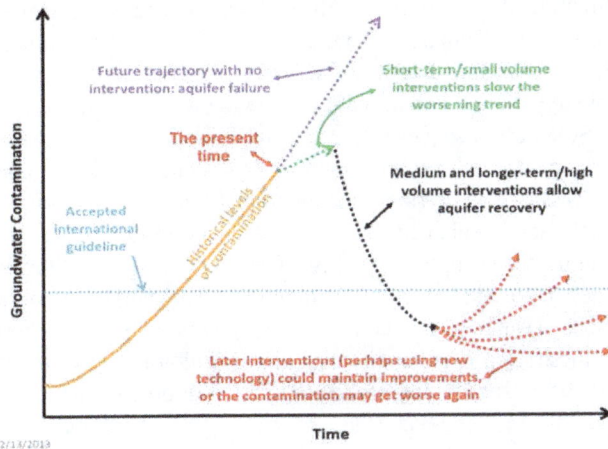

Figure 1. *Problems relating to the groundwater in Gaza Strip*

plants. Wastewater reuse schemes are indispensable option for Palestine in general and Gaza in particular. The Integrated Aquifer Management Program (IAMP) aims to reduce the agricultural pumping from 90 MCM per year to about 50-60 MCM per year. The difference between these two volumes could be supplied by reclaimed water directly from the three regional treatment plants and a small quantity may be extracted from recovery wells in the vicinity of proposed infiltration basins as illustrated in Figure 2.

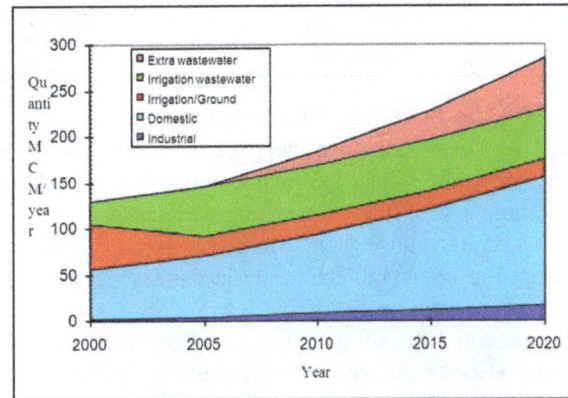

Figure 2. *Planned Utilization of Wastewater in Gaza Strip*

2. Integrated Water Management Plan in GS

Wastewater is becoming a common source for additional water in some water scarce regions and many countries have included wastewater re-use in their water planning. Water demand in the GS is increasing continuously due to economic development and population increase resulting from natural growth and returnees, while the water resources are constant or even decreasing due to urban development. The use of wastewater as a supplemental source of irrigation is inevitable for increased agricultural production in Gaza where irrigation supplies are insufficient to meet crop water needs. Moreover, irrigation with treated waste water is considered a promising practice that helps in minimizing the pollution of the ecosystem subjected to contamination by direct disposal of wastewater (WW)into surface or ground water. In addition, ww is a valuable source for plant nutrients and organic matter needed for maintaining fertility and productivity of arid soils. However, reuse of waste water for irrigation may potentially create environmental problems if not properly treated and managed. It is stated in the National Water Plan (NWP) that wastewater investment costs represents about 37% of the overall Palestinian investment plan for Gaza Strip overall the period time to achieve the strategies and targets outlined in the National Water Plan. The national investment target for wastewater production was set for from a careful consideration of the extent of the sewerage works, and the construction of the wastewater treatment

3. Potential and Perspective of Wastewater Reuse in Gaza Strip

Wastewater generation is around 80% of water use for domestic purposes. The quantities of treated wastewater are expected to reach 180 MCM per year by 2025. This quantity is a good potential to minimize the shortage in water sources in the area. The quality of raw wastewater is considered acceptable for reuse after treatment according to the design criteria of the WWTPs. Effluent is available throughout the year at a fairly constant flow whilst water demand from crops will vary. Demand in winter will be small as the rainfall is marginally sufficient to meet the water demands of all crops. The complexity of attempting to supply effluent in winter is great as demand would fluctuate widely according to the rainfall. The maximum area that can be irrigated by treated wastewater depends on many factors like the distribution of crop patterns, land tenure and on the peak demand of the mix of crops and the flow available at this time. It can be seen that peak monthly water demand for crops occurs in April to extended to September in some cases . Accordingly, the maximum quantity of treated effluent may be used in Gaza Strip –in case of accounting for all the fruits areas (Citrus, Olives and other Fruit Trees)and converting the rain-fed areas to irrigated areas partially , maximally, the quantity of effluent will be used is about35- 40 MCM/year. However, the total needs of the entiretreated effluent is subject to more augmentation in case of climate change impacts and additions of leaching fractions due to the high water salinity. The areas of land that could be irrigated to satisfy peak demand with the effluent flow available during these periods are demonstrated in the table (1).

Table 1. Distribution and Potential Areas of WW reuse in GS. Crops Total area, Dunam Water Quota/dunam Total Water Demand Potential of WW Reuse MCM/Area

Citrus	15,000	1000	15	Yes
Olive	25,000	400	10	Yes
Fruits	28600	500	14.3	Partially
Field crops	40470	100-400	5	Flag
Vegetables	61000	700	42	Excluded
Total	170,448			35- 40 MCM

In the Gaza Strip, pilot wastewater reuse schemes have existed for some years, and there are plans for these to be augmented shortly. The key requirement, however, is for the completion of the four major wastewater treatment plants scattered throughout Gaza, as reuse cannot be introduced at any significant scale in the absence of high-quality wastewater treatment.

The amount of collected, treated wastewater, and that will be reused in the future is a matter of great concernin Gaza Strip and tobe expected to increase substantially with population growth, rapid urbanization, and improvement of sanitation service coverage. But, on the other hand, wastewater utilization should be managed within certain restrictions imposed for environmental protection and to safeguard public health. Availability and utilization of treated wastewater constitutes one of the most factors considered regarding the expansion in irrigated agriculture and the associated agricultural water demand. However, is for the completion of the four major wastewater treatment plants throughout Gaza, as reuse cannot be introduced at any significant scale in the absence of high-quality wastewater treatment as shown in Figure 3.

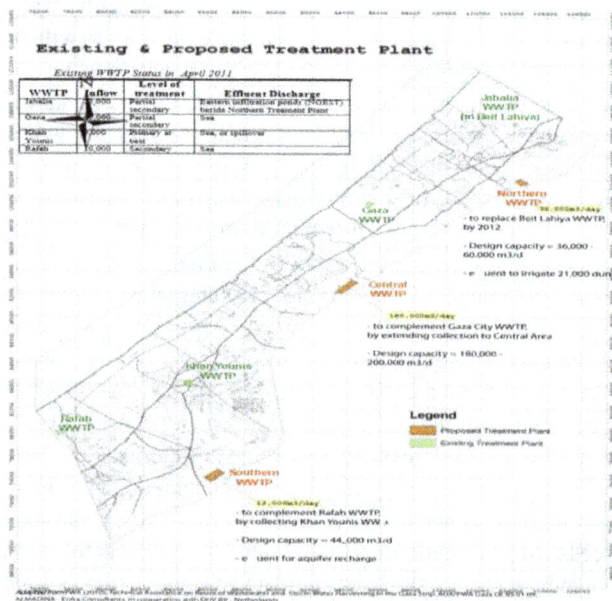

Figure 3. Planned Wastewater Treatment Plants and reuse schemes in GS

4. Interaction between Wastewater Reuse and Climate Change

Climate change is the rise in temperature, is a natural phenomenon that takes place in nature as a result of the release of greenhouse gases (water vapor, carbon dioxide (CO_2), and other NOx gases). Due to the fact that human industrial activities have increased in recent centuries, CO_2 gas emissions and other greenhouse gases have risen dramatically. According to the estimates of the International Panel on Climate Change, the (IPCC), the Middle Eastregion and North Africa will be the region most severely affected by climate change in the coming decades.

Soil carbon sequestration will be an important mitigation strategy to reduce atmospheric CO_2 concentrations. The process of transferring atmospheric CO_2 into soil and biotic pools can enhance soil quality, increase agronomic productivity, improve quality of natural waters, and lower rates of anoxia (decrease in the level of oxygen) or hypoxia (dead water) in coastal ecosystems. Soil carbon sequestration is enhanced through agricultural management practices (such as increased application of organic manures, use of intercrops and green manures, higher shares of perennial grasslands and trees or hedges, etc.), which promote greater soil organic matter (and thus soil organic carbon) content and improve soil structure (see, e.g., Niggli et al. 2008;).Increasing soil organic carbon in agricultural systems has also been pointed out as an important mitigation option by IPCC (2007b). Very rough estimates for the global mitigation potential of Organic matteramount to 3.5–4.8 $GtCO_2$ from carbon sequestration (around 55–80 percent of total global greenhouse gas emissions from agriculture) and a reduction of NO_2 by two-thirds (Niggli et al. 2008). For sound estimates, however, more information on the mitigation potential of OM duly differentiated according to climatic zones, local climatic conditions, soil characteristics, variations in crops and cultivation practices, etc.—is still needed.Two Field experiments wereconductedin South of GS to evaluate the short term effect of irrigation with treated wastewater on the level of soil Organic Matter (OM) and accumulation of heavy metals in the upper soil layer and plants. Soil organic content (OM) significantly increased with wastewater irrigation application and with increasing the period of application which is attributed directly to the contents of the nutrients and organic compounds in the wastewater applied. The soil OM contents accumulated more in the topsoil in all treatments.This increase was the highest in the top soil (0–20 cm) and for the longer period of wastewater application. Several researchers reported accumulation of N, P, and K in the soil with wastewater application which was attributed to the original contents of these nutrients in the wastewater applied O.M% tends to decrease after irrigation by well water, while the opposite trend was obtained with irrigation by treated wastewater as shown in Figure4. This is due to high nutrients result in rich biomass production, showing a benefit to the soil. Because of the soil's organic substance, physical and chemical properties improved. This enabled granular structures to form, and crop growth accelerated and enjoyed nutrient absorption (Wang & Wang, 2005). OM as a mitigation strategy addresses both emissions avoidance and carbon sequestration. It is achieved through:

- lower N2O emissions (due to lower nitrogen input)—it is usually assumed that 1– 2 percent of the nitrogen applied to farming systems is emitted as N2O, irrespective of the form of the nitrogen input. The default value currently used by the IPCC is 1.25 percent, but newer research finds considerably lower values, such as for semi-arid areas [e.g., Barton et al. 2008];
- less CO2 emissions through erosion (due to better soil structure and more plantcover).
- lower CO_2 emissions from farming system inputs (pesticides and fertilizers produced using fossil fuel).

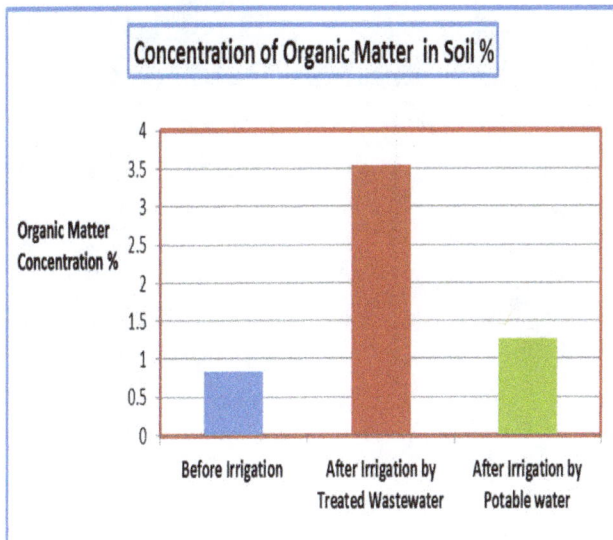

Figure 4. *Concentration of Organic Matter in Soil (%)*

5. Wastewater Reuse and Drought Mitigation

In arid and semi-arid countries, particularly the developing ones, the full utilization and re-use of sewage water is still far from our final goal, i.e. to be used as a water source, in spite of the vital role it could play in reducing the high pressure imposed on the limited available freshwater. Health and environmental problems are the major obstacles restricting the sustainable and safe reuse and recycle of wastewater which require concerted efforts supported by regional and international organizations. Treated water may be substituted for conventional resources and itmay be reused for purposes such as non-agricultural irrigation (parks, green areas, jungles and deserts ,… etc.), and when rainfall is insufficient to meet the increasing needs to feed the future generation.

6. Conclusion

The obvious conclusion and the initial results of the current pilot projects carried out in GS emphasized that a high degree of effluent reuse must be achieved in Gaza in order to reduce the current levels of groundwater withdrawal by the agricultural sector and mitigate the negative environmentally sound impacts. All future collection and treatment strategies should integrate reuse possibilities wherever practical. Reuse of wastewater effluent offers the he release of complementary resources, sustaining the existing and expanding irrigated areas, in addition to the treated wastewater provides a renewable and valuable source for agriculture and free limited water supplies for domestic and industrial purposes. Reuse of sewage water when properly managed, has the benefit of reducing environmental degradation as well as mitigation measure of climate change.

Acknowledgement

The author would like to express his gratitude for PWA, particularly Eng. RebhiAlSheikh, Deputy Chairman-PWA, forhis ever supporting and high appreciation and thanks for International Center ofAgricultural Research in Dry Areas (ICARDA)Amman, Jordanfor theirsupervision , finance and sponsorship for the WW reuse projects in GS.

References

[1]   Abu Nada, Z. M. , 2009 . Long Term Impact of Wastewater Irrigation on Soil and Crop Quality Parameters in Gaza Strip. Islamic University. M.Sc degree thesis.

[2]   AL-Sbaihi,H .Abu Sharekh, Y.,Akelane , S. , 2013.Short Term Effect of Wastewater Reuse on the level of Organic Matter and Accumulation of Heavy Metals on Soil and Zea mays (Corn)in Al-Zaitoun District - Gaza Strip.

[3]   Barton, L., R. Kiese, D. Gatter, K. Butterbach-Bahl, R. Buck, C. Hinz, and D. Murphy. 2008. "Nitrous Oxide Emissions from a Cropped Soil in a Semi-arid Climate," *Global ChangeBiology* 14: 177–92.

[4]   *Climate Change and Food Security: A Framework Document.* Rome:FAO/WHO (Food and Agriculture Organization/World Health Organization).

[5]   French Regional Mission for water and agriculture (MREA), 2004. Partial Results for the season 2003, Technical Report.

[6]   IPCC. 2007a. "Summary for Policy Makers." In *IPCC Fourth Assessment Report*, "Working Group II Report: Impacts, Adaptation, and Vulnerability." See specifically onadaptation, chapter 17; on inter-relationships between adaptation and mitigation, chapter

[7]   KFW, Kreditanstalt fur Wiederaufbau, Sludge and Effluent Reuse Study for Gaza Central Area, February, 2006.

[8]   Metcalf & Eddy. 1991. Wastewater Engineering: Treatment, Disposal and Reuse. New York; McGraw- Hillel Inc.

[9]   Ministry of Agriculture. 2011. Annual Data Report.

[10] Palestinian Central Bureau of Statistics (2007): Population, housing and establishment census 2007, Palestinian National Authority.

[11] Palestinian Water Authority (2001): Coastal Aquifer Management Program (CAMP), Integrated Aquifer Management Plan (Task 25), Gaza.

[12] Palestinian Water authority, 2011. Comparative Study of Option for additional water supply in GS (CSO-G, 2011).

[13] World Bank, 2004. West Bank and Gaza Wastewater Treatment and Reuse Policy Note. Water & Environment Department, the Middle East and North Africa Region.

8

Water grabbing/land grabbing in shared water basins the case of Salween River Hatgyi Dam

E. Zerrouk

GSGES, Kyoto University, Japan

Email address:
emelzerrouk@hotmail.com

Abstract: Land grabbing by foreign governments and international companies is on the rise. Faced by population growth and an ever-decreasing availability of useable/affordable land in populace states, many are looking to buy land where it is available, predominantly for agricultural and industrial purposes. But land alone is not sufficient for either of these uses. The availability of useable water resources is also a prerequisite to each land purchase. To buy land is to own its green water and have access to any blue water available to it. The development of hydropower projects, however, endeavours to buy the use of blue water, and must also come with a purchase/lease of the surrounding lands. Thus, it can also be seen as a type of 'water grabbing'. Where the locally affected, vulnerable, pre-existing stakeholders are against the project and the loss of livelihood and rights it engenders, a hydropower project may be labelled as a vehicle for water and land grabbing. For an international river, a part of a shared basin, the water grabbing affects stakeholders living under various political regimes and with disparate local power relations. The effects of the project on both sides of a border may be the same; however, the manner in which the two governments handle the effects will be different. The Case of the Hatgyi Dam development on the Salween River, a joint project between China, Myanmar and Thailand, is an example of the above. As a controversial dam being built on an international, border river, the Hatgyi Dam case study exemplifies many of the issues to be found in similar developments across the developing world.

Keywords: Hatgyi Dam, Myanmar, Water Grabbing, Land Grabbing

1. Introduction

The 2008 economic crisis, and its run-up, led the way to an increase in worldwide land grabbing and water grabbing. States and powerful interest groups, concerned with food and water security, looked at reducing their reliance on the international market to meet their consumptive needs. With limited resources available within their own borders, and high labour costs, they looked to invest in lands elsewhere that could supply their needs, targeting mostly developing countries eager to attract Foreign Direct Investment (FDI). But land/water grabs can exacerbate complex hegemonic relations between states, and within nations and communities, especially asland and water are being re-appropriated away from vulnerable stakeholders. The 'sale' of water resources, finite andessential, to investors as part of a development project has far-reaching implications for those formerly reliant on them. In the case of a shared water basin, the political and social dimensions of this sale are further complicated. This paper seeks to explore the

concept of water/land grabbing and to apply it to a case study in Myanmar, a least developed country(LDC) that is undergoing a process of rapid investment and reform. The case of the Hatgyi Dam, located in the international Salween River basin, is an example of what can be easily termed 'water grabbing.' Unlike many emerging water grabbing case studies it is concerned with the grabbing of water and land for hydropower development rather than for direct agricultural or industrial purposes. However, an underlying concern for equity and inequality links the Hatgyi Dam case to other land/water grabs occurring within Myanmar and around the globe.

Resource 'grabbing,' land or water, as a concept and not just as an action, names an epistemological framework through which to analyse how control of land and its resources is shifting away from, vulnerable,pre-existing stakeholders[1] and its negative effects. It is concerned with an increasing rate of land acquisition and land use change. The bias exists in this approach that the re-allocation/use of

---

[1]Whether human or other biological dependents.

the land and/or resources is negative, and thus this type of analysis favours subaltern perspectives and risks sidelining larger narratives that may also be present[2]. However, its strength is also in its ability to project the concerns of the 'other'[3].As a narrative framework ithas the ability to simultaneously de-centralize the human aspect of development and investment, by putting the land and its resources front and centre, while also ensuring that the 'humane' aspects are highlighted, i.e. the treatment of, and effects the changes will have on, the human and the environment from a moral and rights standpoint. Thus, in a decade of rapid re-appropriation of land by powerful interest groups, a trend that is detrimentally effecting the livelihood of many in the 'developing' world, the rhetoric of 'land/water grabbing' is gaining popularity and is becoming a useful way in which to communicate the complexities of these issues.

The recent spike in land and water grabbing came out of the 2008 financial crisis and a perceived food crisis. With rising commodity prices a ripple of unease struck those countries that are almost wholly reliant on the market to meet domestic food, and energy, needs. In an attempt to become relatively food secure, and thus further water secure (via virtual water trade[4]), these countries, such as the Gulf States[5], looked to purchase foreign lands on which to produce their own goods. With many developing countries subsequently opening their doors to this brand of FDI, large multi-national and transnational companies also jumped on the opportunity to 'grab' cheap and abundant resources.

In application, those occurrences most obviously labelled as land grabs disproportionately affect those living on the edges of society, such as subsistence farmers, minority groups, and indigenous peoples. This is due to the fact that what remains of a country's 'undeveloped' and accessible land and resources will often be located away from urban centres and/or main infrastructure, such as highways. Thus the 'empty' or 'unused' landrhetoric (these designations discussed later in detail), is easily sold to those detached from the realities of the lives of those on the peripheries and from their customary, traditional attachments to the same land constructed as 'available' for investment.

Investors have been sold the idea that the world still holds vast lands that are 'untapped' and 'unused,' 'marginal,' that can be made productive to the benefit of the nation in which they are located and the investing parties. This green Eldorado is a Trojan horse. It has been created so as to dim the realities and make it appear that these large land grabs are indeed no threat whatsoever to food security

or water security, however at the expense ofpre-existing stakeholders. If the land is vacant then the logic follows, that making it 'productive' is positive development.

A kin to the 'unused' land narrative is the 'inefficient land use' construction. But how do we define an 'efficient' use of land or of water, and for whom? Can it only be defined in commercial terms? The labelling of a land as inefficient implies that it had no purpose in its previous state of use. While a case may be made for industrially degraded lands, the majority of the land-grabbing cases in emerging literature are actually occurring on prime land. This is the case because, in order to develop the land (beyond holding it for speculative purposes), water is required, and as communities and animal life tend to gravitate to freshwater sources, the land is infrequently 'unused'.

The global climate change debate has supported the rise in many alternative, relatively carbon low energy sources. However, the two most important sourcesin terms of the land and water grabissueare biofuels and hydropower.The growth of the biofuel agro-industry has changed the dynamics of land use in producing countries,while a resurge in hydropower development is buoyed by the world's need for 'green energy.'

Many first generation biofuels, such as agrofuels like maize, sugar cane, cassava, are seen as being 'flexi crops,' meaning that their use can be changed dependant on market demands: "soya (feed, food, biodiesel), sugar cane (food, ethanol), oil palm (food, biodiesel,commercial/industrial uses), corn (food, feed, ethanol)" [1]. Yet,when the perceived demands mean that these crops are transformed into fuel, there is considerable distortion of the global food and feed market and the potential for a real increase in food insecurity amongst the global and local population.Another issue with agrofuels, like maize, cassava, sugar cane,etc., is their high water demand. Increasing the use of crops for fuel means that more water resources are being re-directed for their growth, processing and use. According to Philip Woodhouse, "the dimensions of this impact on international agriculture can be gauged from projected increases in land areas dedicated to biofuel production. Estimates of de Fraiture et al. (2008) projected a global total of 42.2 million hectares devoted to biofuels in 2030. White and Dasgupta (2010) cite projections of biofuels accounting for 20% of global arable land by 2050" [2].Evidently, a re-evaluation of this push for biofuel production in the name of energy security is necessary in order to ensure that certain perceived food insecurities do not become real insecurities.

Policies by global powers like the United States (US) and the European Union (EU) are only exacerbating this issue. The US's American Clean Energy and Security Act and the EU's Renewable Energy Directive 20 20 20 create a market for energy that will reduce their overall emissions and diversify their energy portfolio. However, this comes at the cost of the land and water resources of the producing environment.Although,these policies are likely to hold for some time as the desire to be (green) energy secure remains

---

[2]i.e. the securitization of resources and energy from a state perspective.
[3]Being the side-lined land itself, the environment from an ecological perspective, the resources, the marginalized human stakeholders such as ethnic minorities, etc.
[4]See the work of Tony Allan and Arjen Y. Hoekstra.
[5]For many Middle East Countries the ability to be self-sufficient in terms of food production ended in the 1970s, and thus they became heavily dependent on imports and some of the world's largest virtual water importers per capita.

and the push to own the land that can produce this in-demand-energy-source will mean more investment and will likely further land grabs [6] . However, it should be remembered that the energy market is always fluctuating. The recent shale gas and 'clean coal' boom in the US and the oil sands of Canada have added another dimension to the fossil fuel versus renewable energy debate in terms of energy security.

2. Case Study: the Hatgyi

2.1. The Salween's River's First Dam

The Salween River [7] is oneof mainlandSoutheast Asia's last large undammed rivers. Beginning in Tangula Mountain on the Tibetan plateau, China, (known there as the Nu Jiang), 4,000m above sea level [4], it flows through Yunnan Province into Myanmar where it forms 120km of the border between Myanmar and Thailand before finally emptying into the Gulf of Martaban on the Andaman Sea. While in Myanmar, the Salween River runs through the turbulent ethnic minority territories of Shan, Kayah, Karen (Kayin) and Mon States.

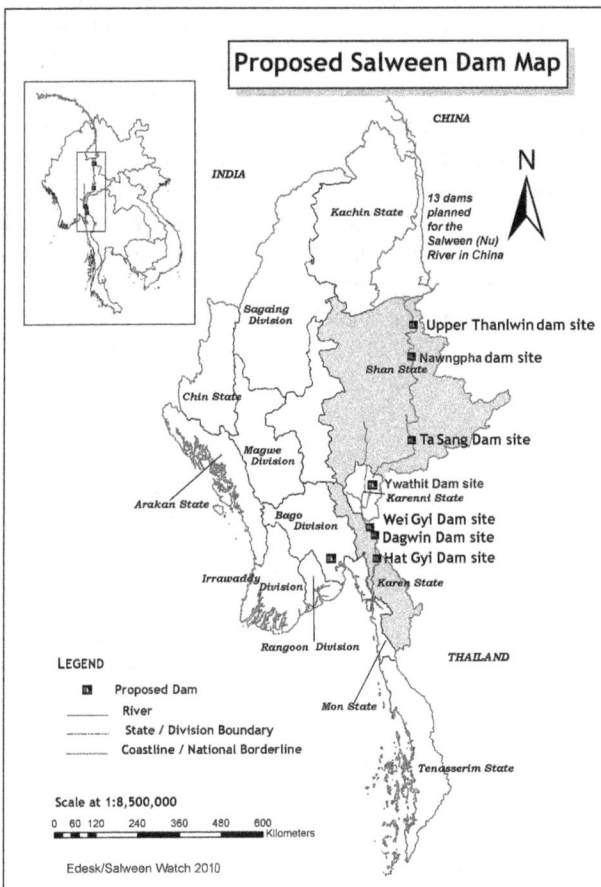

Proposed Salween Dam Map

Edesk/Salween Watch 2010

Approximately 2,413km long, the Salween is,after the Mekong River, the second longest river in Southeast Asia [5]. The river's basin is approximately 271,914sq.kmstretching over three countries [6]. 53% is drained in China, 42% in Myanmar, and 5% in Thailand [Salween Watch]. The Delta,on the bay of Martaban, is the most denselypopulated section of the basin. It is highly fertile and ideal for wet season rice cultivation. The basin supports over 10 million peopleof more than thirteen different ethnicities [5]. Exploited for irrigation, fishing as well as other traditional activities,the river basinhousesimportant religious sites and cultural capitals.

The government of Myanmar's Hydro Electric Power Department has signed Memorandums of Understandings (MoU) with EGAT (Electricity Generating Authority of Thailand), and various Chinese energy companies, to begin the process of building five to eight[8]large hydropower dams on the Salween River mainstream. As these MoUs were signed without the consultation of the generalpublic or the consent of the locals in the construction areas, they have become a highly contentiouspolitical subject in Myanmar and Thailand. Inequity, in terms of benefit sharing and compensation, is an immediate follow-on from this lack of involvement. As a representative of the Ethnic Community Development Forum,Sai Khur Seng, so rightly expressed, "energy projects in [Myanmar] should be for the benefit of the Burmese people and not at their expense".

In Myanmar, dammingin general is a sensitive political issue as unilateral development, on the part of the government, has precipitated further armed conflicts in ethnic minority lands. The internationally infamous Myitsone Dam project is only one other example. Positioned on the headway of the Irrawaddy River,this damming project was set to flood an area in Kachin State equivalent to the size of Singapore [7], and lead to the displacement of thousands (estimated 12,000). The project, now suspended,escalated tensions between local armed resistance groups, e.g. the Kachin Independence Army (KIA) and the central government. Similarly, the Salween River dams are mired in the complex, and often violent, civil conflicts of the Eastern border states. Situated in Myanmar's politically constructed 'terra nullius,' [9] these projects risk exacerbating the insecurities of the local population (in terms of safety, livelihoods, water, and food) and jeopardizing fragile ceasefire agreements.

On the upper part of the Salween River/Nu Jiang, China has been planning its own hydropower generating13 dam cascade. If the entire project were to be implemented it would be expected to generate 21.3GW of electricity (greater than the Three Gorges Dam). However, the project stalled almost ten years ago under Premier Wen Jiabao due to concerns over seismic, environmental and social impacts. However, on January 23rd 2013, the government released a notice related to the implementation of the energy portion

---

[6] However, it should be remembered that the energy market is always fluctuating. The recent shale gas, and 'clean coal,' boom in the US, and the oil sands of Canada, have added another dimension to the fossil fuel vs. renewable debate in terms of energy security.
[7] Known as the Thanlwin in Thailand and the Nujiang in China.

---

[8] Several project plans are still at the feasibility study stage.
[9] 'nobody's land' as opposed to terra nulla which is 'no land.'

of its 12th five-year-plan. It was announced in the notice that at least four dams were to be carried out on the upper Nujiang [8]. The dams are to be located in and/or near the Yunnan Three Rivers UNESCO World Heritage Site, raising further controversy. Besides the immediate impacts the dams will have on the local populations(approximately 50,000 people) and the area's biodiversity, the cascade will have far reaching detrimental effects on China's downstream riparians, Myanmar and Thailand[10].

The planned hydropower projects on the upstream and downstream Salween River threaten delicate ecosystems and the livelihoods ofmillions, the majority of which are from ethnic minority groups. Although the need for lower carbon emitting energy is apparent among the riparian states, these damming projects come at a high price with their benefits being inequitablydistributed among those affected.However, from the government's geopolitical perspective, the Salween River hydropower projects strengthen the Burmese government's position in the region, further establishing itself as a key energy hub. Developing the river as a source of alternative energy aids in Myanmar's regional integrationandbrings the region a step closer towards what Bastien Affeltranger calls the "dream of the ASEAN power grid" [9]. Energysources produced in and exported from Myanmar are already considered as part of the greater Mekong Power Grid, which includes Thailand and China, the region's largest energy consumers. Both countries also purchase electricity from Laos and Vietnam. Thus, the expansion of Myanmar's hydropower export potential would increase its ability to compete with its neighbours as a supplier of 'cheap' energy.

The perceived energy and water security needs of Myanmar's neighbours are what primarily push the hydropower sector to develop the Salween River. The Thai government is heavily dependent (70%) on natural gas for its energy production (around 30% of which it imports from Myanmar)[10]. With an estimated remaining supply of 30years, renewable energy sources such as hydropower are seen asan attractive alternative [10].However, Thailand's growing civil society is staunchly against further hydropower development within their country. Thus, EGAT has been actively promoting hydropower projects in neighbouring states.

EGATis thegenerating body, purchasing company, supplier and distributer of electricityin/to Thailand. It is therefore in their interest to maintain high-energy demandand low-energy costs. Thus, EGAT reputedly forecasts future energy needs well above their actual 'peak load' demands, often with a final discrepancy of as much as 15% [10]. Considering its powerful standing in Thai national politics and its unrelenting push for hydropower and gas development, land and water grabbing in neighbouring countries with weaker civil societies and

conservation legislature islikely to continue.

Myanmar, an LDC, sits between India, China and Thailand, three of the regions fastest growing economies (with India and China being in the top four of the world's energy consumers next to Russia and the US). With growing industrialisation, urbanisation, and a burgeoning middle class, their predicted energy needs now and in the future gives rise to immense competition to control regional energy sources. Myanmar, with gas, oil, and hydropower potential, is a country that many have waited for the opportunity to invest in.

Eager to attract foreign investment, Myanmar's reform process has focused heavily on economic reforms, introducing new legislation on foreign direct investment and on Special Economic Zones (SEZs). Under the constitution "the state owns all natural resources" and land [11&12]. However, environmental regulations and laws are still weak and vulnerable in the face of this fierce development phase. Thus, the first step towards protecting the environment and those who depend on it is to strengthen this branch of law at the national level.With regard to development projects like the Salween River Dams,U Zaw Naing Oo of Resource and Environment Myanmar Ltd (a Burmese consultancy firm)explains that "environmental by-laws have not yet appeared to be able to effectively implement a monitoring plan... if we could have that, it would contain sections such as who should do the monitoring and what kind of groups need to be set up and solve such problems" [13].Until that time, Myanmar's resources, land and water are open to exploitation by unsympathetic investors.Unfortunately, this type of weak environmental regulation is a 'pull' factor for many large extraction and development companies looking to reduce costs by investing in countries where regulations are less demanding in terms of safety and environmental impact regulations.

3. Method

According to the Transnational Institute, there are three traditional groups of thought where the issue of land acquisition and resource grabbing is concerned. The first accepts that the process of land acquisition and land use; change needs to happen but should be carried out within a particular legal framework that regulates investment in land. This group focuses their attention onhow to make would-be land grabs into win-win situations for all impacted stakeholders. The second school of thought is resigned to the fact that the grabs are going to continue as a result of the global market system and thus raise the issues of how to mitigate the damage to be caused by such a trend. A land and water rights based approach is favoured as the best way to ensure that any benefits reach those negatively affected by the grabs. The last group sees little, if anything, to be gained by the proliferation of land grabbing and would like to see an end to the expansion and the down-scaling of the grabs that have already occurred. They argue that such

---

[10] As flows are altered, sedimentation downstream decreases, eutrophication of slow flow areas occurs (increasing spanning grounds for malaria mosquitoes etc.) and fish migration is interrupted.

large-scale and capital-intensive investment is not really necessary for the security of the world in terms of food or climate change. They also argue that, first and foremost, the rights and needs of the pre-existing human and natural stakeholders must be ensured (believing that if this was done, land grabbing would not occur and another solution to these issues would be found). While taking into consideration these various approaches, this paper seeks to objectively project the situation without conforming to any of the three schools of thought. The first step is a thorough understanding of what land and watergrabbing actually means, and to whom, within different contexts [3].

"Water grabbing [is] a situation where powerful actors are able to take control of, or reallocate to their own benefits, water resources already used by local communities or feeding aquatic ecosystems on which their livelihoods are based. This lens demands a focus on how material, discursive, administrative and political power is mobilised to enable such water reallocation and changes to tenure relations as well as the impacts of the latter on local livelihoods, rights, gender, class and other social relations" [1].

The underlying thesis of water grabbing is the same as for land grabbing: control. As a concept, it is concerned with how power is used to achieve that control, usually at the expense of vulnerable primary shareholders (the ecosystem, local peoples, etc). It is impossible not to grab water when grabbing land, and land without an adequate supply of water, for whichever intended purpose, is useless[11]. Water is either essential to facilitate any activity on the land acquired, be it industrial or agricultural, or it is the primary target of a grab, i.e. in the case of hydropower development. Thus, water grabbing is essentially the relocating of domestic, agricultural and industrial water demands to foreign lands perceived to have more abundant water resources. In other words, the investor is buying virtual-water[12] directly instead of trading for products produced by that country.

Water, unlike land, does not obey boundary lines, and thus the activities of an investor on and around their land will likely have much more far-reaching effects than would be apparent at first. Water can even be considered as 'grabbed,' in a broader sense, if it has been polluted to the point that it is no longer safely/easily used for the purposes it was put to previously. In other words, if a development leads to water being un-potable, or unusable for agriculture and watering of livestock, or even to the build-up of contaminants in an ecosystem, then this water has effectively been 'removed' from the cycle of use in that basin.

The most common type of water grabbing is achieved through a commodification of water, the transfer of 'legal' rights of access, and the privatization of this public good and human right, such that powerful interest groups may re-direct/re-distribute it for their own ends. Thus, land grabbing and water grabbing, whether separate goals or overlapping, often occur in countries that have weak legal frameworks, and rights over water and land are unclear. Where there are existing users, there is the problem, as with land rights, of registration which is seen, amongst the commercial market, as the only clear proof of having rights at all. Thus, traditional and communal rights over water and its management are overlooked and unprotected. This can lead to the idea that the water needed by the investor is not in competition with any pre-existing rights holder and that investors are given 'carte blanche' in their utilization of the resources in the purchased/leased area. This myth cannot be sustained, however, in cases where the water/land grabbing is occurring through hydropower development, and most especially in the case of development in an international water basin. The following case study is just such an example. When more than one riparian exists, unilateral or joint development, of the river will have trans-boundary effects, thus the illusion of 'no harm' cannot afford to exist.

4. The HatGyi[13] Dam and Water Transfer Project

The Hatgyi Dam is expected to be the first of the planned Salween River dams to be completed. When built, this gravity dam is expected to stand 33 meterstall with an installed capacity of 1,100-1,500MW, more than triple the initial plans for a 300MW dam outlined by the 1999 feasibility assessment (see Table1)[14]. It is being developed as a joint project between Myanmar Department of Hydro Electric Power, Sinohydro Corp, and EGAT, and will cost over 1 billion USD. However, the Hatgyi Dam site is located in Karen State's Pa-an district, upstream of Pa-an the State capital, and downstream, by 33 kilometres, from the meeting point of the Salween River and its tributary the Moei River [14]. This district, home to the Karen ethnic minority group (among others), is a highly militarized and conflict ridden area of the country. The attempts by the central government to forcefully 'secure' and grab the lands around the projected site has led to much grief amongst the local peoples

Ethnic Issues

"If they go ahead with the dam. We dare not stay in our village because we are close to the dam site – they'll want us to leave.[15]' The Salween is the main artery that pumps life into the local communities on both banks of the border.

---

[11]Except perhaps in some forms of speculation.

[12]Sourced from either green water (of the land) or green & blue water (from rivers and lakes added to the land). See work of Tony Allan and Arjen Y. Hoekstra for further detail.

[13]Sometimes referred to in literature as: Hatgyi, Hat-gyi, Hutgyi, Hut-gyi, Taung Kyar, Hajji, Hatki.

[14]It should be noted that the site has been moved by around 2km from the original site in order to increase the capacity of the dam [Kyaw Thu. Myanmar Times. Issue 334. 18 September 2006].

[15]Naw Eh Paw a villager who lives along the Salween on the Burmese side.

The villagers rely on it for fish, and the animals and plants that inhabit the rich jungles nourished by the river. The lack of infrastructure in the area, especially all-weather roads, means the Salween is the main means of transport for people needing to get to markets to sell their produce or buy supplies" [15].

Myanmar is home to a plethora of different cultures, languages and ethnic groups. The groups that are most directly affected by the Salween River developments are those located on the edges of Shan, Karen[16], Kayah, and Mon States (particularly Karen State for the case of the Hatgyi Dam). Of the four, Karen and Kayah State are the most heavily under government military control [16]. The conflict between the KNU (Karen National Union) and the central government is one of the longest civil conflicts in modern history, starting in 1949 after independence, and ending in September 2012 with a ceasefire agreement. At the time of writing this paper this is the longest holding truce between the two sides since the conflict began. Prior to this point in time it had been the express wish of the Burmese Junta (prior to the 2010 elections), along with its allied DKBA (Democratic Karen Buddhist Army) forces, to consolidate its hold of KNU controlled lands along of the Thai-Burmese border by 2010 [17]. This deadline presumably had much to do with the planned Salween River development projects.

The militarisation of the ethnic borderlands, down the length of Myanmar, was seen as key to ensuring that the central government had a strong foothold in the region bordering its strategic trading partners, China and Thailand. Both investing states seek security over their current and future trans-border investment interests (gas, oil, timber, electricity from hydropower, and water transfer). Thus from a Chinese and Thai perspective, border stability is a key driver in encouraging a Burmese development agenda.

The physical nature of the resource being exploited, i.e. a main, trans-boundary river located in contested spaces, means that the area the government is required to 'secure' is spread out and that development of the river for these purposes cannot be done without interfering with the local population's customary uses and rights over that resource. Conversely, 'divide et impera'(to divide and conquer), is evidently one concept behind the recent liberalization and development of the resources in the Eastern border states of the country. Private land concessions in the ethnic Eastern border regions have been given to China, Thailand, private overseas investors, and Burmese interest groups for lumber, mineral extractions, and development projectslike: the Trans Burma-Yunnan pipelines and Salween River hydropower plants. They are scattered around the ceasefire zones of the Shan, Kayah, and Karen States making a spider's web of loosely connected central government control. This power projection in the ceasefire zones can be seen as, "an explicit postwar [sic] military strategy to govern land and populations to produce regulated, legible,

militarized territory" [12]. The subsequent strengthening of the central government's suzerainty/sovereignty of the Eastern states is maintained by agreements with local leaders and the presence of policing forces around development sites. Thus, any land concessions given to foreign or domestic bodies leads to the extension of government control over the region in general.

A less obvious result of the water and land grabbing, due to hydropower projects along the Salween River, is the re-distribution of the power it causes. Old systems of local authority are altered as new actors develop interests that overlap over the same geographical space. Deals formerly brokered along the Eastern borders between foreign companies and local leaders are now made with the Burmese government, who are now interested in the controlled liberalization of their agriculture and extraction industries, effectively removing local leaders from the negotiations and power play.

5. The Hatgyi Dam in Detail

"We don't want any dams in our area," said Saw Kyaw Phoe from Mae Par village in the upper part of the construction site. "If the dam is built, our village -- the whole area -- including our paddy farms and our gardens will be flooded. I, myself, will have no place to live," he said. "We depend on the Salween to irrigate our farms, but the dam will destroy our livelihood," [18].

The life of the project began in 1998 with the start of a pre-feasibility study entitled the "Preliminary Feasibility Study of Hutgyi Hydropower Project in the Union of Myanmar."The studywas carried out with the involvement of Myanmar Electric Power Enterprise (MEPE), the Japanese Marubeni Corporation, the Italian-Thai Development Plc. Co., Ltd., and NEWJEC Inc., the consulting group and subsidiary of Japan's Kansai Electric Power Company [14]. It recommended the construction of a 300MW dam with a small reservoir and minimal flooding at its highest flow to both the Thai and Burmese side of the project. The project is almost entirely foreign-owned. Distribution of the cost for the dam was originally: Thailand 50%, China 40%, and Myanmar 10%.However, China now owns the majority share [19].

Hatgyi Dam:					
Name	Location	Main Ethnic Groups Effected	Installed Capacity	Energy produced yr	Height
Hatgyi	Thai/Burmese Border Lands in Karen State Pa-an District	Karen	1,100MW – 1,500MW	7,335GhW	33m

Investors:
Myanmar: Ministry of Electric Power,
The International Group of Entrepreneurs Co Ltd. (IGE)
Thai: EGAT
China: Sinohydro Corporation
China Southern Power Grid Co. China Three Gorges Project Corporation

Figure2: Specifications of Hatgyi Dam

---

[16]Formerly known as Kayin.

Table 1: *History of armed resistance in area surrounding dam site.*

	Karen Armed Resistance and Hatgyi Dam
Date: 1949	Karen National Union (KNU) and the Karen National Liberation Army/ Organization (KNLA/O) its military wing formed.
1994	The Democratic Karen Buddhist Army (DKBA), a Buddhist soldiers group from within the KNLA sides with the military junta.
2007 January	KNLA 7th Brigade commander, in charge of Pa-an District, (Hatgyi Dam site), formed the Karen National Union/Karen National Liberation Army Peace Council (KNU/KNLAPC), a separate group.
2007 February	KNU/KNLAPC and State Peace and Development Council (SPDC) announce peace agreement
2007 April	KNU/KNLAPC, DKBA, and SPDC troops attack KNU near Thai border.
2009 June	DKBA and Army start offensive against KNU in Pa-an District
2009 July	KNLA Brigade 7 headquarters 'overrun' in Karen State
2011 Feb - 2012 Feb	DKBA faction fights with central government over becoming Border Guard Force (BGF)

Table 2: *Hatgyi Dam Timeline*

	Hatgyi Dam Timeline:
1998-9	Pre-feasibility study
2004	Survey work begins
05/2004	Two EGAT surveyors killed at site (landmine and grenade fire) – work on hold
12/2005	MoU signed between EGAT and Myanmar Hydro Electric Power Department
06/2006	MoU signed between EGAT and Sinohydro on investment in Hatgyi Dam[17] work starts on power plant and road repair.
02/19/07	EGAT engineer killed in artillery fire – work on Dam on hold
03/2008	Further MoU with Sinohydro Corp as majority shareholder of Hatgyi Dam
07/ 2009	Survey work continued
03/2010	Joint field survey of Hatgyi Dam site. Burmese, Thai and Chinese engineers
24/04/12	New MoA signed with EGAT, Sinohydro, Myanmar Hydro Electric Power Dept. & IGE[18]
2012	Dam and water diversion projects in development

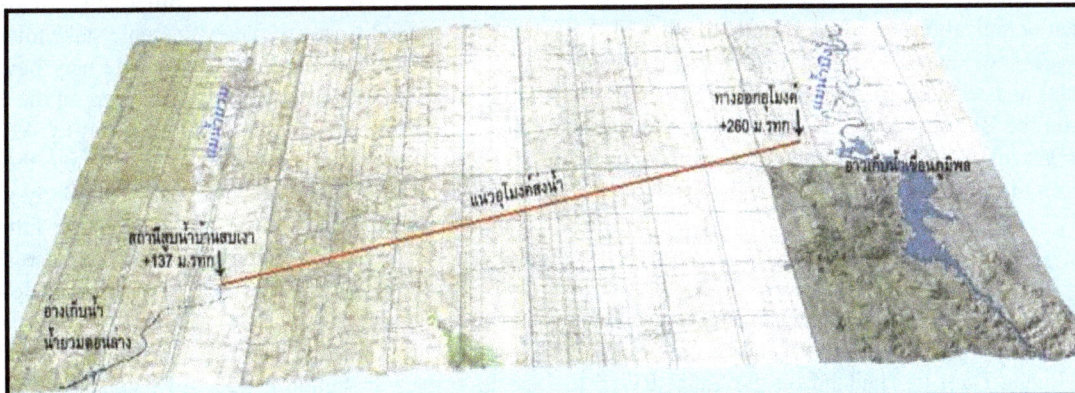

Figure 3: *Proposed Pipeline Path*

---

[17]According to Pornchai, former chairman of EGAT as well as the former chairman for the subcommittee for negotiating power cooperation with neighbouring States, Sinohydro is the majority shareholder as of March 2008 (responsible for securing the loan for the project), and EGAT second, Burmese government minority shareholder in the project [Shu Huaying, 2008; Watcharapong 2010].

[18] The International Group of Entrepreneurs Co Ltd. of Myanmar.

The agreement to begin work on the dam's development was not concluded until 2009. Previous attempts to move forward with the project were complicated by security concerns, as on at least two occasions EGAT staff were injured and/or killed (see Table 2) because of the conflict that has plagued that part of the country since independence. Indeed, the conflict between several of the ethnic minority groups in Karen State and the central government effects the character of the developments on the river, and vice versa, making the Hatgyi Dam project a very clear and unpleasant example of water/land[19] grabbing. The projects are moving ahead against the express wishes of the local populations, and violence, an instrumentused by both sides to press their case. However, with the current prospect of a lasting peace in Karen State, as the largest resistance group the KNU (Karen National Union) agrees to a ceasefire, this case study may not end as traumatically as its past trajectory would indicate. With peace talks taking place, and the impacts of the Hatgyi Dam project a part of the negotiations, there is the hope that local concerns may be heard and some rights respected.

Although the Hatgyi Dam is located squarely in Burmese territory and not on the political borderline, the flood zone of the reservoir is set to inundate Thai lands. Additionally, infrastructure required to transfer the energy generated by the dam, and to maintain it, links the project directly to the Thai side, thus creating a wider border area that cuts across Pa-an District.

The Hatgyi Dam's environmental impact assessment (EIA) has been heavily criticized. Due to a former non-disclosure agreement signed between EGAT and the Myanmar Hydro Electric Power Department, Thailand has not made the EIA public (as it is normally obliged to do by law). What has been made available, under considerable pressure, downplays "the environmental and human impacts" and is disparaged "for making dubious claims about the extent of the opposition to the project by the local ethnic Karen" [20]. The reservoir alone will lead to the flooding of Karen State's wildlife reserves. Non-governmental organizations (NGOs), from both sides of the river, have asked EGAT to carry out a thorough environmental and social impact assessment (EIA/SIA) of the project on the Burmese side, but it has replied that it is unwilling to get involved in what it perceives as a domestic issue (between the Burmese government and citizens).

6. Inter- Basin Water Transfer

At the same time as the Hatgyi Dam area was being assessed (February 1998-1999), the feasibility study for the Mae Lama Luang Dam in Thailand on the Yaum River, a Salween River tributary linked to the Moei River, was conducted by Japan's J-Power group [21]. The Mae Lama

Luang Dam is one the dams that may be built to facilitate the transfer of water from the Salween River Basin to the Bhumipol Dam's reservoir inNorthwest Thailand (one of the country's two largest reservoirs).This bulk water transfer from basin to basin, without consulting, and against desires of,those who are dependent on the river is the most graphic example of water grabbing on this river.

Various routeshave been suggested for the water transfer. Two of theseroutes, the "Salween River-Nam Yuam Upper Dam-Bhumipol Dam's reservoir" and "Salween River-Mae Lama Laung Dam-Bhumipol Dam's reservoir," take water directly from the Salween mainstream by drawing on overflow from the Hatgyi Dam [22].Several other possibilities draw from Salween tributaries. The diverted water will be stored in a second dam, like the Mae Lama Laung (feasibility study done by J-Power February 1998-1999) or Nam Yuam Upper Dam.From there it is transported through a 60+kilometre [20] pipeline to the reservoir of Bhumipol Dam (refer to Figure3) to be added to reserves stored from the Mekong River basin flows[23]. The project, if completed along these lines, will take seven years.

The transfer of water between basinsis a contentious matter. The exchange deprives the people and the environment downstream of their customary flows.In the case of theSalween River-Mae Lama Laung Dam-Bhumipol Dam's reservoir route, water extraction from the Salween River will only exacerbate the damage caused by the Hatgyi dam itself on the downstream (recall that the Hatgyi is meant to be only the first of the Salween Rivers to be completed). The negative physical effects of the dams and the diversion project: disturbance of sediment flows, loss of aquatic diversity and fish stock, coastal seawater intrusion and salinization of delta soils, etc., put pressure on the livelihoods of those communities dependant on the Salween River basin,and further sour central-periphery political relations in Myanmar.

This development affects two nations across an international borderengendering a different set of governance and accountability issues. The vulnerable stakeholders of the water/land grabbing on the Burmese side may have little or no voice in the matter at present. However, on the Thai side, the level of propaganda and promotion that EGAT has been involved in over the last few years to promote the Salween River dams proves that the voices of the similarly vulnerable Thai stakeholders have an influence over the future of the projects. Thus, the dynamic politics involved across borders does not allow one national narrative to dominate the field. Naypidaw may view the lands around the project as an acceptable loss for the sake of the project, but Thailand may be less quick to bulldoze their own villages. Thus, in this case, the international dimension of the project may be a key factor in mitigating the damage caused by, or in stopping, the project.

---

[19]Here 'water' comes first as the primary objective of the grab and land is second as it is simply the means of securing it.

[20]Taking the lowest estimate.

7. Concluding Remarks

A legacy of unsustainable environmental practices has led many of Myanmar's investors to seek water/land grabbing opportunitiesoutside their country. For those States trapped ina situationwith depleted/polluted resources, the options are to rely on the market to provide what they can no longer produce, or 'grab' the resources in a still producing country and be in direct control of supply in order to meet their water/food/energy security needs. Private companies needing to compete on the market are drawn for similar reasons. Land and water grabbing is about control of supply, outside of the restrictions of their country.Thus,Myanmar, with its 'abundant' resources and immature environmental legislature is fertile ground on which to satisfy these geopolitical objectives.

As they revise their investment polices,Naypidaw should demand that investors follow generally accepted principals of social corporate responsibility, such as those laid out by the World Bank. Additionally, the publication of allproject EIAs and SIAs as well as opening the development process to public scrutiny and debate would do much to build confidence.A desire for cheap energy and for quick earnings means that a hydropower project maybe developed solely for the power element and exclude any potential, locally, beneficial by-products, such as irrigation, use of sediments etc. Making the relevant project documents available would mitigate this, and'unnecessary' projects would be contested from the outset[21]. Currently, damming engineering has advanced to the level where not only has the lifetime of dams been increased, but the options for incorporating related damage mitigation to these inexorable projects is available (sediment rotation, fish passage, etc.). However, many of these measures are impossible to incorporate retroactively, thus the emphasis remains on establishing mechanisms that require that environmental and social needs beprivileged.

Investment agreements, such as those linked to hydropower development or agro-industry,need to outline the rights of the investor in terms of water, taking into account the seasonal and yearly variations in water availability and the needs of those in the same basin, i.e. of the level of stress they place on the basin. This clarification is especially sensitive when investments occur in a transboundary basin.

Unlike many other international rivers, like the Danube and the Mekong, the Salween does not have a commission facilitating information sharing and monitoring of activities and dynamics in the river basin. However, according to Sri Su Wan, the co-director of TERRA, a Thailand based environmental NGO, there has been a move made to coordinate the creation of a Salween River commission [24]. If such an organization is created, even if only on the Thai side initially (then including Myanmar and China),

those effected by development projects, on both sides, such as the Hatgyi Dam and water diversion project would have a platform from which to voice their concerns and receive information.Only in an environment of information sharing and consultation can water and land grabbing be prevented.

References

[1] Mehta, L.; Veldwisch, G.J. and Franco, J..Introduction to the Special Issue: Water grabbing? Focus on the (re)appropriation of finite water resources, Water Alternatives 5(2) (2012) 193-207.

[2] Woodhouse, P, Foreign agricultural land acquisition and the visibility of water resource impacts in Sub-Saharan Africa, Water Alternatives, 5(2) (2012) 208-222.

[3] TNI, Transnational Institute,The Global Land Grab, A Primer, (Oct 2012). http://www.tni.org/sites/www.tni.org/files/download/landgrabbingprimer_0.pdf. 25 January 20133.

[4] Salween Watch, Recent Dams and Water Diversion projects,(2011).http://www.salweenwatch.org/index.php?option=com_content&view=article&id=51&Itemid=60.

[5] Wolf, T. Aaron, Salween River, In:Managing and Transforming Water Conflicts, Cambridge: Cambridge University Press, (2009), p236-239.

[6] IUCN, IWMI, Ramsar, WRI. Watersheds of Asia and Oceania AS23 Salween, Water Resources Atlas,http://www.burmariversnetwork.org/images/stories/rivers/eatlassalween.pdf.

[7] Wall Street Journal Staff Reporter, "Tensions Over Dam Projects Shift Myanmar's Politics," The Wall Street Journal, 13th August (2011),http://online.wsj.com/article/SB10001424053111904006104576504123439749638.html.

[8] Office of the State Council, China. 国务院关于印发能源发展"十二五"规划的通知/Notice of the State Council on the energy development of the "12th Five-Year Plan. 23 January 2013. http://www.gov.cn/zwgk/2013-01/23/content_2318554.htm. 1 February 2013

[9] Affeltranger, Bastien. "Inter-basin water transfers as a technico-political option: Thai-Burmese Projects on the Salween River." International Water Security, Domestic Threats and Opportunities. Eds: Pachova, Nakayama, and Jansky. New York/Tokyo: United Nations University, 2008. p 161-179. p163.

[10] Matthews, N. 2012, Water grabbing in the Mekong basin – An analysis of the winners and losers of Thailand's hydropower development in Lao PDR, Water Alternatives 5(2) 392-411.

[11] Ei Ei Toe Lwin. Myitsone should not resume, say experts and locals, A Myanmar Times Special Feature, Energy, August (2012), P9. http://www.burmalibrary.org/docs14/Myanmar_Times-energy2012-red.pdf. [15 October 2012].

[12] Woods, Kevin, Ceasefire capitalism: military–private partnerships, resource concessions and military–state building in the Burma–China borderlands, The Journal of

---

[21] Myanmar became a full member of ICOLD, the International Commission on Large Dams, in 2011. A positive step, as the country now benefits from knowledge sharing with the other 95 member States.

Peasant Studies,38:4(2011), 747-770.

[13] Gaung, Juliet Shwe, Concerns remain over Shwe gas benefits,
A Myanmar Times Special Feature, Energy, August (2012),p4,http://www.burmalibrary.org/docs14/Myanmar_Times-energy2012-red.pdf. 15 October 2012,[January 2013].

[14] TERRA. Status of The Salween (Thanlwin) Dam Plans. November 1999. http://www.ibiblio.org/obl/docs/SW03.htm. 1 Dec 2012.

[15] Burma Rivers Network, Villagers Fear Salween Dam, 8 June (2011),http://www.burmariversnetwork.org/news/11-news/578-villagers-fear-salween-dam.html.[February 2013].

[16] Steinberg, I. David,Burma/Myanmar: What Everyone Needs To Know, Oxford: Oxford University Press, (2010).

[17] Irrawaddy,"KNU to Abandon Bases", 18 June (2009),http://www.irrawaddy.org/article.php?art_id=16078.[February 2013].

[18] KIC, Karen Information Centre, Hatgyi Dam Generates Problems for Karen Farmers, 1 March (2011).http://www.bnionline.net/feature/kic/10171-hatgyi-dam-genrates-problems-for-karen-farmers.html.[February 2013].

[19] Burma Rivers Network, Hatgyi Dam,(2013). http://www.burmariversnetwork.org/dam-projects/salween-dams/hatgyi.html. [24 September 2012].

[20] International Rivers,The Salween River Basin Fact Sheet, 24 May (2012). http://www.internationalrivers.org/resources/the-salween-river-basin-fact-sheet-7481. [January 2013].

[21] J-Power, Mae Lama Luang, Thailand,http://www.jpower.co.jp/english/international/consultation/detail_old/se_as_thailand03.pdf. [1 February 2013].

[22] TERRA, The Salween Water Diversion Project, September (2005). http://www.terraper.org/mainpage/images/keysub/1305711562_th.pdf. [February 2013].

[23] DEDE. Conclusion on Project Features of the Water Supplying Lines ofReservoir at Lower Yuam River – Reservoir at Bhumibol Dam. http://www.dede.go.th/dede/fileadmin/upload/pdf/Page_9.pdf. February 2013.

[24] Sri Su Wan. Phone Interview. 14 February 2013

[25] Anders, Jägerskog et al. Land Acquisitions: How will they Impact Transboundary Waters? Stockholm International Water Institute, SIWI. 2012

[26] Aung Hla Tun. Myanmar state media details new foreign investment law. Reuters. 3 Nov 2012. http://www.reuters.com/article/2012/11/03/us-myanmar-investment-idUSBRE8A204F20121103. 6 Nov 2012.

[27] Bangkok Post. Myanmar's investment law: temper applause with caution. 3 Dec 2012. http://www.bangkokpost.com/business/economics/324282/myanmar-s-investment-law-temper-applause-with-caution. 5 December 2012.

[28] Durham University Archaeology Department. How to Build A Dam and Save Cultural Heritage. Workshop. Durham, July 6th 2012.

[29] Duvail, S.; Médard, C.; Hamerlynck, O. and Nyingi, D.W. 2012. Land and water grabbing in an East African coastal wetland: The case of the Tana delta. Water Alternatives 5(2): 322-343

[30] Earthrights. "Flooding the Future: Hydropower and Cultural Survival in the Salween River Basin." 28 April 2005. http://www.earthrights.org/burma-project/flooding-future-hydropower-and-cultural-survival-salween-river-basin. 15 October 2012.

[31] EGAT, Electricity Generating Authority of Thailand. Main Page. http://www.egat.co.th/ or http://www.egat.co.th/en/. 11 Feb 2013.

[32] ---. Power Development and Future Plan. 2011. http://www.egat.co.th/images/stories/annual/reports/2554/annual2011_eng_p86.pdf. 11 Feb 2013.

[33] ---. Key Statistical Data, EGAT Annual Report. 2011. http://www.egat.co.th/images/stories/annual/reports/2554/annual2011_eng_p113.pdf. 11 Feb 2013.

[34] FAO. Foreign direct investment –win-win or land grab?ftp://ftp.fao.org/docrep/fao/meeting/018/k6358e.pdf. Nov 2012.

[35] ---. Salween Dams. 2011. http://www.internationalrivers.org/node/3231. January 2013.

[36] Karen Rivers Watch. Damming at Gunpoint. November 2004. http://www.freewebs.com/krw_reports/Dam%20english.pdf. January 2013.

[37] Kay, Sylvia and Franco, Jenny. Transnational Institute. Global Water Grab, A Primer. 13 March 2012. http://www.tni.org/sites/www.tni.org/files/download/watergrabbingprimer-altcover2.pdf. 5 Feb 2013.

[38] Mizzima. Push to build big dams undermines peace process in Karen State. Mizzima News. 14 March 2012. http://www.mizzima.com/news/inside-burma/6758-push-tobuild-big-dams-undermines-peace-process-in-karen-state.html. 1 Feb 2013.

[39] Myint, Soe. The geopolitical challenge for Myanmar's energy sector. A Myanmar Times Special Feature, Energy. August 2012. P2. http://www.burmalibrary.org/docs14/Myanmar_Times-energy2012-red.pdf. 15 October 2012. January 2013.

[40] Naing, Saw Yan. "KNU Allows EGAT to Survey Salween River Dam." The Irrawaddy. 29th August 2007. http://www.irrawaddy.org/article.php?art_id=84166. 24th October 2012.

[41] Resource and Environment Myanmar Ltd. Projects. 2012. http://www.enviromyanmar.net/projects.html. 12 Feb 2013.

[42] Robinson,Gwen. Myanmar to delay foreign investment law. Financial Times. 17 September 2012. http://www.ft.com/intl/cms/s/0/6245e878-00b3-11e2-8197-00144feabdc0.html#axzz28DZvsZz4. 3 October 2012.

[43] Salween Watch, The Salween River. http://www.salweenwatch.org/index.php?option=com_content&view=article&id=50&Itemid=59>. 5 February 2013.

[44] ---. Recent Dams and Water Diversion projects. 201).

http://www.salweenwatch.org/index.php?option=com_conte nt&view=article&id=51&Itemid=60. 5 February 2013.

[45] Saw U, Alan. "Reflections on Confidence-building and Cooperation amond Ethnic Groups in Myanmar: A Karen Case Study." P219-235 In: Ganesen, N and Hlaing, Yin Kyaw eds. Myanmar: State, Society and Ethnicity.

[46] Shu Huaying. Sinohydro Corp News Release. 11 March 2008.
http://www.sinohydro.com/english/portlet?pm_pl_id=7&pm _pp_id=19&COLUMNID=11424148920001&CHCOLUM NID=11424149100001&ARTICLEID=12054551070001. February 2013.

[47] TERRA, (Towards Ecological Recovery and Regional Alliance). Calling for the Hatgyi Dam to be Stopped Immediately. 7 February 2011. http://www.terraper.org/mainpage/media_detail.php?mid=10 44. 5 January 2013.

[48] Watcharapong Thongrung and Chularat Saengpassa. Controversial Hatgyi Dam to Go Ahead. Nation. 16 February 2010. http://www.nationmultimedia.com/2010/02/16/business/busi ness_30122674.php. 1 December 2012.

[49] Wolf, T. Aaron. Salween River. p236-239. In Managing and Transforming Water Conflicts. Cambridge: Cambridge University Press, 2009.

[50] World Bank. Thailand. 2011. http://data.worldbank.org/country/thailand. August 2012.

[51] U Moe Myint. I expect to see rapid change for the better. A Myanmar Times Special Feature, Energy. August 2012. P6-7. http://www.burmalibrary.org/docs14/Myanmar_Times-energy2012-red.pdf. 15 October 2012.

[52] UNCITRAL. 1958 – Convention on the Recognition and Enforcement of Foreign Arbitral Awards - Status. 2013. http://www.uncitral.org/uncitral/en/uncitral_texts/arbitration /NYConvention_status.html. 11 Feb 2013.

[53] UNCTAD. Investment Policy Framework for Sustainable Development. 12 June 2012. http://www.unctad-docs.org/files/UNCTAD_IPFSD_2012.pdf. 11 October 2012.

[54] ---. World Investment Report 2012 .5 July 2012. http://www.unctad-docs.org/files/UNCTAD-WIR2012-Full-en.pdff. 12 October 2012.

[55] ---. Country fact sheet: Myanmar, World Investment Report 2012 .5 July 2012. http://unctad.org/sections/dite_dir/docs/wir12_fs_mm_en.pd f. January 2013.

[56] ---. World Investment Prospects Survey2012–2014. http://unctad.org/en/PublicationsLibrary/webdiaeia2012d21 _en.pdf. 21 October 2012.

[57] UNESCAP. Game-Changing Resolutions for Inclusive & Sustainable Transformation in Asia and the Pacific – Closing Statements. 23 May 2012. http://www.unescap.org/speeches/game-changing-resolutions-inclusive-sustainable-transformation-asia-and-pacificc. 20 October 2012.

[58] ---. Member States. http://www.unescap.org/about/member-statess. 20 October 2012.

9

Biological treatment of textile wastewater and its re-use in irrigation: Encouraging water efficiency and sustainable development

S. Senthil Kumar, Mohamed Jaabir

PG and Research Department of Biotechnology, National College (Autonomous), Tiruchirappalli-620001, Tamil Nadu, India

Email address:

envsenthil@gmail.com(S. S. Kumar), senthil@nct.ac.in(S. S. Kumar),mohamedjaabir@nct.ac.in(M. Jaabir)

Abstract: The present study focused on the isolation of potential bacteria from contaminated soil of textile industries and subsequent employment of those organisms in treatment of textile waste-water. Wastewater was treated by novel isolates and the biologically treated wastewater was used for the irrigation (phytotoxicity evaluation) of two important edible crop plants (*Brassica nigra* and *Cyamopsis tetragonolobus*). For this, plants were grouped as I, II, III and IV that received the tap water, raw effluent, chemically treated and biologically treated wastewater respectively. 46 bacterial isolates were obtained and optimization of parameters revealed that one strain, namely UBL-27 (*Comamonas sp. UBL 27)* decolorized the wastewater to a max. of 80% in static (anoxic) condition at pH 8 in 24 hours at 32°C. There was a remarkable performance in the germination percentage under biologically-treated wastewater to about 83.6% when compared to that of Control Group producing 92.9%. In contrast to this, the germination % was significantly too low ($p \leq 0.05$) in the other cases with the raw wastewater and chemically treated wastewater. The wastewater had marked effect on the growth of the *Brassica nigra*, the height of the plant was higher in the biologically treated effluent (11.2 ± 0.4 cm) and control group (12.1 ± 0.2) than Group II ($8.9 \pm .17$ cm) and Group III (9 ± 0.2 cm). Weight of the plant was 1.95 ± 0.35 g and 1.68 ± 0.47 g in Group I and Group IV. It was significantly lower in case of Group II and Group III. In *Cyamopsis tetragonolobus*, heights of the plant among the four groups at the end of 80 days were 102.3 ± 3.4, 52 ± 7.6, 45.3 ± 4.9 and 92.8 ± 5 cm respectively. Similarly, no. of leaves/plant among the four groups was 49.2 ± 3.2, 26.8 ± 4.5, 32 ± 2.4 and 47 ± 4.5. Total yield of the plant under the experimental area for Group I was 3.15 ± 0.09 kg while that of the Group IV was 2.92 ± 0.09 kg. The yield was significantly lower in the Group II and III such as 1.67 ± 0.17 kg and 2.06 ± 0.22 kg respectively. To consolidate, the raw effluent has decreased the yield by more than 45% ($p \leq 0.05$) while that of the chemically treated group by more than 30%. Though, biologically treated wastewater may not be absolutely fit for drinking purposes or for recycling in dyeing processes, it is proved from this, that the eco-friendly alternative can be used for the irrigation purposes beside abatement of water and soil pollution.

Keywords: Textile Wastewater, Biodegradation, Comamonas Sp., Water Efficiency, Phytotoxicity, Textile Wastewater.

1. Introduction

India faces the serious problem of natural resource scarcity, especially that of water in view of population growth and economic development. Groundwater is the major and also an important source of water for the agricultural and industrial sectors in India. During the post-liberalization period of the Indian economy, the textile industries grew rapidly due to the availability of cheap water and labor. The cheap water that is rapidly pumped from underground aquifers is a major factor in the success of textile industries in contributing to India's economic growth. For instance, the textile valley of India, 'Tirupur-city' in the southern state of Tamil Nadu, is one of the largest and fastest growing urban agglomerations, also the textile hub of India and a vast generator of employment, income and foreign exchange - is gaining universal recognition as the leading source of hosiery, knitted garments, casual wear and sportswear.

Despite the recent rapid growth in industrial production, agriculture is still an integral part of India's economy and society. There is a constant competition over water between farming families and industrialists seeking to commodify the resource base for commercial gain. According to the Ministry of Water Resources, industrial water use in India stands at about nearly 6 per cent of total freshwater abstraction and the demand is expected to increase dramatically. On the other hand, Indian agriculture sector claims 90% of total water resources and has caused groundwater depletion. Analyses warn that the water demand in India will overshoot the current supply in the near future. Another major source of India's water supply is surface water, which is highly monsoon dependent. Only 48% of rainfall ends up in Indian rivers, especially in the peninsular rivers like 'Cauvery' in South India. Unfortunately, due to lack of storage and crumbling infrastructure, only 18% of the water can be utilized making water scarcity a critical problem (UNICEF, 2002). Climate change also has an effect on rainfall pattern and is exacerbating the depleting supply of water.

Invariably, ground water and surface water pollution due to unfettered economic growth and poor wastewater management practices of textile industries further worsen the issue of water scarcity. This acute water scarcity crisis in India along with pollution gives reason for concern and the need for appropriate water management practices focusing attention on water conservation, efficiency in water use, wastewater treatment and its recycling. Regardless of the many benefits from a thriving economy, considering both volumes discharged and wastewater composition, the wastewater generated by the textile industry is rated as the most polluting among all industrial sectors (Asgher et al, 2009; Lopez et al, 2006; Vanndevivera et al, 1998).Growing awareness on the impact of pollution in groundwater and surface water, textile wastewater treatment is now receiving greater attention. During the last few years, new and tighter regulations coupled with increased enforcement concerning wastewater discharges have been established in many countries including India. This new legislation, in conjunction with international trade pressures such as increasing competition and the introduction of 'ecolabels' for textile products on the European and US markets, is threatening the very survival of the Indian textile industry.

Considering the vast potential of textile industries in contributing to the Indian economy and supporting millions of people for employment, it becomes mandatory to save the industries from shutting down due to constraints of the Pollution Control Board of the country. At the same time, environmental deterioration also needs immediate attention for preventing any further damage and also to find a viable alternative to the problem. Though physico-chemical methods for wastewater treatment have been accepted and followed by textile industries, they are not cost-effective and involve large volumes of sludge disposal. Accumulation of such concentrated sludge poses a lot of practical difficulties in thedisposal of wastewater. Presently, sludge is being deposited into the lands owned by the

textile industries converting them into waste lands (Anjaneyulu et al, 2005; Robinson et al, 2001).

On the other hand, biological treatment could achieve greater efficiencies in the decolorization and detoxification of textile wastewater by making use of the native microorganisms. These microorganisms that are present in the soil of the textile industrial areas, where the dumped dye wastewater contains a lot of synthetic compounds, adapt themselves to the adversity of their micro-environment over the ages due to their persistency. There are a lot of reports suggesting the use of native bacterial isolates that have the capacity to decolorize the dyes that are commonly used in the respective textile industries (Asad et al, 2007; Ali, 2010; Jadhav et al, 2010; Saratale et al, 2010).Since many synthetic dyes and their derivatives are carcinogenic and mutagenic, the process of bioremediation requires not only the decolorization of the water containing the dye but also the complete degradation or detoxification of it (Couto, 2009; Forss and Welander, 2009; Kaushik and Malik, 2009). The re-use of bioremediated textile wastewater for agricultural irrigation is often viewed as a positive means of recycling water due to the potential large volumes of water that can be used.

The present study was an attempt in finding an economically and environmentally (eco) friendly solution for the prevailing textile wastewater pollution. It is focused on the isolation of potential microorganisms (especially bacteria) from the contaminated soils of the industrial areas of Tirupur district, Tamil Nadu, South India, and the subsequent employment of those organisms in decolorization, degradation and detoxification of textile wastewater, while re-using the water for irrigation thereby reducing the demand–supply gap on water, encouraging water efficiency and sustainable development. The remainder of this paper is organized into three sections: section one describes the methodology adopted for the biological treatment of the textile wastewater and phytotoxicity studies, section two presents the results and discussion and the last section provides conclusions and policy implications.

2. Materials and Methods

Textile Wastewater Collection: The raw discharge (Raw wastewater/ wastewater) of the dyeing unit of United Bleacher's (UBL) Pvt. Ltd, Mettuppalayam, Tamil Nadu, India, was collected in barrels and transported to the laboratory within 24 hours.

Isolation and screening of textile waste water decolorizing microorganisms: Soil samples were collected from five different sites around the industry including the drains. All the soil samples were mixed and used for the isolation of dye-decolorizing microorganisms due to the decades-long usage of the location since the establishment of the industry. The soil samples were serially diluted by following the standard protocol and the dilution series from 10^{-2} to 10^{-7} was plated in Nutrient Agar medium (Elisangela et al, 2009). Each dilution was maintained in triplicates. All the plates were incubated at 37°C for 24 h.Pure cultures were raised and maintained on nutrient agar slants at 4°C.

*Decolorization experiments:*A loopful of each isolated bacterial culture was inoculated into a separate 250 ml Erlenmeyer flask containing the raw wastewater in nutrient broth and incubated for 24 h at 37°C for the initial screening of the ability of the isolates to decolorize the wastewater. Aliquots of the culture (3 ml) were withdrawn at different time intervals and centrifuged at 5000 rpm for 15 min to separate the bacterial cell mass. Decolorization was determined by measuring the absorbance of the decolorization medium at absorption maxima, and the percentage of decolorization was calculated as follows: (%) Decolorization = {(Initial absorbance – Observed absorbance) / Initial absorbance} X 100. All decolorization experiments were performed in triplicates (Elisangela,et al, 2009; Saratale,et al, 2006).

*Growth medium:*The growth medium used for this study was nutrient broth. However, to evaluate the nutritional requirements and to optimize the decolorization process of the isolate, the experiments were conducted with minimal media (mg L^{-1}) containing Glucose 1800; $MgSO_4.7H_2O$ 250; KH_2PO_4 2,310; K_2HPO_4 5,550; $(NH_4)_2SO_4$ 1,980. All chemicals used were of highest purity available and of analytical grade.

*16s rDNA sequencing:*The chromosomal DNA of the strains with the best decolorization potential was isolated according to the procedure described earlier (Rainey,1996). A partial DNA sequence for 16S rRNA gene was amplified by using 5' - ATG GAT CCG GGG GTT TGA TCC TGG CTC AGG-3' (forward primer) and 5'-TAT CTG CAG TGG TGT GAC GGG GGG TGG-3' (reverse primer) (Jing et al, 2004; Senthil Kumar et al, 2010). The nucleotide sequence analysis of the sequence was done at Blast-n site at   NCBI   server (http://www.ncbi.nlm.nih.gov/BLAST).*Phytotoxicity studies:*To assess the toxicity of the above biological treatment and to compare it with that of the chemically treated wastewater (physico-chemical treatment followed in the industry) and untreated (raw effluent) wastewater, black mustard (*Brassica nigra* L.) and cluster beans (*Cyamopsis tetragonolobus* L.) were chosen.

*Germination index and morphological studies:*Emergence of shoot to 0.5cm in length or more was considered germination of the seeds and the germination percentage was determined (Kaushik, 2005). *Cyamopsis tetragonolobus L.* (cluster beans) was chosen for the field trial in which the plants were grouped as I, II, III and IV receiving tap water, raw wastewater (untreated), chemically treated wastewater and biologically treated wastewater, respectively (10 x 15 square feet area each). The morphometric parameters (shoot and root length; dry and wet weight of the whole plant, root and shoot; no. of leaves, nodes, pods, seeds per pod and pod length); yield parameters (weight of a pod, total yield of a plant); and biochemical parameters (chlorophyll, total carbohydrate, protein and reducing sugar) were compared among the treatment groups.

*Statistical analysis:*The data recorded in all the experiments were subjected to Two-Way Analysis of Variance (ANOVA) using MS-Excel (Office 2007). Readings were considered significant when P was $p \leq 0.05$.

3. Results and Discussion

From the soil sample taken at the United Bleacher's Pvt. Limited, Mettuppalayam, Tamil Nadu, bacterial cultures were raised in the laboratory and 46 bacterial isolates were identified based on their distinct colony morphology and was named from UBL 01 to UBL 46. Among the 46 bacterial isolates tested for decolorizing the wastewater in the nutrient medium, only six bacterial cultures demonstrated promising decolorizing activity with over 45% on an average, in shaking condition within 72 h when optically measured at 598nm (data not shown). Those six isolates namely UBL02, 03, 04, 23, 24 and 27, were subjected to 16S rDNA sequencing method of identification and found to be *Enterococcus faecalis* (HM451428), *Bacillus thurungiensis* (HM451439), *Bacillus sp.* (HM45431), *Bacillus megaterium* (HM451443), *Bacillus flexus* (HM451429) and *Comamonas sp.*(HM451426) respectively. All the 46 isolates were also tested in static condition (microaerophilic) in nutrient broth. In this experiment, bacterial strain UBL 27 decolorized the wastewater to a maximum of 57.67±0.61 %. This isolate was one among the six isolates that performed well in shaking condition. The other five isolates UBL02, 03, 04, 23 and 24 produced decolorization to similar levels in that of the shaking condition. Based on the findings, UBL-27 was chosen for optimization studies. Experiments revealed that UBL 27 was the best decolorizer for the wastewater (73.2 %) and worked best at pH 8, 30°C in static condition (data not shown).

The difficulties encountered in the wastewater treatment resulting from dyeing operations lies in the wide variability of the dyes used and in the excessive color of the wastewater/effluents (Machado et al, 2006). Thus, in spite of the high decolorization efficiency of some strains, decolorizing a real industrial effluent is quite troublesome (Wesenberg et al, 2003). This is evident in the present study. Bacterial isolates such as UBL01, 02, 03, 43 and 45 that demonstrated significantly high capacityto decolorize the disperse group of dyes (data not shown), could not show similar efficiency in decolorizing the real-time textile wastewater/effluent from the same industry (United Bleachers Pvt. Ltd, Mettuppalayam, Tamil Nadu, India). However, decolorization using fungal culture has greater disadvantages due to 'blanket of biomass' (Overgrowth of mycelial growth) and in further downstream processing of the wastewater. For the convenience and effective management of the treatment plant, use of bacteria is greatly admired.

Physicochemical status of the wastewater samples of

United Bleachers Pvt. Ltd revealed a reasonably high load of pollution indicators compared to the prescribed standards of Pollution board (Banat et al, 1996) (data not shown). There was a gradual change in the color from dark brown to light brown of the textile wastewater from the source (collection point at the textile industry) to the 'Lab-sink' (a temporary storage tank for textile wastewater transport and usage in the laboratory) indicating sign of decolorization. The decreasing color intensity of the wastewater has been related to absorption / chemical transformation of dyes (including metal complex by biotic or abiotic components of the wastewater/effluent) (Blanquez et al, 2004). The increasing bacterial count at the 'lab-sink' might have been responsible for such color change in the present study.

Initially the temperature of the wastewater generated from UBL was considerably high, however, declined to mesophilic status (30°C) at the 'lab-sink', which ultimately has favored biologically mediated remediation of effluent. This was supported by the finding through optimization (data not shown) where 30°C was found to be the optimum temperature for maximum decolorization of the isolates. This is in consistency with the findings of Asgher et al, (2009); Muhammad et al,(2009); and Swamy and Ramsay, (2007). The trend in decolorization decreased above and below 30⁰C. Incubation temperature is a very critical process parameter which varies from organism to organism and slight changes in temperature may affect its growth and ultimately its enzyme production. Higher temperatures may inhibit the growth of organism and enzyme formation which is responsible decolorization (Asgher et al,2009). Bioremediation at higher temperature (40°C) reduces solubility of gases in water that ultimately express as high BOD/COD. This increase in temperature reduced the biodecolorization by almost 10% at 40°C than that at 30°C. High values of BOD/COD as observed in present case demands significant amount of dissolved oxygen for enhanced intrinsic remediation of wastewater. Generally alkaline pH of textile wastewater/effluent is associated with the process of bleaching (Banat et al, 1996) and it is extremely undesirable in a water ecosystem. Both chemically and biologically mediated adsorption/reduction of dyes are initiated with decreasing pH level under redox-mediating compounds (Van der Zee et al, 2003). Decrease in pH i.e., from 10.2 to 8.0 of the effluent significantly improved bacterial count and thereby associated remediation. This is consistent with the findings of Naeem et al, (2009). Total Suspended Solids and Total Dissolved Solids in effluents correspond to filterable and non filterable residues, respectively. Reduction in pH for bioremediated favored microbial growth and the lattereventually resulted in increased in flocculation contributing to the rise as TSS. Microbial community (both aerobic and anaerobic) establishes itself in granulated floc as activated sludge plays a vital role in biodecolorization /bioremediation of wastewater (Lin and Liu, 1994; Lin and Peng, 1996). In the present study, among the 46 bacterial isolates screened for the bioremediation process, only a few isolates show potential decolorizing abilities though of varying degrees under shaking and non shaking conditions. A detailed physiological understanding of such microbes is much needed for bioremediation technology in future.

There has been a strong global awakening during the last few decades regardingthe proper management of existing natural resources. Among them, irrigationwater is one which is becoming costlier due to increasing demands of humanpopulation. Simultaneously the demand for food is also increasing, which has brought moreand more land under cultivation and focused the attention on fertilizer and irrigationwater. With these certain limitations, one has to turn to non-conventional recourses tomeet the irrigation water demand. Among others, one of the most important irrigation aswell as nutrient resources is industrial wastewater, which consists of about 95% waterand the rest as organic and inorganic nutrients. Since, its disposal is a big problem in urbanareas, applying the textile wastewater to agricultural fields instead of disposing off in lakes

and rivers can make crops grow better due to the presence of various nutrients like N, P, Ca, Mgetc. (Kannan et al,2005 and Khan et al, 2003).Use of untreated and treated textile wastewater for agriculture purpose has direct impact on the fertility of soil (Jadhav et al, 2010). Therefore, it is of concern to assess the phytotoxicity of the textile dye effluent before and after degradation by any mode of treatment. Seed germination and plant growth bioassays are the most common techniques used to evaluate the phytotoxicity (Saratale et al,2010). Two important edible crop plants (*Brassica nigra L.* and *Cyamopsis tetragonolobus L.)* have been tested for phytotoxicity of the biologically treated effluents in comparison to the raw wastewater/effluent and chemically treated wastewater.

There was a remarkable performance in the germination percentage of mustard and cluster bean seeds under biologically treated wastewater to about 83.66±0.5 % and 81.31±3.24 % when compared to that of the Control (Group I) on day 4 and day 7, respectively (Table 1). From the results, it was obvious that the raw wastewater (Group II) and the chemically treated wastewater (Group III) reduced the seed germination and early growth in both the test plants (Table 1, 2 & 3). Further, biologically treated textile wastewater did not show any inhibitory effect on seed germination. These findings are in accordance with the others in similar studies. Mohammad and Khan (1985) found that industrial wastewater reduced the germination percentage of kidney bean (*Phaseolus aureus*) and lady's finger (*Abelmoschus esculentus*). Dayama (1987), while working with *Cicer arietinum*, reported that even highly diluted industrial wastewater (5% of industrial wastewater) adversely reduced the seed germination as in *Sesamum indicum* (Neelam and Sahai, 1988), *Holchus lanatus* and *Agrostiss tolonifer* (Amzallag, 1999). The reduction in germination percentage in untreated textile wastewater might have been due to presence of high concentration of Ni^{2+}, Cd^{2+}, Pb^{2+}, and other toxic organic compounds that cause a range of cellular toxicities (Kadar and Kastori, 2003). Although the osmotic potential of the wastewater was not recorded, it is also possible that the presence of high amount of salts and organic compounds in untreated textile wastewater reduces the availability of water thereby resulting in reduced germination. This aspect is further supported by the fact that the presence of high salts in water or soil reduces the germination and early growth of plants by salt-induced osmotic stress that varies from species to species as reported by Ashraf (2004). Few years back, Ramana et al. (2002) found that the osmotic potential of the distillery wastewater is higher at higher concentrations, which retards germination of different vegetable crops.

Slight reduction in growth of the vegetable crops irrigated with 100% treated textile wastewater(Table 2 & 3) might have been due to the persistent levels of some heavy metals (Ramana, 2002). This argument supports the findings of Srivastava and Sahai (1987) who reported that irrigation with distillery wastewater reduced the growth of *Cicer arietinum* (Srivastava and Sahai, 1987). Overall, this inhibition in growth and germination may have been due to the presence of heavy metals such as Ni^{2+} and Pb^{2+} that cause toxicity at cellular as well as at the whole plant level (Kadar and Kastori, 2003). Furthermore, presence of heavy metals in the growth medium also causes reduction in uptake of other essential nutrients thereby resulting in reduced growth (Table 3). For example, while working with tomato seedlings, Palacios et al. (1998) concluded that the presence of elevated concentration of Ni in the rooting medium may cause disturbances and imbalances in different essential mineral elements (Palacios et al, 1998).

Photosynthetic pigments such as chlorophyll 'a', 'b' and total chlorophyll were decreased in the Group II and Group III plants compared to that of Control and Group IV (data not shown). A decrease in chlorophyll content may be due to either the inhibition of chlorophyll synthesis or its destruction or replacement of Mg ions (Barcelo and Gunse, 1985; Chandra et al, 2009). Sahai et al. (1983) reported similar observation when *Phaselous radiatious* was treated with distillery wastewater (Sahai et al, 1983). The more adverse effect of raw textile wastewater and chemically treated wastewater on photosynthetic pigments in *Brassica nigra* and *Cyamopsis tetragonologus* can also be due to the damage done by the presence of heavy metals like Ni and Pb to the photosynthetic apparatus, which possesses both chlorophyll a and b (Seregin and Kozhevnikova, 2006). Similar results were found by Krupa et al. (1993) and Sheoranet al. (1990) who observed a reduction in growth and chlorophyll concentration of bean and pigeon pea, respectively (Krupa et al, 1993; Sheoran et al, 1990).Consequently the two crops irrigated with biologically treated wastewater have demonstrated better growth and yield when compared to untreated (Group II) and chemically treated wastewaters (Group III) (Table 4 & 5). These results can be explained in view of the arguments of different scientists that heavy metals such as lead, nickel

and cadmium are highly toxic though at relatively low concentration. They interfere in enzyme action by replacing metals ions from metallo-enzymes and inhibit different physiological processes of plants resulting in poor growth and yield parameters (Kadar and Kastori, 2003; Palacios et al, 1998; Seregin and Kozhevnikova, 2006). However, biological treatment revealed better growth and yield parameters due to accumulation of such heavy metal ions by microbial uptake mechanisms (Ahluwalia and Goyal, 2007). Though, chemical treatment is being currently carried out in the industries that are under pressure to reduce pollutant load, the water that is drained into the streams is not significantly improved in qualities, in terms of toxicity. It is evident in the present study that the chemically treated wastewater is toxic to the plants. However, in biological treatment, though the clarity of the treated water is not to the desired re-use standard due to the biomass accumulation, the detoxification of the treated water is revealed by the phytotoxicity and yield parameters in the plants under study.Since water is a commodity of high demand in the country, with these limitations, one has to turn to non-conventional resources to meet the irrigation-water demand. Since the disposal of untreated textile wastewater is a major problem in urban areas, applying the bioremediation to textile wastewater can make crops grow better due to presence of various nutrients such as N, P, Ca and Mg (Kannan et al, 2005; Khan et al, 2003). Based on the above results, treatment of wastewater can be considered as an effective method in reusing the wastewater from industry for irrigation purposes.

4. Conclusion

In times gone by, staple food production has been dependent on irrigation, and irrigated production is estimated to account for 60% of the world-wide agricultural output. This also holds for India where the Green Revolution was responsible for countering the country's food deficit and was largely been successful due to ground water irrigation. However, currently, effects of overdrafts, like premature failure of wells and monsoon, decline in ground water yield and lowering water tables are apparent. This situation is expected to further worsen due to population growth and the increase in the effective demand for ground water by intensive agricultural production. Within this context, improving water use efficiency in face of the increased water deficit in agriculture requires an integrated approach with the strong commitment of all stakeholders (e.g. textile-industrialists, farmers, plant breeding industry and technology developers). The multiple issues related to water and agriculture are too often hampered by the lack of coordination and exchange of information. On the other hand, the enormous volume of water expelled out of the textile industry may help in encountering the water crisis. For this, treatment of wastewater water and elimination of pollutants is crucial for human health and environmental welfare. The proposed integrated approach of using adapted-bacteria in the detoxification and the re-use of wastewater in irrigation allows farmers, who are unable to access the ground water / unable to make necessary investments in tube wells, to meet their irrigation–water demand. The findings in our study are important to guide the government policy towards providing a solution for textile wastewater treatments and improved water recycling for agriculture. Government should facilitate policies encouraging wastewater treatment and re-use by developing a legal framework. The proposed integrated approach would help textile industries to comply with the norms of the Pollution Control Board; maintain the existing industrial infrastructure contributing to the Indian economy; reduce or eliminate solid/chemical waste deposition in the form of sludge out of chemical treatment and finally encourage industries to develop and maintain a greener environment by diverting biologically treated wastewater water into fields for irrigation, ensuring sustainability of the ground water. There is still plenty of scope for research in this area to improve and optimize the current methods of wastewater treatment. The increased attention to this topic will improve the health, economic and agricultural factors of a sustainably developing community.

Table 1. *Percentage of germination in Brassica nigra and Cyamopsis tetragonolobus in different treatment groups.*

Germination Index (%)	Group I	Group II	Group III	Group IV
Brassica nigra	91.66±2.51[a]	36.33±1.52[b]	29±3[c]	83.66±5.03[a]
Cyamopsis tetragonolobus	87.27±1.52 [a]	32.27±2.17 [b]	30.12±0.67[b]	81.31±3.24 [a]

Values followed by same letters in a column are not significantly different (p ≤0.05) (n=3, mean ± SD)

Table 2. *Mean values of total height; root and shoot length; wet and dry weight in Brassica nigra on day 12.*

Groups/ Parameters	Group I	Group II	Group III	Group IV
Total Height (cm)	13.3±0.5[a]	5.9±0.25[b]	4.93±0.2[c]	11.23±0.37[d]
Root Length (cm)	1.98±0.37[a]	1.47±0.15[a]	1.37±0.26[a]	1.71±0.25[a]
Shoot Length (cm)	11±0.1[a]	4.15±0.05[b]	3.33±0.23[c]	9.1±0.2[d]
Wet Wt. (g)	2.1±0.15[a]	0.92±0.035[b]	0.78±0.056[c]	1.52±0.04[e]
Dry Wt. (g)	0.28±0.32[a]	0.13±0.015[b]	0.09±0.005[c]	0.19±0.02[d]

Values followed by same letters in a column are not significantly different (p ≤0.05) (n=3, mean ± SD)

Table 3. *Mean values of the root and shoot length; height; wet weight in Cyamopsis tetragonolobus under different treatments.*

Groups/ Parameters	Group I	Group II	Group III	Group V
Total height (cm)	123±4.58[a]	75.6±3.05[b]	70.3±1.52[b]	102.6±5.03[c]
Root Length (cm)	26±2[a]	19.3±1.52[b]	18.6±1.52[b]	21±2.64[a,b]
Shoot Length (cm)	97±2.64[a]	56.3±1.52[b]	51±2[c]	81.6±2.51[d]
Wet Wt. of Plant (g)	260.3±5.13[a]	130±5[b]	124.3±4.04[b]	227.6±2.51[c]

Values followed by same letters in a column are not significantly different (p ≤0.05) (n=3, mean ± SD)

Table 4. *Length and Weight of individual pods, Yield/plant in Cyamopsis tetragonolobus among treatment groups*

Groups/ Parameter	Group I	Group II	Group III	Group IV
Pod length(cm)	16.3±0.258[a]	13.1±0.1[b]	12.9±0.15[b]	15.7±0.32[d,a]
Weight of individual pod (g)	5.97±0.08[a]	4.04±0.14[b]	3.73±0.19[b]	5.58±0.24[c,a]
Yield /plant (g)	181.57±3.75[a]	85.9±1.71[b]	75.1±4.57[c]	173.98±8.12[e,a]

Values followed by same letters in a column are not significantly different (p ≤0.05) (n=3, mean ± SD)

Table 5. Gross yield in Cyamopsis tetragonolobus among various treatment groups in the field study

Groups/ Parameter	Group I	Group II	Group III	Group IV
Total Yield (Kg)	3.15 ± 0.09^a	1.84 ± 0.06^b	1.64 ± 0.07^c	2.92 ± 0.09^d

Values followed by same letters in a column are not significantly different (p ≤0.05) (n=3, mean ± SD)

Acknowledgement

The authors are thankful and would like to acknowledge the Science and Engineering Research Board (SERB), Department of Science and Technology, Govt. of India, New Delhi for funding the research project. The authors are grateful to the Principal, Dr. K. Anbarasu and Shri. K. Ragunathan, the Secretary of National College (Autonomous), Tiruchirappalli, India, for all their encouragement in the pursuit of this project.

References

[1] Ahluwalia S S, Goyal D (2007) Microbial and plant derived biomass for removal of heavy metals from wastewater. Bioresource Technology 98: 2243–2257.

[2] Ali H (2010) Biodegradation of Synthetic Dyes - A Review. Water Air Soil Pollution 213: 251-273.

[3] Amzallag G N (1999). Regulation of Growth: the Meristem Network Approach. Plant Cell Environment, 22: 483–493.

[4] Anjaneyulu Y, Sreedhara Chary N, Raj D S S (2005) Decolorization of industrial wastewaters – available methods and emerging technologies – A review. Review of Environmental Science Biotechnology 4: 245–273.

[5] Asad S, Amoozegar M A, Pourbabaee A A, Sarbolouki M N, Dastgheib S M M (2007). Decolorization of textile azo dyes by newly isolated halophilic and halotolerant bacteria. Bioresource Technology 98: 2082–2088.

[6] Asgher M, Azim N, Bhatti H N (2009) Decolorization of practical textile industry wastewaters by white rot fungus Coriolus versicolor IBL-04. Biochemical Engineering Journal 47: 61–65.

[7] Ashraf M (2004) Some important physiological selection criteria for salt tolerance in plants. Flora 199: 361–376.

[8] Banat I M, Nigam P, Singh D, Marchant R (1996) Microbial decolorization of textile dye containing effluents, a review. Bioresour. Technol. 58: 217–227.

[9] Barcelo J, Gunse B C (1985) Poshenrieder: Effect of Cr (VI) on mineral element composition of bush beans. Journal of Plant Nutrition 8: 211-21.

[10] Blanquez P, Casas N, Font Z, Gabarrell X, Sarra M, Caminal G, Vicent T (2004) Mechanism of textile metal dye biotransformation by Trametes versicolor. Water. Res. 38(8): 2166–2172.

[11] Chandra R, Bhargava R N, Yadav S, Mohan D (2009) Accumulation and distribution of toxic metals in wheat (Triticum aestivum L.) and Indian mustard (Brassica compestries L.) irrigated with distillary and tannery wastewater. Journal of Hazardous Material 162: 1514 -1521.

[12] Couto S R (2009) Dye removal by immobilized fungi. Biotechnology Advances 27: 227– 235.

[13] Dayama O P (1987) Influence of dyeing and textile water pollution on nodulation and germination of gram (Cicer Arietinum). Acta Ecologia 9(2) : 34–37.

[14] Elisangela F, Andrea Z, Fabio D G, Ragagnin de Menezes Cristiano Regina D L, Artur C P (2009) Biodegradation of textile azo dyes by a facultative Staphylococcus arlettae strain VN-11 using a sequential microaerophilic/ aerobic process. International Biodeterioration and Biodegradation 63: 280–288.

[15] Forss J, Welander U (2009) Decolorization of reactive azo dyes with microorganisms growing on soft wood chips. International Biodeterioration and Biodegradation 63: 752– 758.

[16] Jadhav J P, Kalyani D C, Telke A A, Phugare S S, Govindwar S P (2010) Evaluation of the efficacy of a bacterial consortium for the removal of color reduction of heavy metals and toxicity of textile dye wastewater. Bioresource Technology 101: 165-173.

[17] Jing X, Shaorong G, Yong T, Ping G, Kun L, Shigui L (2004) Antagonism and molecular identification of an antiobiotic bacterium BS04 against phytopathogenic fungi and bacteria. High Technology Letters 10 (3): 47-51.

[18] Kadar I, Kastori R (2003) Mikroelem-terhelés hatása a repcére karbonátos csernozjom talajon. Agrokemiaes Talajtan 52: 331–346.

[19] Kannan V, Ramesh R, Sasikumar C (2005) Study on Ground water characteristics and the effects of discharged wastewaters fro textile units at Karur District. Journal of Environmental Biology 26(2): 269-272.

[20] Kaushik P, Garg V K, Singh B (2005) Effect of textile wastewater on different cultivar of wheat. Bioresource Technology 96:1189-1193.

[21] Kaushik P, Malik A (2009) Fungal dye decolorization: Recent advances and future potential. Environment International 35: 127–141.

[22] Khan N A, Gupta L, Javid S, Singh S, Khan M, Inam A, Samiullah (2003) Effects of sewage wastewater on morphophysiology and yield of Spinacia and Trigonella. IndianJournal of Plant Physiology 8(1): 74 -78.

[23] Krupa Z, Siedlecka A, Maksymiec W, Baszynski T (1993) In vivo response of photosynthetic apparatus of Phaseolus vulgaris L. to nickel toxicity. Journal of Plant Physiology 142: 664–668.

[24] Lin S H, Liu W Y (1994) Continuous treatment of textile water by ozonation and coagulation. J. Environ. Eng. 120 (2): 437–446.

[25] Lin S H, Peng C F (1996) Continuous treatment of textile wastewater by combined coagulation, electrochemical oxidation and activated sludge. Water Research. 30: 587–592.

[26] Lopez M J, Guisado G, Vargas-Garcia M C, Estrella F S, Moreno J (2006) Decolorization of industrial dyes by ligninolytic microorganisms isolated from compositing environment. Enzyme Microbial Technology 40: 42–4532.

[27] Machado K M G, Compart L C A, Morais R O, Rosa L H, Santos M H (2006) Biodegradation of reactive textile dyes by basidiomycetous fungi from Brazilian ecosystems. Brazilian Journal of Microbiology. 37: 481–487.

[28] Mohammad A, Khan A U (1985) Effect of a textile factory wastewater on soil and crop plants. Environmental Pollution 37: 131–148.

[29] Muhammad Asghar, Naseema Azim, Haq Nawaz Bhatti (2009) Decolorization of practical textile industry effluents by white rot fungus Coriolus versicolor IBL-04. Biochemical Engineering Journal. 47: 61-65.

[30] Neelam Sahai R (1988) Effect of textile wastewater on seed germination seedling growth pigment content and biomass of Sesamum indicum. Linnean. Journal of Environmental Biology 9: 45–50.

[31] Palacios G, Gomez I, Carbonell-Barrachina A, Pedreno J N, Mataix J (1998) Effect of nickel concentration on tomato plant nutrition and dry matter yield. Journal of Plant Nutrition 21: 2179–2191.

[32] Rainey F A, Ward Rainey N, Kroppenstedt R M, Stackebrandt E (1996) The genus Nocardiopsis represents a phylogenetically coherent taxon and a distinct actinomycetes lineage: proposal of Nocardiopsaceae fam. nov. International Journal of Systematic Bacteriology 46: 1088–1092.

[33] Ramana S, Biswas A K, Kundu S, Saha J K, Yadava R B R (2002) Effect of distillery wastewater on seed germination in some vegetable crops. Bioresource Technology 82: 273–275.

[34] Robinson T, McMullan G, Marchant R, Nigam P (2001) Remediation of dyes in textile wastewater: a critical review on current treatment technologies with a proposed alternative. Bioresource Technology 77: 247–55.

[35] Sahai R, Shukla N, Jabeen S, Saxena P K (1983) Pollution effect of distillery waste on the growth behaviour of Phaseolus radiatus L. Environment Pollution 37: 245–253.

[36] Saratale G D, Kalme S D, Govindwar S P (2006) Decolorization of textile dyes by Aspergillus ochraceus. Indian Journal of Biotechnology 5: 407–410.

[37] Saratale R G, Saratale G D, Chang J S, Govindwar S P (2010) Decolorization and biodegradation of reactive dyes and dye wastewater by a developed bacterial consortium. Biodegradation 21(6): 999-1015.

[38] Senthil Kumar S, Shariq Afsar T, Mohamed Yasar M, Mansoor Hussain A, Mohamed Jaabir M S (2010) A Study on Fungal Antagonism by Chitinolytic Bacterial Isolates from Prawn Culture Farms of Ramanathapuram District Tamil Nadu. Journal of Pure and Applied Microbiology 4(1): 429-432.

[39] Seregin I V, Kozhevnikova A D (2006) Physiological role of nickel and its toxic effects on higher plants. Russian Journal of Modern Phytophysiology 53: 257–277.

[40] Sheoran I S, Singal H R, Singh R (1990) Effect of cadmium and nickel on photosynthesis and the enzymes of the photosynthetic carbon reduction cycle in pigeon pea (Cajanus cajan L.). Photosynthesis Research 23: 345–351.

[41] Srivastava N, Sahai R (1987) Effect of distillery wastewater on the performance of Cicer arietinum (L). Environmental Pollution 43: 91–102.

[42] Swamy J, Ramsay J A (2007) The evaluation of white rot fungi in the decolorization of textile dyes. Enzymes and Microbial Technology. 24: 130-137.

[43] Van der Zee F P, Bisschops I A, Blanchard V G, Bouwman R H, Lettinga G, Field J A (2003) The contribution of biotic and abiotic processes during azo dye reduction in anaerobic sludge. Water. Res. 37 (13): 3098–3109.

[44] Vanndevivera P C, Bianchi R, Verstraete W (1998) Treatment and reuse of wastewater from the textile wet-processing industry: review of emerging technologies. Journal of Chemical Technology and Biotechnology 72: 289–302.

[45] Wesenberg D, Kyriakides I, Agathos S N (2003) White-rot fungi and their enzymes for the treatment of industrial dye effluents. Biotechnology Advances. 22: 161–187.

10

An anthropological approach to HEPPs in Eastern Anatolia: The case of Aksu Valley

Pervin Yanikkaya Aydemir

Master's Degree Program, Yeditepe University, Anthropology Department, Istanbul-Turkey

Email address:
pya.aydemir@gmail.com

Abstract: Both water and development have very important functions in human life. Throughout the history, people have designed and constructed dams, reservoirs and irrigation systems to supply agricultural lands with water as well as converting water into energy as part of development projects. While water resources development projects are mostly preferred as they are cheaper and clean compared to other alternatives, impacts of such projects on people, their livelihoods and nature have been particularly devastating in many parts of the world such as Asia, Africa and Latin America. Recently, with an argument of increasing energy demand and reduction in dependence on imported energy, Turkish government has initiated some sort of "mobilization" for small hydroelectric power plants (HEPPs) to be run by private companies, particulary in the Eastern Anatolia and Black Sea regions. Despite recent initiatives, there is no established water policy in Turkey. Outsourcing control over free-flowing streams out of local representational structures into the hands of private companies has resulted in social movements and protests against these projects. I conducted a fieldwork in one of the valleys in Eastern Anatolia where two HEPPs have been constructed. Methods used during the 8-week fieldwork included participant observation, focus group studies and in-depth interviews. Privatization of the water resource in the Aksu Valley (formerly Salaçor) not only gave the entire control of water to the contractor company for 49 years, but also left all the public services in the valley to the mercy of the company while use of water has been historically well-managed by the local community, who was in control and distribution of the water. This paper discusses outcomes of the HEPP project in daily life of the local people in Aksu Valley, asserting that users of water resources should have been considered as participants in water management, planning, and decision-making of development projects. *A drop of water is a sea to an ant*, Afghan proverb

Keywords: Anthropology, Water, Development, HEPPs, Eastern Anatolia

1. Introduction

Development is indispensable, so is water. It is apodeictic that both have very important functions in human life in relation to natural, cultural, spatial, temporal, physical, symbolic, artistic and ideational landscapes. However, the mutual relationship between development and water may have both negative and positive outcomes depending on their convergence. Lundquist and Gleick indicate that "Lack of water is a barrier to sustainable socioeconomic development; lack of development is a barrier to solving water problems. Because water integrates so many aspects of life, it must be given primer consideration in the context of development objectives. This includes the day-to-day management of water, decisions about allocations for socioeconomic activities, and the preservation of natural resource capital" [1].

At present, globalization and development are usually used interchangeably. As a post-World-War II phenomenon, globalization is associated with the emergence of development, where both have their roots in colonial histories [2]. Since many of the definitions of globalization feature mainly economic aspects, Ted Lewellen, with an attempt to make an anthropological definition of globalization, proposes that "Contemporary globalization is the increasing flow of trade, finance, culture, ideas, and people brought about by the sophisticated technology of communications and travel and by the worldwide spread of neoliberal capitalism and it is the local and regional adaptations to and resistances against these flows." [3]. Even though anthropologists did not see development as one of their areas of major interest, they often made assumptions in their ethnographies. Accordingly, Lewellen argues that globalization represents a significant break with traditional anthropology since "traditional anthropology looked at bounded cultures and communities" whereas "globalization theorists are more likely to be interested in transnationals, diasporas, nations that are scattered in many countries, and deterritorialized ethnicities.".Anthropology is not about isolated people living in distant and exotic places anymore, but about plurality and diversity of people who are made to be part of a universal world as bounded to each other in a dynamic and interactive way by means of various globalization processes. Accordingly, as countries of the world and their people have been tied into the global economy since 1960s, anthropologists as well as other social scientists, are becoming more concerned with the impact of globalization in social and environmental policies [4].While optimization of human adaptation and maintenance of ecosytems were primary concerns of cultural practices, new ecological or environmental anthropology blends theory with political awareness and policy concerns[5].The political and economic interconnectedness in the global world alters local ethnoecologies by challenging, transforming, and replacing them.

2. Environment and Anthropology

Since the earliest days of the discipline, anthropologists are interested in questioning how people modify, symbolize and adapt to their immediate environment. According to Edelman and Haugerud,in the 1970s anthropologists were under the influence of dependency and world-system theories, peasant studies and feminism, and their focus was centered on culture-political economy [6]. By the mid-80s, there was an important shift and anthropologists started to avoid systematic analyses of political economy and the new economic liberalism which favors fragmentary attacks on economic reductionism and cultural essentialism. In the 1990s, there were only a few anthropologists who resisted neoliberal arguments, favoring free markets in decision-making rather than governments while the opposite was also untrue, and preference of the late 20th century anthropology to focus on "flux and fragmentation rather than powerful economic actors perhaps reflected anthropology's traditional focus on small-scale phenomena". Similarly, economic policies and everyday politics are not among the points of interest of anthropological studies of the environment and resource conservation, which mainly focus on indigenous rights and social aspects of the nature. However, nature has always been one of the central concerns of anthropology, "whether in the field of folk-sciences and cultural ecology or in the study of myths and rituals linked to the environment and subsistence systems" [7].

Okongwu and Mencher argue that anthropologists have written mainly for other anthropologists, "not for those who have the power to change the world", and in reaction to this trend, some anthropologists pointed out the importance of anthropology moving to shape public policy and assist in formulating the critical issues of the society by proposing solutions that meet the desires and needs of local people, and creating a synergy between theory and practice [4]. Political positioning is inevitable in the production of anthropological knowledge since we became "aware of the profound political implications of seemingly objective forms of knowledge" [8]. In the meantime, in response to changing circumstances, a new ethnoecological model has emerged: sustainable development, which aims at culturally appropriate, ecologically sensitive, self-regenerating change. It became a very popular approach in planning and development projects, however there are only few examples of successful sustainable development projects [5] due to several reasons such as multiple definitions and interpretations of sustainability, lack of an integrated approach, lack of community involvement, lack of effective monitoring and evaluation, co-operative governance unsupportive of effective sustainable development, and unsustainable rising levels of natural resource abuse as outlined by Nealer and Naude [9]. Each of these problems experienced with sustainable development eventually intersects with multiple domains of anthropology,in one way or another. And, water is one of the main components of sustainable development and environmental issues.

3. Water and Anthropology

Water, as a basic human right, has become a compelling theme in anthropology due to scarcity all over the world. For anthropologists, it is not only a resource, but a substance connecting many realms of social life to study different forms of valuing water, unequal distribution of water, rules and institutions that govern water use and shape water politics and knowledge systems of water [10]. However, there is a need for a broader understanding of water as suggested by Blatter et al. [11] who emphasized the significance of 'renewed awareness of natural imperatives' and 'multiple meanings of water bound to various cultures' as an expansion to legal, technical and economic definitions of water. Focusing on connectivity and materiality aspects of water, Orlove and Caton suggest five principal themes for studying connectivity of water by anthropology: value (natural resources and human rights, equity (access and distribution), governance (organization and rules), politics (discourse and conflict) and knowledge (local/indigenous and scientific systems), and they conclude that 'waterworlds' must be studied ethnographically including all components such as 'waterscapes', 'watersheds' and 'water regimes' in connection to science and technology studies as well as maintaining connections with political ecology and material culture studies [10], particularly given the degeneration of sustainability of water into a political and ecological chaos due to its potentiality as a hydropower.

As a hydropower, water has been one of the essential

renewable energy resources. Throughout the history, human kind have designed and constructed dams, reservoirs and irrigation systems to supply agricultural lands with water as well as converting water into energy as part of development projects for modernization. The water resources development (WRD) projects are favored because they produce relatively clean energy besides being a cheaper option to invest and operate. However, these projects have pros and cons: on one hand, urban dwellers enjoy the availability of electricity; industries increase their productivity and create more jobs for workers; and farmers cultivate double or more crops resulting in increased agricultural production, on the other hand they create socioeconomic, cultural, psychological, ecological, even health problems in people living in the immediate environment of the constructions [12]. In modern times, we are all familiar with the most dramatic effect of large dams that were built all over the world resulting in relocation of 40-80 million people, mostly indigenous, tribal and peasant communities, from their lands in the past six decades as indicated by the World Commission on Dams [13]. The impacts of dam building on people and livelihoods have been particularly devastating in Asia, Africa and Latin America.

Unfortunately, through globalization and implementation of neoliberal policies, privatization of water resources takes the control of water from local communities and gives it to a small group of people and does not include all legitimate actors in decision-making processes although it has been recognized that "involvement of users and sharing of responsibilities and management tasks is a prerequisite for proper choice of technological and organizational approaches" [1]. Management of water not only as an economic entity, but a social entity, requires social participation. This participation should be democratized in the short and long runs, where democratization of participation will aim, in the short run, "to improve information flow and introduce realizations about the political dimensions of social participation" while in the long run, democratization of participation will aim "to promoteregulation of asymmetrical powers" [14]. The most effective water policies and institutions are those who involve users of water as participants in water management, planning, and decision-making. With experience, it has become clear that major decisions made without involving local communities and those affected by these decisions are more likely to fail. For example, Veronica Strang points out that "the process of centralisation, enlargement and alienation, compounded by the recent privatisation of the industry, has perceptually severed the ties between most people and water resources", leading to anxiety and anger of local people in Stour Valley, Dorset "since they feel that 'their' most vital substance is in the hands of a highly untrustworthy 'other" [15]. The impact of globalization and privatization of natural resources has been extensively studied and selected (16, 6, 2, 17). Outsourcing the resource control to the 'highly untrustworthy other' while

excluding local communities from the regulatory processes severely damage the notion of water as a collective good that has been well-managed so far by social agreements, and unmask the political actors acting on behalf of the international capital. The experience is no different in Turkey.

4. Water and Turkey

Turkey has a surface water gross production of 193 billion m^3. Fifty-eight percent of this gross potential represents usable water potential, where annually usable amount of water per capita is 1430 m^3making Turkey one of those countries with water stress. It is a known fact that with an estimated population of 100 million in 2030, Turkey will become one of those water-poor countries, with an annually usable amount of water per capita of 1000 m^3 [18]. Turkey, surrounded by waters on three sides, has a coastline of 8300 km, and a total land area of 779452 km^2, of which 98.17% comprising land, and 1.83% water. At present, there are 230 finished dams in Turkey. Consumption of water for drinking, irrigation and industry is 15%, 75% and 10%, respectively.

Although water is a public propertyall over the world and water resources are mainly managed by public institutions (99% in Asian, 97% in African, 96% in Central and East European, 95% in North American and 80% in Western European countries) [18], at present Turkey has adopted an aggressive commodification and privatization process for water services. The public aspect of water has been lost to a private and profitable commodity through legal mechanisms, which gradually render national and local organizations legally responsible from drinking, utility and irrigation waters into inactive entities such as General Directorate of State Hydraulic Works (DSİ), General Directorate of Rural Services (KHGM, already shut down), General Directorate of Electric Power Resources Survey and Development Administration (EIE), and the Bank of Provinces. However, it has been shown that decreasing investment, operation and control power of the public sector in favour of private sector leads to several problems in planning, control and environment [19]. Lack of effective planning and appropriate mechanisms for monitoring and surveillance presents further legal and environmental issues. Tekiner Kaya argues that it is, in fact, the lack of legal environmental regulations and inappropriate implementation of law for nature protection, which are directly reflected in both operational and initial investment costs that make hydroelectric energy cheaper in Turkey [20].

Turkey's recent energy policy is established around two main arguments; "increased energy demand" and "dependence on imported energy". With a focus mainly on management of supply rather than management of demand, the policy encourages energy production as much as possible while efficient use of energy is considered to be of secondary importance [21]. Like many other developing countries, Turkey also wants to make use of its energy resources at best in order to meet increasing demands of a rapidly rising population as well as reducing reliance on external financing and energy sources. Following the enactment of the Electricity Market Law No. 4628 on 20.02.2001, the Energy Market Regulatory Board of Turkey (EPDK) allowed private sector companies to build, generate, operate and distribute electricity by liberalising the market. This law paved the way for many national and international energy or irrelevant companies to scramble, particularly for renewable energy investments. Due to a

potential hydroelectric power capacity of 130 billion kWh/year [22], use of water for many streams in Anatolia has been contracted to private companies for 49 years, rendering it into a benefit-loss economy. According to the 2012 Report of Union of Chambers of Turkish Engineers and Architects [23] on Hydroelectric Power Plants (HEPPs), at present there is a total of 1941 projects for river-type HEPPs; 205 already operating, 514 being constructed, and 1222 with a ready research and master plan (Table 1). Furthermore, projects for 10000 micro power plants have been in progress (Fig. 1).

Table 1. Development of Hydroelectric Power Production in Turkey

Status	Quantity	Installed Capacity (MW)
In Operation	205	14.405,24
Construction Phase	514	14.098,52
Provincial Administration Survey, Master Plan, Planning and Final Design Ready	1222	47.067,34
Grand Total	1.941	75.571,10
Developed by Legal Entities Inside Grand Total	1215	5.360
Projects Developed by DSİ and EİE According to Article 4628 inside Grand Total	259	4.857

Source: TMMOB Hydroelectric Power Plants Report 2011

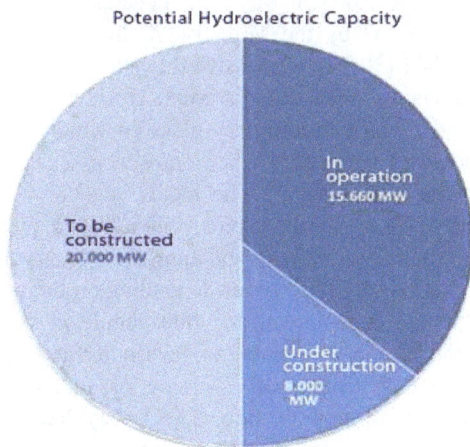

Figure 1. Potential Hydroelectric Capacity, Source: TMOBB Report 2011

5. Management of Hydroelectric Power Plant Projects in Turkey

Mostly, inelaborate feasibility, planning and localization of HEPPs result in juridical cancellation of many projects in most environmentally vulnerable regions. One of the major issues is flow requirements. Most of the projects promote release of only 10% of the average instream flow. The 2001 TMMOB Report indicates that EIA reports have serious scientific and technical problems as they are mostly prepared by 'copy-and-paste' technique. Even engineers who are in support of HEPP projects criticize planning and surveillance stages of these plants [24]. Social participation is not encouraged, and for projects under 25 MW, local

people are totally ignored while meetings held by the contractors for projects over 25 MW are usually non-functional [21]. In a survey with 90 people in Artvin for the present and planned dams and HEPPs on the Çoruh River, major negative outcomes listed by the local people included damage to green life, agricultural products, and culture by 50%, 33%,and 33% respectively while positive outcomes included changes in the regional climate, new job opportunities for local people and economic recovery by 22%, 15% and 8%, respectively [25]. As concluded in a parallel report by several NGOs for submission to the UN Commission, "The dams in the Çoruh Valley violate the right to an adequate standard of living in many respects." [26].

It has been reported by Özalp *et al.* that the government planned to build 15 large dams on the main channel of the Çoruh River Watershed, and 116 small HEPPs on the tributaries of the river, which are likely to disturb forests, water and soil during construction works due to steepness and roughness of the terrain in the region as well as serious damaging in riverine habitats, riparian zone vegetation, accelerated soil erosion and sedimentation problems due to changes and derivations of the river channel's water regime [27].

Inevitably, privatization of water resources and intrusions in many unheard valleys led to social movements and protests against these projects. Local people from these valleys gathered around a movement called "*Anadoluyu Vermiyoruz/We Will Not Surrender Anatolia*", and organized a nation-wide walk in January 2011 towards Ankara, the capital city of Turkey from all geographic regions under the threat of HEPP projects, but the walk was blocked by security forces, and they were not allowed to enter the city and visit the National Assembly [28]. While these protests are still going on mainly on local-basis, and rarely in big cities with the participation of locals and activist groups, both the local protestor and the activist-protestor struggle against the same water policy, i.e. the global capitalism, that, in fact, makes it a global class struggle, and thus these protests shouldn't be considered only as local ecological movements [29]. However, the contractors do not give up, and either through illegal ways or exploitation of legal gaps, they continue their construction activities with an 'intimate help' from the government by enactment of an expedited expropriation for energy projects.

Although it is believed that construction of a small HEPP does not necessarily require any change in the structure of the community, it, in fact, has a strong impact in power relations, daily-life settings, gendered economic activities, social activities, interpersonal relationships, etc. While administrators and some people from the local community may see this construction as a service and benefit to their daily life, some others may not favour such a radical change and disturbance in their daily life. This was exactly what happened in the Aksu Valley. While the site offers a variety of topics, each to be discussed elsewhere in detail such as 'extermination' of peasantry and subsistence-agriculture, intervention in a well-functioning hydrosocial circulation, impact of privatization on natural resources and destruction of ecosystems, strong formation of national and ethnic identity with respect to geography, social and cultural changes, role of religion and hegemonic power in social change, gendered activities, local architecture, toponymy, foods, etc., below you will find a story about how a valley has been appropriated politically by construction of two HEPPs to legitimize a variety of neoliberal practices in the name of energy development as a vivid example of 'neo-colonialism'.

6. Aksu Valley

Aksu Valley is located in the Eastern Anatolia. It is within the borders of Ispir, one of the northern districts of Erzurum. It has a transitional climate between the Eastern Black Sea Region and the East Anatolia. It is 163 km far from the Province of Erzurum, and 20 km from the District of Ispir. It is a green valley starting from the 20th km of Ispir - Artvin road that runs alongside the Çoruh River, and extending to the borderline of Hemşin District of Rize Province up toan altitude of approximately 1000 m. After the 35th km, it rises to an altitude of 3711m where the Verçenik Hill is one of the peaks of the renowned Kaçkar Mountains [30].

There are four villages in the Aksu Valley: Aksu, Yedigöl, Çatakkaya, and Yıldıztepe. Aksu Village has 13 quarters scattered around both banks of the streambed.The houses typical to this region with metal roofs were built on the rock slopes. Aksu, formerly Salaçor, is a very narrow valley

breached by the Aksu Stream, and the major basis for subsistence in the valley is horticultural products, particularly 'İspir bean' grown in an area of ca. 950.000 m^2 wheremost of it have been created out of nothing by local people through terracing, filled with soil carried on their back. So, the soil is not thick enough to retain water, that's why it needs frequent irrigation. Another major basis for subsistence is mulberries. Most of the people earn money from mulberry molasses. Apiculture is also a common source of subsistence with a production of approximately 20 tons/year. It is an organic valley since all crops are raised using animal manure and green manure not as part of a current trend, but traditional agricultural methods maintained for hundreds of years. Animal husbandry and dairy products are the main source of income for villages and quarters in the higher parts of the valley. With an ideal climate and water, the valley is also suitable for pisciculture. There is only one trout farm at present.

It is under the influence of Black Sea region climate. The weather is not too warm in the summer or too cold in the winter. It is home to various endangered animal species; mainly eurasian otter, chamois, black spotted wildcat, red speckled trout, wild goat, and grizzly bear. Organically grown apples, wild cherries, wild pears are among the fruits that grow in the area as well as walnut, hazelnut, hawthorn, hibiscus and linden trees. It was announced as a "Wild Life Development Area" by the Ministry of Environment and Forestry in 2004 [30]. To emphasize the natural uniqueness of their homeland, locals make a wonderful description of the valley; "It is an extraordinary place in which cherries blossom down in the valley when snowdrops (galanthus) flourish in the higher parts". The locals, men or women, even children know all the plants, animals and physical resources in their environment, and appreciate their value.

The population of valley varies depending on the season, from 450 people in winter up to 2000 people in summer. All of the inhabitants are related to each other, in one way or another, either from mother's side or father's side, sometimes even from both side. The valley had a self-sufficient population until 1970s when some men went to big cities to seek for a job because of sources that fell short with increasing population. After a while, some were followed by their immediate families. Now, most of the families have two homelands; 'place of birth' and 'place of breadwinning'. Majority of the people live in big cities, mostly engaged in bakery. The District of İspir is renown with its bakers across the country, so through relations they have easy access to such jobs. Some locals work at seasonal jobs, and return to the valley for summer. There is an in-betweenness of village and urban life reflecting itself in a complex relationship to modernity.

I conducted a fieldwork in the valley in the summer of 2012 for eight weeks to collect data for my master's thesis, which is about the impact of HEPP projects on daily life of local people living nearby. During my stay, in addition to participant observation, I conducted focus group studies and in-depth interviews with locals. I was very nicely welcome by the villagers, and they seemed to be happy to have someone living with them and sharing their experience. They had a lot to say, but they had serious trust issues, so did I. A few avoided participating in focus groups or interviews while some others seemed very enthusiastic about talking. This paper aims to give an overall picture of their experiences in relation to changes in daily life as a result of HEPP construction in their valley.

7. Hydroelectric Power Plants and their Impacts in the Valley

At present, there are two hydroelectric power plants in the 35-km long valley, and three other projects are on the way, as well as a dam project, which will flood the Dereağzı quarter located at the convergence of Aksu Stream with the Çoruh River. The two plants have been operated by Borusan EnBW Energy, which is a partnership between Borusan Group, one of the leading Turkish holdings engaged in steel, distributorship (BMW), logistics and energy industries, and EnBW Holding Ag, one of the leading energy companies in Germany. Two plants have an installed capacity of 50 MW, composed of 2 regulators, 2 tunnels, 2 powerhouses and switchyard facilities [31]. The Yedigöl plant started to operate in October 2011 while Aksu plant was not operating when I was in the valley. The immediate impact area of these two plants included four quarters, namely Kadıbağı, Massırdap (Yolbilen), Gılok (Otluca) and Dereağzı.

As well said by Jamie Linton, it is difficult to write, talk, and even think about water without involving people in the story, and "the state of water always reflects, in one way or another, the state of society." [32]. For local people, the Aksu Stream was the source of life and joy until huge construction machines and crowded workforce appeared at the confluence of their stream with the Çoruh River five years ago. Nobody knew why they came for. Neither official authorities nor the contractor company had organized a public consultation meeting before initiating construction activities. At first, the villagers were told that they were there to enlarge the road, and thus they were gladly welcome. However, after a while, it turned out that these machines and workforce were, in fact, part of a different agenda, which became evident day by day with dynamite explosions, thousands of cut trees, day-and-night traffic of hundreds of trucks and land-rovers, thick depositions of dust on trees, growing plants and crops, windows and interior of houses, hundreds of dying red speckled trouts, alterations to the stream bed, cloudy stream water and irrigation water, discharge of waste waters directly to the stream, towering piles of debris on the lands or in the stream, damages to the road and overall landscape, damages to the graveyards, rolling rocks from the hills, gradual elevations in the road, and making wild animals, particularly wild goats easier targets for hunters with a few unnatural pathways at certain points, etc. Such negative

events experienced in a cascading manner during the construction process were immediately reflected in inhabitants' quality of life, putting them into a sociological and psychological stress. The sociopsychological benefits they enjoyed living in a stable and isolated environment such as certainty, familiarity and comfort, and most importantly, their vital source, the Aksu Stream, and its environs suddenly had become under the threat of an energy project. They were about to lose the sound of the stream they were born into! Things were not what they used to be anymore in their landscape.

The threat induced by bulldozers into their life space and scenery resulted in complaints by the villagerswho developed anxiety and anger against the idea of losing their homeland, but they found nobody to direct their complaints. The official authorities did not pay attention to their questions while the contractor turned to a deaf ear. As a result, a few individual protests turned into a community action against the construction activities with an aim to protect and defend their living space. As Clifford Geertz argues, environment is an active and central factor in shaping social life, and "an established society is the end point of such a long history of adaptation to its environment that it has, as it were, made of that environment a dimension of itself' [33]. It was very unusual for these native people, who call themselves "Dadaş" to resist because they are renown with their strong nationalist feelings, and being Dadaş is identified with braveness, courageousness, fairness and chivalrousness. While they still feel uncomfortable to have shown disobedience and resistance to the authority, they say that they are always proud of being loyal to the state and obeying the authority, but, due to the threats to their living space accompanied with fear of loss of their land and possibility of relocation, they had nothing to do, but resist. One of the most common characteristics is that they are proud of never being a burden on the state, and they all tell the story about their grandparents who did not abandon their land and fought against the Russian army when Ottoman soldiers retreated from the region at the beginning of World War I. At that time, there was also an Armenian community living in the valley. Although officially changed, local people still use Armenian names of some quarters.

They also talk proudly about being able to collectively finance all public services by their own resources for years. For example, they indicate that they used their own collective resources to build the potable water pipeline from springs feeding the Aksu Stream, mosques in some of the quarters, the schools, the village clinic and a house for the midwife, the leisure rooms in each quarter, side roads to quarters, bridges over the stream and many other public services. They indicate the state was only involved in the construction of the once-a-stabilized road running alongside the stream, which was completed within a period of 13 years during governance of different governments.

8. Protests and Disrupted Social Order

Their resistance and protests became organized with the help of the Wildlife Conservation Society of Aksu Valley, which became active after arrival of the contractor in the valley. The activities of the Society representing all inhabitants of the valley are financially supported by the rich 'expatriate' people of the valley who live in big cities as well as membership fees. For a while later, their protests called attention of environmental activists, who supported protests against the HEPP projects across the country. The Society organized events in and outside the valley with their participation. Some of these events had widespread media coverage. For example, in a protest organized in İstanbul in November 2010, protestors involving local people of Aksu (both from 'place of birth' and 'place of breadwinning') and activitists wore masks of *Boldoroz* as a cultural symbol of a water monster narrated by the local people to keep their children away from the stream, and held a demonstration in front of a concert house where Borusan Quartet was giving a concert [34]. They had banners like; "Borusan, a friend of art and nature, destroys the nature of Aksu Valley", "Killer Borusan, get out of Aksu!", and "They dry out our streams, they kill our fishes; revolt!" accompanied with photographs showing the damages to the nature and graves in the valley. The company rejected these claims, and declared that they were acting within legal limits of their contract, respecting the life and expectations of the local people. Initially such protests had a broad repercussion in written and visual media. Many national and regional TV channels, reporters, and academicians visited the site to make the voice of Aksu people heard across the country and support their reactions. However, after a while the mainstream media stopped broadcasting about the protests against HEPP projects throughout the country as if the public dissent was squelched by invisible forces.

Despite so many efforts, nothing changed in the Aksu Valley. Whenever local people protested and tried to prevent contractor's activities, they found gendarmerie standing against them whereas they were expecting the gendarmerie to be on their side against improper working conditions of the contractor that threatened and changed their daily life. During interviews, they all mentioned that this was a real disappointment for them, particularly given the fact that gendarmerie has a particular place in daily life of rural people because their 'sons' also become gendarmerie when they serve in the army, so each *Mehmetçik* is everybody's son. Frequent confrontations with gendarmerie and their negative approach to the demands of the local people resulted in loss of trust and respect in this institution as well as the state itself.

In the meantime, the villagers became aware of the fact that the contractor started to win some of the villagers and protestors over, sometimes with money, sometimes by giving away some construction material like pipes, cement, etc., and sometimes by hiring them, which unsurprisingly resulted in conflicts among the local people. From then on, they started to engage in never ending discussions, arguments, and disputes, sometimes in loud outcries. Finally, when they found themselves alone with the contractor as 'orphans', in their words, feeling deserted by the state, some regretted to have been resisted to the contractor and the government by actively participating in protests, and believed that if they would have started negotiating with the contractor at the beginning, they might have been getting some benefit, while some others felt very disappointed with the outcome, and had two minds as to whether or not to move from the valley. At present, local authorities point the contractor for any public request made by the villagers, impelling them to become good negotiators with the contractor. When I was in the valley, the construction work was over, and the contractor was running one of the plants with a work force of 40 people, where 22 of them were hired from the Aksu Valley. The villagers told that these people preferred to work for the contractor although wealthy people of the valley offered them jobs with better conditions in other cities while the workers justify their preference stating that although they earn less, they are staying with their family, and taking care of their house, children, and crops. Although ultimately everybody made peace with each other, protestors still feel resent, and try to develop empathy with these 'traitors'.

No immediate impact was observed in the irrigation system. At present, the irrigation is carried out by conveying water from the stream through a number of ditches and using these ditches in a collective way in order. When someone finishes irrigating his/her (it is usually women doing these works) piece of land, s(he) puts the stone back to its place by the ditch to block the flow of water, and the neighbour starts irrigating his/her land. Since they usually know when and who is irrigating and see each other all the time, there is no flaw in the system. No disputes and no fights. Everybody knows what and when to do! The water is socially well controlled and regulated. Practice of collaborative use of these ditches forms sort ofa 'hydrosocial cycle' [35,32] within the valley, water shaping the culture and vice versa. In addition to being a source of their irrigation system, water is an inherent part of their daily life. Children (only boys) play, swim and catch fish in the stream; women and girls spend almost every day outside nearby the stream, irrigating, gardening, picking vegetables and fruits (particularly mulberries every morning from June to September), chatting, sometimes eating, and setting fire in their courtyard or near the stream under cauldrons to cook mulberry molasses ('pekmez'), which is a main source of subsistence for many of them; and retired and older men sitting all day longin the gazeboslocated mostly on the streamside, accompanied with young and working men during their resting hours, chatting and drinking tea, even eating while enjoying the view.

However, the contractor designed and installed a new irrigation system in the valley, which was on trial when I was there. The jerry-built system did not look promising when relatively complex use of physically demanding pipeline valves is considered. Claimed to be a favour to the people of valley by the contractor, it rather seems to be a sign of complete control of the stream in the future. Most probably, none of the villagers will be able to use existing ditches and/or excavate new ones to divert water from the stream to his/her land anymore. Although the contractor argues that the new system, which replaces ditches with contractor's pipes, will be more productive and efficient, everybody I talked had hesitations, expecting potential problems and disputes over irrigation next summer. This hesitation, in fact, is doubled by another fact, that is, the ecological amount of water that will be left to the stream for living organisms and irrigation ('cansuyu' in Turkish). When they went to the court, the villagers were only able to increase it from an average of 200 lt/sec to 830 lt/sec. They are very much concerned about the potential outcomes of reduced quantity of water including damage to their livelihoods, environment, biodiversity, natural food sources such as fish (already drastically reduced), crops, etc. The disturbed integrity of the stream is very likely to vindicate their concerns.

To their surprise, one issue that was well received and accepted by the contractor was placement of the poles for energy transmission line higher than it was planned. Interestingly, in the project the valley was shown as a moor without any settlement and inhabitants like many other valleys with a prospective HEPP project. It seems that what hurt local people most was to be regarded as "no one" in the official documents, but they remind that politicians will remember their existence when it comes to elections. This can be regarded as a sign of political awareness, in a way, referring to the strong relationship between water and power because now through the construction of a HEPP and an embedded intrusive contractor, the state became visible in their isolated living space, and acquired reality in their daily life whereas up until now it was mostly something to be confronted only when they went to İspir, the district center, to deal with a couple of official formalities, and when they used voting boxes.

It was upsetting to see that relationships among neighbours, relatives, and people living in other quarters had been all corrupted mainly because the contractor played them off against each other. Lack of trust to local and foreigner others was evident. Although almost all of them are related to each other, some people who had some benefit from the contractor seemed to have been ignored and isolated. Backbiting was common. Everybody said something behind others' back regarding their relationship with the contractor. I was also approached very cautiously by some locals unlike well-known traditional Turkish hospitality because of their preoccupation with what if I was a disguised contractor employee or a government official. I was explicitly asked a few times if I had some sort of a connection with the contractor.

Be it a woman or a man, they were very eager to talk about water whenever I raised a question. First thing they usually said was a passive reaction; "They took our water away", and "They poisoned our water". Most of the women were concerned about use of chemicals that may lead to cancer. During my stay, stomach upsets, diarrhoea and eye infections were common among villagers, most probably due to discharge of untreated sewage and solid wastes into the stream from some quarters, which was in fact a conflicting situation while they were complaining about contractor's polluting their water. Some already lost their land. The money paid by the contractor was spent shortly. Since the land was small, they were paid small. Now they are left with nothing. Some try to continue their subsistence by cultivating some relatives' land and share the earnings. Some women seemed to have been psychologically more affected, and obsessively repeated the same thing, e.g. "They took our water away". Even tough it seems unlikely in the near future, potential threat of resettlement makes them very depressed due to a strong feeling of belongingness to Aksu, and anxiety to lose homeland since their living space is, now, a market place.

In conclusion, control of the water is an extension of the control of local people and nature. In Aksu, subsistence agriculture is disappearing and being forced to become a part of the system, accompanied with a vanishing village life and culture. Local people are forced to break from their surrounding nature, and access to their own water is restricted.Discontinuation of a sustainable use and communal management of water putstheir cultural and social life at stake. With increasing number of examples, we see that gains from dams and hydroelectric plants to generate power, supply to urban life, and large scale irrigation projects are at the cost of social, cultural and ecological losses since these water development projects are usually devoid of its social content. In the Aksu Valley, an improperly managed and privatized energy development project, justifying itself for public good have incurred local people to exploitation of all natural and social resources by the private capital. Unfortunately, the state has sacrificed its citizens in favour of global corporate powers. Once there remains no need to remove the stone in front of the ditch to allow flow of water into a land, it is inevitable for all of us to see its reverberation as part of a global hydrosocial circle of waves.

References

[1]   J. Lundquist and P. Gleick, Sustaining Waters into the 21st Century. Background paper to the Comprehensive Assessment of Freshwater Resources of the World, Stockholm: Stockholm Environment Institute (1997).

[2]   S.B. Banerjee, C.M.C. Vanessa and M. Raza, The Imperial formations of globalization, in: Organizations, Markets and Imperial Formations: Towards and Anthropology of Globalization, Edward Elgar Publishing, Inc., Cheltenham, 2009, pp.3-16.

[3]   T.C. Lewellen, The Anthropology of Globalization: Cultural Anthropology Enters the 21st Century, Bergin &Garvey, Westport, 2002.

[4]   A.F. Okongwu and J.P. Mencher, The Anthropology of Public Policy: Shifting Terrains, Annual Review of Anthropology, Vol. 29 (2000) 107-127.

[5]   C.P. Kottak, The New Ecological Anthropology,in: N. Haenn and R. Wilk(Eds.), The Environment in Anthropology: A Reader in Ecology, Culture and Sustainable Living, New York University Press, New York, 2006, pp. 40-52.

[6]   M. Edelman and A. Haugerud, The Antropology of Development and Globalisation, in: M. Edelman and A. Haugerud(Eds.) The Anthropology of Development: From Classical Political Economy and Contemporary Neoliberalism, Blackwell Publishing, Oxford, 2005, pp. 1-74.

[7]   P. Descola and G. Pálsson, Introduction, in: P. Descola and G. Pálsson (Eds.), Nature and Society, Routledge, London-New York, edition published in Taylor&Francis E-Library, 2004, pp. 1-21.

[8]   S. Narotzky, The Production of Knowledge and The Production of Hegemony: Anthropological Theory and Political Struggles in Spain, in: G.L. Riberio and A. Escobar (Eds.), World Anthropologies: Disciplinary Formations within Systems of Power, Berg, Oxford and New York, 2006, pp.133-156.

[9]   E.J. Nealer and M. Naude, Integrated co-operative governance in the context of sustainable development, TD The Journal for Transdisciplinary Research in Southern Africa, 7(1) July (2011) 105-118.

[10] B. Orlove and S.C. Caton, Water Sustainability: Anthropological Approaches and Prospects, Annual Review of Anthropology, 39 (2010) 401-415.

[11] J. Blatter, H. Ingram and S.L. Levesque, Introduction, in Expanding Perspectives on Transboundary Water in:J. Blatter and H. Ingram (Eds.) Reflections on Water: New Approaches to Transboundary Conflicts and Cooperation, The MIT Press, Massachusetts and London, 2001, pp. 31-53.

[12] S.Sornmani and K. Okanurak, Socioeconomic and health impacts of water resources devleopment in Thailand, in: Brian H. Kay (Ed.)Water Resources: Health, Environment and Development, E&FN SPON, London and New Yok, 1999, pp. 229-245.

[13] World Commission on Dams. Dams and Development: A New Framework for Decision-Making: The Report of the World Commission on Dams. London and Sterling: Earthscan Publications (2000).

[14] K.A. Berry and E. Mollard, Introduction, in: K.A. Berry and E. Mollard (Eds.) Social Participation in Water Governance and Management: Critical and Global Perspectives, VA: Earthscan, London and Sterling, 2010, pp. xx-xxvii.

[15] V. Strang, The Meaning of Water, Berg, Oxford-New York, 2004.

[16] R.B. Johnston, L. Hiwasaki, I.J. Klaver, A.R. Castillo and V. Strang (Eds.), Water, Cultural Diversity and Global Environmental Change: Emerging Trends, Sustainable Futures,UNESCO and Springer, 2012.

[17] H.G. Brauch, P. Kameri-Mbote, U.O. Spring, N.C. Behera, J. Grin, B. Chorou, C. Mesjasz and H. Krummenacher in:H.G. Brauch (Ed.) Facing Global Environmental Change: Environmental, Human, Energy, Food, Health and Water Security Concepts. Hexagon Series on Human and Enviroment Security and Peace, Vol.4, Springer, Heiderberg, 2009.

[18] TMMOB Su Raporu: Küresel Su Politikaları ve Türkiye. Ankara, 2009. Information on www.tmmob.org.tr.

[19] F.H. Topçu, Hidroelektrik Santrallarında Kamu ve Özel Sektörün Rolünün Değişimi ve Yarattığı Sorunlar, Uluslararası Alanya İşletme Fakültesi Dergisi 3/1 (2011) 223-242.

[20] T. Kaya, Türkiye'de Su Gücü ve Küçük Hidroelektrik Santraller,Nevşehir Üniversitesi Sosyal Bilimler Enstitüsü Dergisi, 1(2011)207:238.

[21] O. Ürker and N. Çobanoğlu, Türkiye'de Hidroelektrik Santrallerin Durumu (HES'ler) ve Çevre Politikaları Bağlamında Değerlendirilmesi, Ankara Üniversitesi Sosyal Bilimler Enstitüsü Dergisi, 3(2) (2012) 65-88.

[22] Information on http://www.euas.gov.tr/Sayfalar/YillikRaporlar.aspx, last access on 15.07.2013

[23] TMMOB Hidroelektrik Santraller Raporu. Ankara, Ekim 2011, Information on www.tmmob.org.tr

[24] M. Hamsici, Dereler ve İsyanlar: "HES, HES, HES! Hadi be sen de! Kes, kes, kes! Beni iyi dinle!, NotaBene Yayınları, Ankara, 2011.

[25] R. Sever and Ö. Ulukalın, Artvin İlinde Yapılan/Yapılmakta Olan Barajlar Hakkında Artvin Halkının Bazı Görüşleri, Doğu Coğrafya Dergisi, 23 (2010) 65-80.

[26] Parallel report in response to the Initial Report by the Republic of Turkey on the Implementation of the International Covenant on Economic, Social and Cultural Rights, submitted on 14 March 2011 by CounterCurrent – GegenStrömung to the UN Committe on Economic, Social and Cultural Rights for its 46th Session, 2-20 May (2011).

[27] M. Özalp, O. Kurdoğlu, E.E. Yüksel and S. Yıldırımer, Artvin'de Nehir Tipi Hidroelektrik Santrallerin Olduğu/Olacağı Ekolojik ve Sosyal Sorunlar, Cilt:II. III. Karadeniz Ormancılık Kongresi, 20-22 Mayıs (2010) 677-687.

[28] Information on http://www.yesilgazete.org/page/3/?s=Buyuk+Anadolu+Yuruyusu&x=0&y=0, last access on 17.02.2012

[29] Ö.S. Işıl, Toplumsal Hareket Örneği: HES Karşıtı Mücadeleler, Sosyolojik Tartışmalar, 3(2012) 8-24.

[30] Information on http://www.necatiaksu.net/aksuvadisi/genel.htm, last access on 17.02.2013.

[31] Information on http://www.borusanenbw.com.tr/en/ProjectAndInvestments/ExistingPlants.aspx, last access on 17.02.2013.

[32] J. Linton, What is Water? The History of a Modern Abstraction, UBC Press,Vancouver, 2010.

[33] C. Geertz, The Wet and the Dry: Traditional Irrigation in

Bali and Morocco, Human Ecology, 1 (1972) 34-39.

[34] Information on http://bianet.org/bianet/bianet/125816-konser-oncesi-borusan-in-hes-insaatina-protesto, last access on 17.02.2013

[35] E. Swyngedouw, Social Power and the Urbanization of Water: Flows of Power, Oxford University Press, New York, 2004.

11

Assessment of the opportunities and threats of Maggaia-Lamido River Basin communities in Sokoto, Nigeria

Muhammad Muktar Namadi[1, *], Mohammed Yau[1], Faruruwa Mohammed Dahiru[1], Manu Haruna Isa[2], Katsina Sani Mamman[3]

[1]Chemistry Department, Nigerian Defence Academy, Kaduna, Nigeria
[2]RightLinks Integrated Services Limited Kaduna, Nigeria
[3]Desertification Control Dept. Federal Ministry of Environment Abuja, Nigeria

Email address:
ammimuktar@yahoo.com(Muhammad, M. N.)

Abstract: This paper investigates the environmental challenges associated with the management and utilization of Maggaia-Lamido River Basin, Sokoto State, Nigeria. It also identified the spatial distribution of communities whose means of livelihood depends heavily on the river basin under study. The paper identified that the most important environmental challenge is how to maintain the land base resource on which the economies of the local people are founded. The land base itself is challenged by desertification and behind this ecological problem stands increasing human pressures, such as, overgrazing, felling of trees for fuel, over cultivation of marginal lands etc. The results are escalating tension and conflict between villages and communities and increasing poverty, which further undermine the prospects for addressing the problems. Without significant change, environmental degradation will continue to step up within the communities like Kalmalo, Gidan Kaura and Tajaye- huchi and the natural resource base on which people's livelihood depend on will continue to decline. Equally, this paper identified the prospects of Maggaia-Lamido river basin and proposed a paradigm shift to an innovative community based Integrated Water Resource Management (IWRM) strategy to serve as a catalyst for poverty reduction and increase societal resilience to the impact of climate change.

Keywords: Problems, Prospects, Water Resource Management, River Basin

1. Introduction

Water is life and essential resource that supports existence of both animals and plants. It plays a key role in the metabolic breakdown of essential molecules such as protein and carbohydrates. Of the five basic human needs (water, food, health, education, peace) water is a common factor to the other four. Water supply is central to life and civilization (CSD, 2000). While the efficiency of food production depends on water availability, most society's socio-economic activities largely rely on both quantity and quality of water. Though, water is present everywhere, its use has always been constrained in terms of availability, quantity and quality (Biswas, 2004). Water has been a very important factor in settlement development in the country where it usually serves as human settlement boundaries (FGN, 2004). Water is constantly in motion, passing from one location to another, which makes its rational planning and management a very complex and difficult task. According to the Food and Agricultural Organization, out of the global available water resource, only 0.3% of it is freshwater (UN-HABITAT, 2008).

Freshwater is inland water that is fit for both agricultural and industrial uses and for human consumption. One of the sources of freshwater is river. A river may pass through several settlements and hence adjacent communities could benefit from the various services it provides such as drinking water, fishing, irrigation, domestic use etc. River Basin referred to as drainage system or basin, represents a system of interconnected system of water tributaries that flow towards single outlet (Longe et al., 2010). It combines the natural processes of precipitation, surface and ground water runoff with man-made features such as dams and reservoirs and hydro-power projects, diversions and irrigation schemes, industrial and residential water and environmental and cultural protection services (Global Water Partnership, 2003). Therefore, the aim of this paper is to assess the problems and prospects of Maggaia-Lamido River Basin and make recommendations in line with Community Based Integrated Water Resource Management principles.

2. Methodology

The methodology of this study involved review of available literatures and field survey. The desk review entailed searching national and international publications on community practices on natural resources utilization in the dry land ecosystems. Each practice area was theoretically assessed with respect to its capacity to promote sustainability in water resources management, management of vegetative cover, dry land farming, natural resource management, drought preparedness and coping capacity, pastoral development and management. Such an appraisal of existing research and coping mechanisms with environmental challenges in dry lands, permits the development of a conceptual framework for the identification of key issues that needed to be explored or clarify during the field survey that centered on consultation with local communities, policy and decision makers, researchers and scientists, on land use and conservation issues. Working within the broad spectrum of the Integrated Water Resource Management, this study employed the exploratory and topical Rapid Rural Appraisal (RRA) in the interview and field survey, using semi-structured group and individual interviews.

Fig 1: Map of Nigeria showing Sokoto state.

3. Overview of Maggia-Lamido River Basin

The Maggia River takes its source in Niger Republic and after 140 km, the river enters Nigeria via Birnin-Konni in Sokoto State Nigeria, after which the river is called Lamido (FMEnv 2002). After about 8 km, the river flows into Lake Kalmalo (see figure 2). The basin has a total land area of 4,138 km^2. While about 2,119 km^2 (51%) of the water basin is in Niger Republic, Nigeria accounts for the balance 2,019km^2 representing 49%. Essentially, the river contains

water only between July and October with annual discharge varying between 3 to 92mm per year (Agunbiade, 2002). A significant part of the river is captured in ponds and does not reach Lake Kalmalo. In Nigeria the water from Lake Kalmalo is used for drinking, irrigation and fishing. The Maggaia-Lamido River Basin cuts across Illela, Gwadabawa, Wurno, Goranyo, Gada, Sabon-Birni Local Government Areas of Sokoto State.

Fig 2: Map of Maggaia-Lamido River Basin

Table 1: *GPS location of settlements around the River Basin*

Settlement	GPS Coordinates
Kalmalo	05^0 14' 37.35E: 13^0 43'20.58N
Munwadata	5^0 16' 2.52E: 13^0 41' 37.98N
Runi	05^0 12' 27.53E: 13^0 41' 53.23N
Jema	05^0 10' 54.66E: 13^0 43' 9.78N
Gidan Kaura	05^0 44.15E: 13^0 35'.17N
Kaddi	05^0 43' 50.57E: 13^0 36' 8.31N
GidanGyado	05^0 43' 21.15E: 13^0 37' 37. 31N
GidanKashim	05^0 44' 8.74E: 13^0 38' 43.87N
Huchi	05^0 26' 44.53E: 13^0 22' 5.95N
Tajaye	05^0 26' 37.95E: 13^0 22' 18.08N
Salame	05^0 24 34.81E: 13^0 22' 26.92N

4. Biophysical Characteristics of Communities around the Basin

Climate: The communities lie within the semi arid region of northern Nigeria characterized by long dry season and very short rainy season. Average rainfall within the region is quite low, at roughly 250mm per year. The rainy season starts usually in June and ends in September. The dry season on the other hand stretches from October to May. Temperatures are high throughout the year with annual average of 35°C. However during the cool harmattan seasons (November – February), the temperatures range between 20°C – 25°C. The relative humidity ranges between 28% in the dry season to 68% in the rainy season. The average wind speed ranges between 145km/h. – 230km/h (Water Resources, 2006). Topography and Drainage: The area is of gently sloping terrain (1^0 – 3^0), drained by the seasonal Kalmalo stream. Lake Kalmalo that used to be perennial and the fadama areas have completely dried up. The ground water depth is more than 60m (ICRISAT, NA).

Soil: there are basically two soil types across the length and breadth of the six local government areas, namely Red Acid Sands and Yellow and Brown Acid Sands (UNDP/FAO, 1969). The Red Acid Sands are derived from eolian and fluviated deposits of Sangiwa, Tureta, Sokoto and Illela formations. The soils occur on flat to gently undulating terrains and the external drainage is good. The A horizons vary in thickness and are usually brown to reddish brown, fine to medium sands. They are virtually structure less or weak sub angular blocky, always very fragile and loose. The underlying horizons consist of yellowish red, loamy sands with weak sub angular blocky structure, very friable when moist, soft when dry and extend to a considerable depth. The PH values of the top layers show high acidity at less than 5 but there is a gradual increase in PH with depth, though it seldom exceeds 6 even in the lowest layers. The organic carbon content in the A horizon varies from 0.15 to 0.30%. Due to coarse texture, both permeability and infiltration rates are very rapid at 5 to 10m/hour or more and water holding capacity is low to very low, often less than 2m of moisture in the top 2ft of soil.

Yellow and Brown Acid Sands: The A horizon vary in thickness from 6 – 25in and consist of brown to pale brown fine sands with little or no structure. The consistence is friable to very friable when moist, slightly hard when dry. The underlying horizons consist of yellowish to pale brown or strongly fine sands to loamy sands, friable or very friable when moist to slightly hard to hard when dry. The structure is porous massive or weak sub angular blocky. The organic carbon content in the A horizon varies from 0.2 – 0.5% and the pH values show wide variation from 4 – 6.7. Both permeability and infiltration rates are very rapid and water holding capacity are very low. The soils with good external drainage and occurring on flat to gently undulating terrains are derived from eolian or fluviatile sediments of the Sangiwa, Sokoto, Tureta, Rabah, Zazagawa and Argungu formation (Agunbiade, 2002).

Vegetation/Forest & Grazing Reserves: The vegetation is of Sudan savannah type comprising of short grasses, shrubs and scattered trees. A forest reserve of about 60km^2 consisting of mostly 'neem trees' has been established in the area. A grazing reserve covering an area of about 10km^2 has also been established to serve as an important rainy season grazing area for transhumant pastoralists as well as for the herds of the sedentary farmers. The vegetation equally consists of scattered grasses and drought resistant trees. Common trees and shrubs found are Acacia albida (Gawo), Acacia nlotica (Bagaruwa), Acacia seyal (Farin Kaya), Balanite etc. Common Pest/Weed: Common pests in the community that present threat to agricultural production include quela birds, locust, Termites and rodents (Agunbiade, 2002).

5. Socio–Economic Characteristics, Environmental Challenges and Traditional Responses

Demography: Based on 2006 population census, the six local government areas through which the Maggaia-Lamido river basin traverse have the total population of one million one hundred and eighteen thousand three hundred and sixteen people (1, 182, 316) as shown in table 2.

Table2: population of the six Local governments

Local Govt. Area	Male	Female	Total
Illela	66112	84377	150489
Gwadabawa	118150	113208	231358
Wurno	83343	78964	162307
Goronyo	91694	90602	182296
Gada	122844	25423	248267
SabonBirni	99247	108352	207599
Grand Total	**581390**	**500926**	**1182316**

Source: NPC, 2006

Farming is the major economic activity in the area. Majority of the people are subsistence farmers engaging in both rain fed farming and fadama farming. Crops grown in the dry season under fadama farming include onion, tobacco, rice and maize. Crops grown under rain fed farming on the other hand include millet, sorghum, maize, and cowpea. Kalmalo is one of the major areas producing large quantity of onion within the basin and Sokoto state at large. Animal rearing is another important economic activity in Kalmalo area. Animals particularly cattle, sheep and goats are kept by both the Fulani pastoralists and sedentary Hausa farmers as symbol of wealth and source of income. Fishing used to be important in the community, but this has completely stopped due to the drying up of Lake Kalmalo. The major constraints to farming in the area are, drought, active sand dunes burying prime farmlands, weeds, pests, lack of farm implements and poor means of transportation. Similarly, the major constraints to livestock rearing in the area include; inadequate grazing reserve to accommodate the growing animal population, scarcity of dry season fodder, inadequate water points for the animals and lack of vet nary facilities.

Fig. 3: Onion Market - Kalmalo

Fig. 4: Fulani pastoralist Searching for Pasture - Gidan Kaura

Figure 5: Hand dug well in Kalmalo Community

Figure 6: Focus group discussion with the community members

Sources of Energy: Fuel wood and animal dung are the major sources of energy for domestic use in the communities. Cutting down of trees contributes to the process of desertification and land degradation in the area, while the use of animal dung deprives the soils of organic materials. Lack of alternative energy sources, poverty, level of illiteracy and increasing population has exacerbated the problem. There is also complete absent of either Government or civil society intervention to address these challenges.

Drinking Water Supply: Water shortage particularly in the dry season has been identified as the most critical challenge facing most of the communities. Between the months of June – September, the river serves the communities with drinking water. While in the dry season, the entire community in most cases depends on a single hand dug well as source of water for drinking and other domestic uses (see figure 5 & 6). The water drown from well or River is used directly without treatment. This has resulted to serious health problems in most communities around the basin.

Opportunities and Threats of the River Basin The persistent ecological challenges facing the communities around the river basin over the years has resulted in devising local adaptation strategies which increase their resilience to the impacts of draught and climate change. This community based approaches ranges from water resource management to land use, farming practices and energy supply. The community responses to the recurrent environmental challenges are summarized in table 3.

Table 3: Environmental Challenges and Traditional Responses

S/No	Environmental Challenge	Impact	Response
1	Active sand dunes	Burying of farmlands, settlements, roads and other facilities Reduced agricultural production Forced migration of people and animals	Use of palm leaves and straws to stabilize the some dunes
2	Drought	Declined crop yield Famine Acute water shortage Drying up of water bodies Loss of livestock Increased poverty Forced migration	Planting of drought resistant and short maturing crops. Engagement in off – farm activities Sales of stored grains and animals Migration out of the area
3	Wind Erosion	Land degradation	Tree planting
4	Declined soil fertility	Reduced crop production	Organic fertilizer (manure) Mixed cropping
4	Pests	Destruction of crops	Use of pesticides
5	Drying up of Kalmalo Lake	Reduced farming and fishing activities Lack of water for people and animals Migration of people and livestock out of the area Increased unemployment Reduced household income	No response

The Magga-Lamido river basin provides some vital services to the communities around it, some of these services include; drinking water and fishes. Table 4 presents some highlights of the prospects and challenges of the basin.

Table 4: Prospect and Challenges of Maggaia River Basin

Strengths	• Fishing • Irrigation • Water supply scheme • Aquatic life • Agriculture
Weaknesses	• Flooding • Scarcity of dry season fodder • Inadequate water supply • Drying up of Lake Kalmalo which affects fadama cultivation • Recurrent drought • Erratic rainfall • Weeds and pests • No institutional presence • Illiteracy of the local communities • Lack of facilities, utilities and services
Opportunities	• IWRM • Transportation • Recreation • Environmental sustainability • Economic efficiency • Ecological balance
Threats	• Climate change • Drop in water level/shrinking of water bodies • Dead of aquatic life • Deforestation • Communal conflict • Land degradation • Siltation • Pollution of water from the use of fertilizers

6. Integrated Water Resource Management

Integrated Water Resources Management (IWRM) refers to a process that promotes the coordinated development and management of water, land and related resources, in order to maximize the resultant economic and social welfare in an equitable manner without compromising the sustainability of vital ecosystems (ODA, 1996). It is also defined as a participatory planning and implementation process, based on sound science, which brings together stakeholders to determine how to meet society's long-term needs for water and coastal resources while maintaining essential ecological services and economic benefits (IUCN, 2006; Agishi, 2002). Generally, IWRM seeks to protect the environment, promote economic growth and engender sustainable agricultural development, support democratic participation in governance, and improve human health.

The key elements which IWRM tends to uphold are; coordinated process that brings together stakeholders,

focuses on both economic and social welfare and equity as well as protecting ecosystems, uses scientific tools and data to provide sound base for judgment and emphasizes proper governance involving democratic participation (World Bank Institute, 2006). The implementation of IWRM involved a logical sequence of phases that is supported by continuous events some of which are; initial planning process, vision statement that guide the future direction of the management process. Other events include situational analysis that elucidates the types of solutions deem necessary based on the water needs of the primary beneficiaries. It equally identifies the strengths and weaknesses in water resource management, points out the aspects that should be addressed in order to improve the situation and guides the path for attainment of stated vision. Selecting the suitable strategy is another important event in the implementation of IWRM. Taking a cue from the vision, the situation analysis, and the water resources strategy, an IWRM plan may be prepared. Its note worthy that IWRM is a participatory approach to water resources management where major stakeholders, communities and policy makers are brought together to come up with solutions that are environmentally sustainable, economically efficient and socially inclusive so as to maintain ecological balance of river basins.

It is against this background and our strong believe that community based natural resource management which springs from genuine community demand can nurture enterprises that both generate considerable income and improve the state of local ecosystems. These enterprises could be scale up to achieve a significant poverty reduction effect. Such stride could also be achieved through the implementation of community based approaches that agrees with the principle of IWRM. In the case of Maggaia-Lamido river basin therefore, the paper seeks to propose the followings;

a. Community Based Integrated Water Resource Management: The introduction of community based solutions calls for awareness building measures as well as organizational and technical support. Community based organization (CBO & NGO) and community leaders may provide essential inputs towards building community capacity for water resource management. This is in view of the fact that Community-based water arrangements should be allowed to play their full roles paramount to the sustainability of the people's immediate environment. Indeed, both the public sector and community-based water arrangements have their strengths and weaknesses as assessed , and the key question is not which one is best, but rather which combination is most likely appropriate to address needs in specific areas and in particular for those most at risk and the vulnerable rural poor women and men of Maggaia –Lamido River Basin communities.

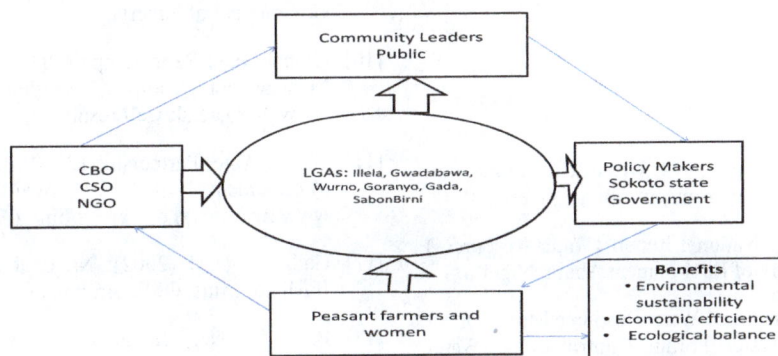

Fig 7: Proposed Community Based Integrated Water Resource Management Framework.

Task of each of the above proposed stakeholders:

LGAs: the six local governments through which the Maggaia-Lamido river basin pass through should provide logistics and counterpart funding and equally apply pressure on the Sokoto state to put in place necessary policy options

CBOs, CSO, NGO: these stakeholders would articulate the positions of the communities and apply pressure on the local government departments responsible for works and water supply.

Policy Makers: The state government through the ministries of Environment and Water resources should initiate an Integrated Water Resource Management action plan with clear vision, mission and objectives with definite timeline.

Community leaders and the general public: in line with bottom up approach to planning issues, the community should be engaged in thorough consultation processes to identify areas of felt needs and available sustainable options. This is with the view to accentuate what is working for the local communities.

Peasant farmers and women: these are the primary beneficiaries of any local based project.

b. Water harvesting: Climate change is to a great extent water change. Water is the primary medium through which climate change impacts will be felt by humans and the environment. The Impact of Climate change on water cuts across all sectors; therefore there is urgent need to embark on water harvesting measures so as to ensure full utilization of rain water.

The paper also proposed interventions based on Identified Priority Needs and weaknesses of the Communities along Maggaia-Lamido River basin (see table 5). Implementation of these interventions would potentially improve the livelihood of the communities and address poverty.

Table 5: Identified Priority Needs of the Community

Environmental Rehabilitation & Conservation	Alternative Energy	Infrastructure	Agriculture	Water Management/ Supply	Poverty Alleviation
1.Rehabilitation of abandoned gypsum mining sites 2.Afforestation including agro forestry 3.Establishment of rangelands to accommodate the growing livestock population 4. Rehabilitation of existing grazing land and	1.Woodlot establishment 2.Solar Energy 3. Biogas 4. Provision of Efficient wood stove	1. Road 2. Skill development center 3. Health centre 4. School 6 Community centre	1.Provision of fertilizer 2. Provision of credit facilities 3. provision of tractor hire services	1. Provision of boreholes 2.Construction of dam 3. Rain water harvesting facility	1 Skill acquisition centre 2 Credit facilities

7. Conclusion

In the final analysis, Maggaia-Lamido River Basin has indeed its strengths and weaknesses and the adjoining communities are the direct beneficiaries of these strengths and equally contribute to its deteriorations and current climate change issues have further worsen the already appalling situations. It is envisioned that Maggaia-Lamido River Basin could serve as a toolkit against poverty and environmental degradations evident in the communities along the basin only if the Authorities accentuate on what is working for the people. Integrated Water Resource Management is all encompassing, for its put into considerations community input, policy makers, and other stakeholders in the designing of water management processes. It is by so doing that felt needs could be identified and alternative means can be integrated to promote environmental sustainability and check the devastating effect of climate change on both human and the environment.

Reference

[1]   Agishi E (2002). National Report (Livestock Resources) on IEM Federal Ministry of Environment Abuja, Nigeria.

[2]   Agunbiade O.E (2002). National Report (Water Resources) on IEM Federal Ministry of Environment Abuja, Nigeria.

[3]   Biswas A.K (2004). Integrated Water Resource Management: A Reassessment of Water Forum Contributions. Water International; 29(2)

[4]   CSD, (2000). Commission on Sustainable Development : Integrated planning and management of land resources. Decision 8/3.

[5]   Encarta (2006), *Water.* Encarta Microsoft ltd

[6]   FMEnv., (2002). National Report on Integrated Management of Natural Resources in the Transboundary Areas of Nigeria & Niger

[7]   FGN, (2004). National Water Policy. Available online: www.waterresources.gov.ng/assets/ (Accessed May, 2013)

[8]   Longe E.O., Omole D.O., Adewumi I.K. and Ogbiye A.S., (2010). Water Resources Use, Abuse and Regulations In Nigeria. Journal of Sustainable Development in Africa; 12(.2).

[9]   NPC (2006): National Population Census, Federal Government of Nigeria.

[10] Global Water Partnership (2003): Integrated Water Resource Management Toolbox, version 2. Available online: www.iboro.ac.uk/well/resource/publ. (Accessed: 29/5/2013)

[11] Global Water Partnership (2000): Integrated Water Resource Management. Available online: www.iwawawiki.org/xwiki/bin/ (Accessed: 28/5/2013).

[12] Gadzama N.M (2002): National Report (Biodiversity) on IEMFed. Mins. Of Environment

[13] ICRISAT (NA). Integrated Ecosystem Management in the Trans-boundary Areas between Nigeria and Niger.

[14] IUCN (2006).Catchment Management Plan for Integrated Natural Resources Management of Komadugu – Yobe Basin

[15] Kolawale A (2002). National Report on Socio – Economic & Cultural Characteristics Federal Ministry of Environment Abuja, Nigeria.

[16] ODA (1996). Renewable Natural Resources Profile – NigeriaNatural Resources Institute

[17] UNDP/FAO (1969). Soil & Water Resources Survey of the Sokoto Valley; Final Report. 5(1).

[18] UN-HABITAT. 2008. State of the World's Cities 2008-2009.

[19] Water Resources, (2006). Nigeria Support to the Federal Ministry of Water Resources: Water Resources Management and Policy: Lot No 2; COWI A/S.

12

Adaptation of the submerged pumps intended for the irrigation in the arid regions

CHEBIHI Lakhdar, KHODJET KESBA Omar

Laboratory of Mobilization and valorization of the water resources Ecole Nationale Supérieure de l'Hydraulique, RN n 29, BP 31, Soumaâ, Blida 09000, Algeria

Email address:
thegreen_only@yahoo.fr (C. Lakhdar), okhodjet@yahoo.fr (K. K. Omar)

Abstract: The first area of work is to study only submersible pumps PUVAL "Pumps Valves" of Berrouaghia. While tracing the effects of abrasion on submersible pumps installed in the shelters in drilling and suffered the consequences are for irrigation in arid areas. The second line stain work to answer the following position: "Should we fight against the causes and not against the consequences?" The result of the proposals will help manufacturers to pump Algerian, one with the best hydraulic performance for the chosen material. Certainly, there have been abrasion tests on samples, but determining the duration of wear and the wear rate of the types of materials is always a line of news. Increasing the life of the pump while remaining within the proper range of operation.

Keywords: Optimization, Energy, Wheels, Abrasion, Performances

1. Introduction

In Algeria, and more particularly in the Sahara, the independent source of satisfaction of the demand for water it is the subterranean water, because of the aridity of the climate and the relatively easy exploitation of this underground resource.

The population growth, drillings albian, and the modernization of agriculture involve a major problem either by the reduction in the quantity or by the deterioration of the quality of this underground source by causing an unfavorable process of this subterranean water evolutionary in the space and the time, which present a current problem and remain not easily controllable.

Our zone of study, namely the wilaya of Ghardaïa, is located in an area Saharan, the repetitive dryness's and the limitation of the water resources did not allow the valorization of the enormous efforts authorized as regards agricultural hydro installations from where the development of the exploitation of the tablecloth.

2. General Information on the Wilaya of Ghardaïa

Ghardaïa is located at the center of the Northern part of the Sahara Algerian, It counts 13 communes distributed out of 03 daïras and covers a surface of 86,560 km ². [15]

The valley of M'Zab, founded in XI century, is a true museum with open sky located in full desert and covering a surface of 50 km ² (20 X 2,5 km). It is classified like national heritage in 1971 and inheritance of humanity by UNESCO since 1982.

Table 1. Surfaces of the communes of the wilaya [15].

Communes	Superficies (Km²)
Ghardaïa	300
El-Menia	27 000
Daya	2 175
Berriane	2 250
Metlili	7 300
Guerrara	2 900
EL-Atteuf	750
Zalfana	2 220
Sebseb	5 640
Bounoura	810
Hassi El-F'hel	6 715
Hassi El-Gara	22 000
Mansoura	6 500
Total	86 560

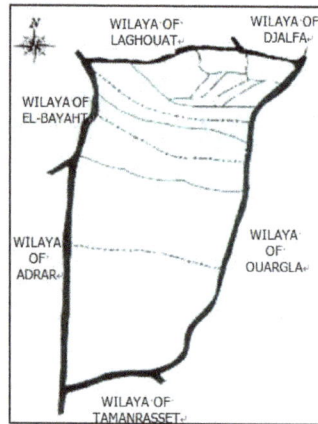

Figure 1. *Localization of the wilaya.*

2.1. Systems of Collecting of Subterranean Water

The subterranean water is preferred because it allows the space extension of the farms and because the works of collecting occupy of reduced spaces. It is presented by drillings of well going beyond the 50 m of depth and the realization to this level of galleries of collecting of source and storage. [13]

According to the monograph of the wilaya, the hydraulic systems are the following:

✓ Systems of collecting of subterranean water by the drilling of well exceeding 50m depths, and the realization of galleries of collecting of source and storage: kind of system of Foggara. [15]

✓ The systems storages of the water of believed by the realization of the dams of reserves in the level of palmerais and small tanks in each garden.

2.2. Resource of Subterranean Water

2.2.1. The Final Complex

The tablecloth of the Final Complex is contained in the various permeable horizons of the higher Cretaceous and the Tertiary sector. It extends in Algeria on a vast territory going from the dorsal from M'zab in the West until the hamada from Tinhert in the South. [12]

2.2.2. The Continental Guide

In geology, the Continental Guide indicates the continental formations which settled between the marine cycle of Paleozoic closes by orogenesis hercynienne and the marine invasion of Cénomanien. It covers a broad period of Sorted with the albian.

2.3. Drillings

The first drillings albian carried out in 1948 and 1950. The gushing impressing of subterranean water revealed the importance of the tablecloth contained in the sandstones of the Continental guide. [04] [14]

From this period, the number of drilling did not cease increasing. The wilaya account currently more than 288 drillings, which arise as follows:

75 for drinking water supply DWS ; 108 for the irrigation IRR ; 5 for the water supply for industry WSI ; 24 in mixed matter DWS and IRR ; 76 drillings not exploited for various reasons (equipment, not electrified…) [15]

3. Abrasive Wear

Abrasive wear is defined as being wear by displacement of matter transported by hard particles or hard protuberances. Thus abrasive wear is the result of the sliding friction by tilling or plastic deformation. Wear can be done with two bodies, directly starting from the solids in contact or with three bodies. [5] [10]

Figure 2. *Pump housing immersed damaged by abrasion.*

3.1. Diagrams of the Mechanisms of Abrasive Wear

According to the severity of the contact, several transformations of abraded surface occur. They depend on the mechanical properties and the geometry of materials in contact. If the surface of the solid solicited is regarded as perfectly planes, the following damages can be observed: cross, if the abrasive is sufficiently sharpened;

✓ embossing of the matter on the surface;

✓ fracturing, if the solid is fragile;

✓ Removal of grains, if the material is insufficiently homogeneous.

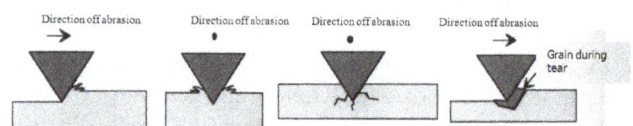

Figure 3. *Diagrams of the mechanisms of abrasive wear [10].*

3.2. Symptoms of Wear

The symptoms of wear are caused so much by the presence of these bubbles, like by their later collapse or of the implosion. [1]

The symptoms include/understand: A sudden fall of the discharge pressure, Popcorn noise of bursting, vibrations, sometimes violent, a deaf noise, Loss of great listening, a wear fish scales on the back face of the frequent need and turbine blades to adjust packing tree. [7]

3.3. Factors Influencing Wear

The depression during which a given pump starts to use depends on several factors: quantity of sand; the quality of sand; the total head total of the pump; the type of wheel; vibration (frequency and amplitude); penetration depth; carrying fluid (natural, circulation terms); tool (natural of material, forms and dimensions); part (natural of material, forms to realize). [11]

3.4. Parameters Influence Abrasive Wear

The quantity of the metal of the abrasive furrow eliminated from surface in the form of particles of wear compared to that affected by the plastic deformation is relatively weak, approximately 10% to 20% on average. This removed quantity depends on many parameters and more particularly by hardness and the form on the particles abrasives.

3.4.1. Hardness

When the hardness of striped surface approaches that of the abrasive, this one blunts, decreasing its aptitude for the cut. [6] [8]

3.4.2. The form

The form of the active part of the abrasive, if it presents sharp angles, suitably directed compared to the direction of slip, abrasion will be similar to a process of cut and the left trace will be indicated like a stripe of micro cut. [3]

4. Methods and Diagnoses

In order to achieve our goals, the followed step consists of prospections of several forms:

- Documentary prospections: it is carried out near the various services of DHW, AOW, NOC, NAHR and NOID. These prospections aim to characterize the zone of studies and to determine the problems well.
- Prospections on the ground near the farmers, of the suppliers of material of pumping and the companies of drilling. The prospections were carried out in the racks Hassi El-F'hel (F1, F2, F3 and F4) most developed as regards exploitation of subterranean water.

Work within the framework of the project has enabled to us to approach the socio-economic and technical context submerged pumps used for the management of subterranean water thanks to the training courses organized in the perimeters.

4.1. The Sample

With an aim of having a sample of the submerged pumps used by the farmers who exploit the tablecloth, it is necessary to leave on the site and carries out an inventory which locates at least the intake points (drillings of the private farmers).

Measurements of the flows for subterranean water were taken without any indication of the surfaces in addition to the coordinates which do not correspond inevitably to the data of the farmers.

4.1.1. Evolution of the Outputs

The farmers distinguish three successive periods:

- ✓ Period before the dryness's characterized by the availability of water and the acceptable outputs
- ✓ Period during the dryness's characterized by the fall of the outputs;
- ✓ Period of implementation of the pumping which made it possible to make up the deficit out of water and a certain improvement of the farming techniques

4.1.2. Oncern

To supplement the information received from the farmers, it was necessary to consult the company's suppliers of the equipment and the services. Main concerns which interest us are:

- ✓ To characterize the sector of private pumping and the methods of intervention in decision making and for the installation;
- ✓ To define the characteristics of the material of pumping.

4.2. Tests of Pumping

The tests of pumping constitute a preliminary stage before the exploitation of a given work, they are necessary to avoid the interferences between nearby works and to plan the flows and the schedules of operation of pumping for a whole long-term area in order to prevent the causes of wears for the pump. [2]

They consist in pumping during a certain time until the stability of the water level to the permanent mode in the objective to optimize the exploitation of the work (flow, time of pumping and the optimal position of the pump) and to consider the characteristics hydrological of the tablecloth to be exploited.

4.3. Methods of Analysis of the Data

The results of the prospections were individually treated then in a total way to detect the relations which exist to reach the ideal output with the use of these submerged pumps from point of view lasted life (operation), better output at end as much as possible to avoid the risks of wears generated by the particles of sands.

We tried in this study to integrate economic factors and organizational affecting the decisions of the farmers for better determining the problems.

The economic study was carried out while being based on the technical routes of each culture such as they are practiced by the farmers and by consulting some companies sometimes. The breeding's are not treated in this study considering unreliability of the results of the prospections.

4.4. Diagnosis of the Operation of the Pump

The THG (Total Height Gauge) is calculated under the most underprivileged conditions and by taking account of the pressure losses caused by the various bodies of the installation.

One calculation outputs mechanical $\eta_m = P_m/P$ and the total output $\eta_{global} = P_n/P$

Table 2. *Calculation of the total outputs of the pumping plants.*

drillings	HMT (m)	Q (m³/s)	P (Kw)	P$_{moyen}$ (Kw)	□$_{global}$ (%)
F.1	37,95	0,02	5,73	31,32	18
F.2	33,15	0,01	2,68	31,32	9
F.3	40,14	0,01	4,73	10	47
F.4	36,43	0,02	6,85	41,76	16

The following remarks arise from table 2:

- The total output generally lower than the average output of each pump is immersed;
- The report R$_{global}$ /R$_{normal}$ measures the effectiveness of the material, more it moves away from the unit plus the output of the material concerned is abnormal. This calculation deduced distinguished two groups, the first includes the F.2 pump having reports/ratios lower than 20%, and the second relates to the remainder of the pumps which have reports/ratios beyond 47%. the first group is in a situation of abnormal operation compared to the second;
- The installations fed starting from the electrical communication have the best outputs; the remainder of the exploitations can be classified according to the technique of adopted irrigation.
- The output of the oldest installations is weakest.

The bad dimensioning of the material of pumping noted in certain visited installations is the origin of poor yield. The bad condition of the material, rare talks and repairs (do-it-yourself) which are carried out only when the breakdowns are serious contribute in a way important to reduce the outputs and to increase the cost of pumping.

5. Profitability of the Submerged Pump

5.1. Loads

Pumped volumes are estimated on the basis of technical route of each culture such as they are practiced by the farmers. These routes made it possible to evaluate the frequency of the irrigations and the operating hours of motor-driven pump group for each culture at the majority of the farmers. If the farmer is not able to remember the details, it puts estimates using planning's of the irrigations; if not, a total volume can be estimated starting from the loads of energies, the exploitation in this case is not taken into account in detailed calculations.

Table 3. *The calculation of cost of pumping (fixed costs).*

Drillings	Basic investment				Fixed costs			
	Digging of wells (DZD)	Pumps purchase (DZD)	Shelter (DZD)	C1 (DZD)	Repair (DZD)	Maintenance (DZD)	C2 (DZD)	C1+C2 (DZD)
D.1	144 000	152 000	20 000	15 600	31 000	8 240	39 240	54 840
D.2	41 000	212 000	40 000	16 832	34 800	15 600	50 400	67 232
D.3	220 000	200 000	0	20 668	0	10 000	10 000	30 668
D.4	289 800	280 000	16 000	28 860	24 000	14 880	38 880	67 740

6. Loads of Production

They include the loads relating to the purchase of the seeds; with the purchase of fertilizers; with mechanized work; and with labor. The table recapitulates the results

Table 4. *loads of production.*

cultures	Drillings	Total loads DZD	% additional expenditure related to the irrigation by Loads report/ratio of irrigation	% Loads of irrigation compared to the total loads	% Fresh of pumping compared to the total ones
Durum wheat	D.1	6 914	15	48	83
	D.2	2 721	30	3	56
	D.3	31 538	23	58	71
	D.4	41 196	21	31	73
Barley	D.1	1 074	13	7	85
	D.2	23 152	5	25	96
	D.4	28 287	20	21	75
Tomato's	D.4	11 896	8	9	92
But; fodder	D.1	6 541	13	45	85
	D.2	68 272	2	72	95
	D.4	52 070	5	39	94

According to this calculation, the following elements arise:

- **Durum wheat**

The loads of irrigation vary from 15% of which 83% are consisted the expenses of pumping, with 30% whose pumping occupies 56%.

- **Barley**

The loads of irrigation are weaker even if they have priority. The loads of irrigation occupy from 07 to 25% whose expenses of pumping represent from 75% to 96%.

- **Cultures of summer**

The other loads of the market gardening dominated the loads of irrigation thus the contribution of the irrigation varied between 09% and 72%. For corn, the irrigation could reach 72% of the total loads including 95% only consisted the expenses of pumping. The additional expenditure related to the irrigation is weaker and do not exceed 13% of the loads of irrigation.

7. Conclusion

The discussions with the farmers showed that fodder and the corns are the principal cultures practiced for a long time with the introduction from time to time of the market gardening on more or less limited surfaces. Generally, one can speak about a single strategy consisting with the culture of the same speculations in spite of the recourse to pumping, in other term, generally pumping is only one means to restore their cultivation methods before the dryness, are added to the two criteria of priority (corns and fodder), the control of the practices of a culture (truck farming) which encourages the farmers to specialize.

The intervention of the costs of pumped water are more dominant compared to the other loads of irrigation, therefore the control of device of pumping such as the installation, the chock, maintenance and repair are the essentiaux factors such as wear by abrasion of the wheels by the crucial factor which is fine sand which intervenes directly if one speaks about the profitability of the submerged pump. New processes invented to reduce the quantity of sand entering a submerged pump in a drilling, is done either by a system of convergent cone perforated along the strainer casing and submerged pump, or by a very new system `cyclone' to avoid any quantity of sand returning inside the immersed suction filter. In order to make profitable the pump from the consumption point of view of electrical energy, better output, therefore improvement of the lifespan of the pump.

Nomenclatures

D.W.S: Drinking Water Supply
W.S.I: Water Supply for Industry
AOW: Algerian Of Water
NAHR: National Agency of the Hydraulic Resources
C_1: the sum of the loads of basic investment.
C_2: the sum of the fixed charges.
DHW: Direction of Hydraulics of Wilaya
DZD: Algerian Dinars
F: Force applied [N]
THG: total height gauge [m]
IRR: Irrigation
NOC: National Office of the Cleansing
NOID: National Office of the Irrigation and the Drainage

P_m: Weight [%]
P_v: Volume [%]
PUVAL: PUmp Valve Algerian
Q: Flow [$m^{3.s-1}$]

References

[1]   WHITE, K.B; TAYLOR, S.E; "Pumping low yielding wells with conventional submersible pumps", Ground water monitoring and remediation (USA), (1995).

[2]   GENETIER B; "La pratique des pompages d'essai en hydrogéologie", Manuels et méthodes du B.R.G.M, N° 9, France, (1984).

[3]   CEMAGREF; "Les stations de pompage individuelles pour l'irrigation", (1996).

[4]   MABILLOT, A ; "Le forage d'eau", Guide pratique, (1988).

[5]   JEANMARIE GEORGE ; "Frottement, usure et lubrification", Eyrolles, (2000).

[6]   GODET, M ; "Les fondements mécaniques de la tribologie", Mécanique Matériaux Électricité, (1972).

[7]   DERRIEN, JACQUES ; "Surface des solides, propriétés électroniques", Paris, Techniques de l'Ingénieur, (1990).

[8]   JEAN DHERS; "Usure, avaries et corrosion", Facteur de destruction des matériels industriels, (1978).

[9]   J. AYEL ; "Les lubrifiants, moteur et pertes par frottement et usure", Institut Français du Pétrole, (1979).

[10] JEAN BLOUET; "Usure et frottement", T. ing. A 3139.

[11] R. BUTIN, M. PINOT; "Fabrications mécaniques", Technologie, tome 1, (1981).

[12] UNESCO; "Etude des Ressources en Eau du Sahara Septentrional", Rapport sur les résultats du Projet REG-100, UNESCO, Paris, (1972).

[13] OSS ; "Système Aquifère du Sahara Septentrional", Rapport interne. Annexes. Tunis, Tunisie, 229p, (2003a et b).

[14] BRL ingénierie ; "Etude du Plan Directeur Général de Développement des Régions Sahariennes", Lot I. France, 94p, (1999).

[15] Projet DELTA ; "Développement Des Systèmes Culturels Territoriaux, Plan d'action, cas de Ghardaïa", (2003).

[16] http://www.google.com/earth/

[17] http://en.wikipedia.org

[18] http://www.hardide.com/applications/coatings-for-pumps/

[19] http://www.poval.com.dz/.

13

Validation of terrestrial water storage change estimates using hydrologic simulation

Sang-Il Lee[1], Jae Young Seo[1], Sang Ki Lee[2]

[1]Department of Civil and Environmental Engineering, Dongguk University, Seoul, South Korea
[2]Department of Civil Engineering, University of Idaho, Boise, ID, USA

Email address:

islee@dongguk.edu (S. –I. Lee), dabbi2011@naver.com (J. Y. Seo), sklee81@hotmail.com (S. K. Lee)

Abstract: New methods estimating the amount of water storage on Earth have evolved over the years. One of them utilizes the gravitational field variation observed from the GRACE satellite. Compared to conventional methods such as water balance analysis, the method makes it simple and straightforward to obtain the terrestrial water storage change (TWSC). Previous studies show that there is a discrepancy between GRACE-based and water balance-based estimates especially in wet periods. Along with precipitation and evapotranspiration, it is common that runoff data needed for the water balance analysis are obtained from GLDAS (Global Land Data Assimilation System). In this study, GLDAS runoff data are replaced with hydrologic simulation results with such anticipation that local geomorphologic and hydrologic characteristics can be better incorporated. In an application to a relatively small basin during a wet period, GLDAS- and simulation-based TWSCs showed values 2.73~3.58 times higher than the GRACE-based estimate. It implies that the GRACE-approach underestimates TWSC during wet periods. It also suggests the need for correction factors to adjust the GRACE-based estimates in the rainy season.

Keywords: GRACE, Terrestrial Water Storage, GLDAS, Hydrologic Simulation

1. Introduction

GRACE (Gravity Recovery And Climate Experiment), twin satellites launched by NASA in 2002, measures minute variations in Earth's gravity field. A gravity field map obtained by GRACE shows how the Earth's mass varies from place to place. Therefore, by converting the gravitational field data into mass data, information on the distribution of water across the Earth's surface can be obtained [1].

Numerous studies have been conducted in the fields of geodesy, glaciology, hydrology, oceanography and solid Earth sciences, utilizing the data provided by GRACE. In the water resources engineering, studies mainly focus on the understanding of water relocation (e.g., on the basin scale estimation of evapotranspiration [2]; for the estimation of groundwater storage changes [3, 4]), and the improvement in the estimation accuracy [5, 6]. GRACE's ability to detect water storage changes at varied spatial scales over different parts of the globe has been demonstrated: Some examples include the Congo River Basin [7], India [8], East African lakes [9], California

Central Valley [10] and Turkey [11].

Lee et al. [12] was the first to attempt to check the applicability of the GRACE-based method to the water resources research of the Korean Peninsula. In their study, terrestrial water storage change (TWSC) was estimated and compared to the value calculated from a water balance analysis. Hydrogeologic data from GLDAS (Global Land Data Assimilation System) [13] and WAMIS (Water Management Information System) [14] were used for the analysis. Except for relatively wide discrepancies in the rainy season, the GRACE-based TWSC estimates showed a good agreement with the ones from the water balance analysis, suggesting a high potential of the GRACE-based technique.

In this study, the accuracy issue of the GRACE-based TWSC in the rainy season is further investigated. Since the big cell size ($1° \times 1°$) of GLDAS was considered to be one of the main sources of uncertainty in the water balance analysis, we replace the runoff data from GLDAS with ones from hydrologic simulation. By doing so, we expect detail hydrogeologic conditions of the basin are reflected and it will lead to the enhancement of the accuracy of the technique based on the gravitational field variation.

2. Methodology

GRACE measures the long wavelength component of the Earth's gravity at an altitude of about 500 km. The GRACE Level-2 data are created after compensating the tidal effect, and the non-tidal effect resulting from the variability of atmosphere and ocean.

Since the change in the terrestrial water storage, which consists of surface water, soil moisture, groundwater, etc., will change the gravitational field, the monthly gravitational field data can be converted into the liquid water equivalent thickness [15]. TWSC can also be computed using a monthly basin-scale terrestrial water balance which can be approximated as follows

$$\left[\frac{\Delta S}{\Delta t}\right]_N = \sum_{N-1}^{N} P - \sum_{N-1}^{N} R - \sum_{N-1}^{N} E \quad (1)$$

Where, S is the TWS, t is time, N represents month, P is precipitation, R is the net surface/subsurface runoff, and E is evapotranspiration.

Lee et al. [12] estimated the TWSC of the Korean Peninsula using land-based WAMIS data for precipitation and evapotranspiration, along with GLDAS data for runoff [13]. GLDAS provides meteorological satellite observation data such as precipitation, runoff, snow water equivalent,

soil moisture, etc. at the grid interval of $1° \times 1°$ (111.1 km $\times$ 88.8 km) which is the same as GRACE (Fig. 1). The mean TWSC of the Korean Peninsula was estimated to be 0.986 cm/month during the period of 2002~2010.

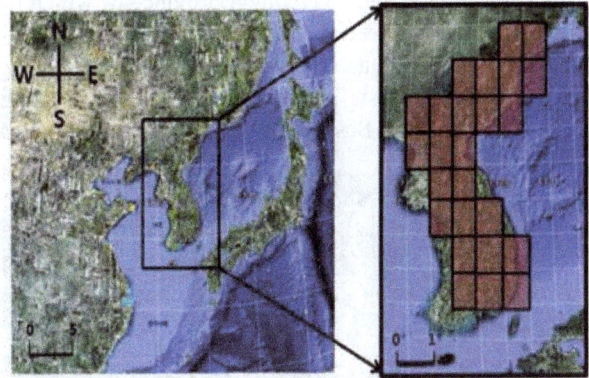

Figure 1. Cells for the Korean Peninsula from GRACE Tellus [12].

In order to reflect the geomorphological and hydrological characteristics in detail, here we substitute the runoff data from GLDAS with simulated ones from a hydrological model, HEC-HMS [16]. Our objective is to compare three different approaches; GRACE-based, Land/GLDAS-based, and Land/Simulation-based technique in estimating the TWSC (Fig. 2). Table 1 shows the spatial resolution and sources of hydrologic components used in different approaches.

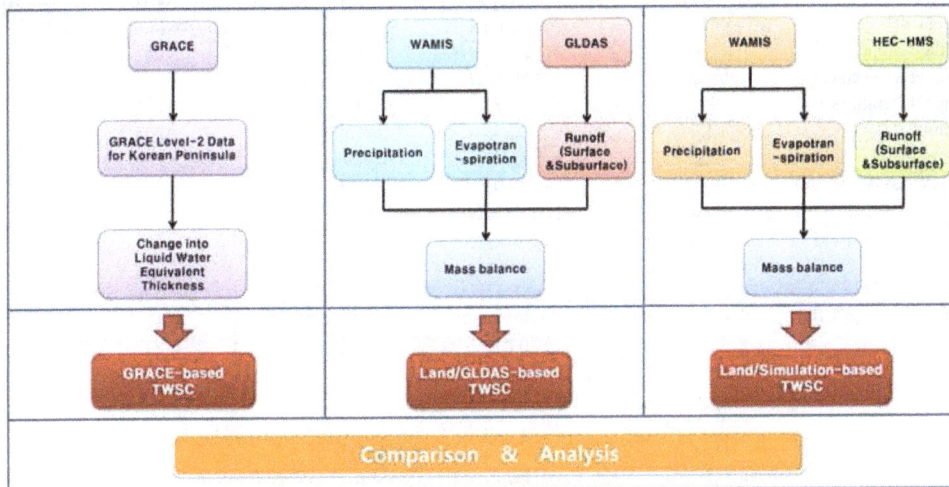

Figure 2. Three Different Approaches of the Study.

Table 1. Spatial Resolution and Sources of Hydrologic Components.

Hydrologic Component	Spatial Extent	Spatial Resolution	Time Span	Source
Precipitation Evapotranspiration	126°E ~ 137°E	Local	Jul 15, 2003	WAMIS
Runoff [1] (surface& subsurface)		1° × 1°	~	GLDAS
Runoff [2] (surface& subsurface)	37°N~38°N	Local	Aug 14, 2003	HEC-HMS

1) Land/GLDAS-based approach, 2) Land/Simulation-based approach

3. Hydrologic Simulation

3.1. Study Area

The study area is Seoul/Gyeonggi province of which the basic information needed for hydrologic analysis is quantitatively and qualitatively superior to other areas (Fig. 3(a)). The area consists of four basins, (A) Imjin River Basin, (B) Han River Basin, (C) Ganghwa Basin, and (D) Banwol River Basin (Fig. 3(b)). High rate of urbanization of this area results in the high ratio of impervious area, which leads to intensified runoff concentration during the rainfall.

Geomorphological data were obtained from the digital maps provided by National Geographic Information Institute (NGII) [16], and hydrogeological data such as precipitation, river stage, soil and land use were provided by WAMIS [17].

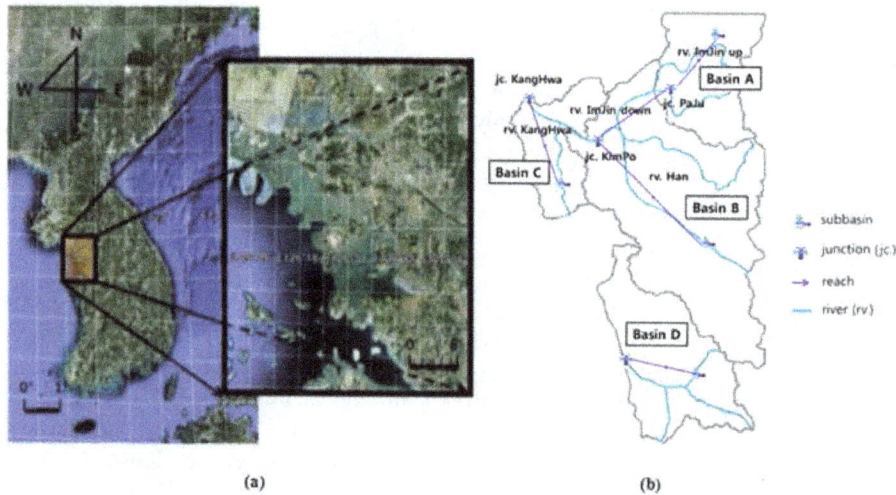

(a)   (b)

Figure 3. *Study Area and Four Basins.*

Table 2. *Model Parameters.*

		Basin A	Basin B	Basin C	Basin D	Remark
Basin Information	Area (km^2)	461.8	734.6	322.8	628.3	
	Average Slope (%)	2.89	2.07	0.59	2.08	
	Main River Element	Imjin	Han	Seokjeong	Banwol	
	Channel Length(km)	29.89	74.73	14.27	31.98	
Initial Abstraction (AMC-III)	I_a(mm)	8.13	10.85	13.34	10.04	NRCS Curve Number Method [1]
	CN (-)	86.2	82.4	79.2	83.5	
	Impervious Area (%)	24.99	29.08	6.19	33.43	
Unit Hydrograph	T_c (hr)	2.37 (1.01)*	5.93 (8.96)	1.32	2.54	Clark Unit Hydrograph [2]
	K (hr)	2.56 (1.09)	4.90 (7.42)	1.68	4.00	
Baseflow	Initial Runoff(m^3/s/km^2)	0.11	0.11	0.11	0.11	Exponential Recession [3]
	R (-)	0.844	0.815	0.551	0.779	
	Peak Rate (-)	0.035 (0.080)	0.035 (0.036)	0.035	0.035	
Stream Routing	K (hr)	up: 5.887 Down: 1.427	4.902	5.922	4.001	Muskingum Method [4]
	x (-)	0.2	0.2	0.2	0.2	
	NSTPS	1	1	2	2	

1) $I_a = 0.2S$, $S = \frac{25400}{CN} - 254$ (I_a: Initial abstraction, S: Potential maximum retention of rainfall)

2) $T_c = 0.2778L/V$ (T_c: Time of concentration, L: Channel length, V: Mean velocity), $K = \frac{T_c}{1.46 - 0.0867\frac{L^2}{A}}$ (K: Storage constant, L: Channel length, A: Basin area)

3) $R = e^{-1/K}$ (R: Recession constant, K: Storage constant)

4) $K = L/V_w$ (K: Travel time, L: Distant, V_w: Flood wave velocity), x: weighting factor, NSTPS: Number of routing steps

* ( ) is the optimized value.

3.2. Model Parameters

Hydrologic simulation using HEC-HMS requires a series of selection for process-describing methods and related model parameters. To represent the initial abstraction, for example, the NRCS Curve number method was used. Since the simulation period corresponds to the rainy spell in summer (July 15 ~ August 14, 2003), the soil condition was set to be AMC-III. The CN-value was calculated from soil- and land use maps. As for the unit hydrograph, the Clark unit hydrograph model was used. It is known to be suitable for a small/medium basin and widely applied to urban and natural basins. For baseflow calculation, the exponential recession model was used. The Muskingum method was used for stream routing. The main parameters and their initial values required for the analysis are shown in Table 2.

The parameter optimization program provided in HEC-HMS was used to reduce the difference between simulated and observed runoff. Optimized values are shown in the parenthesis of Table 2. The result of hydrologic analysis at Kimpo Junction is shown in Fig. 4. It shows the improved performance when optimized parameters are used in the simulation. The error after the optimization turned out to be 29% lesser than before.

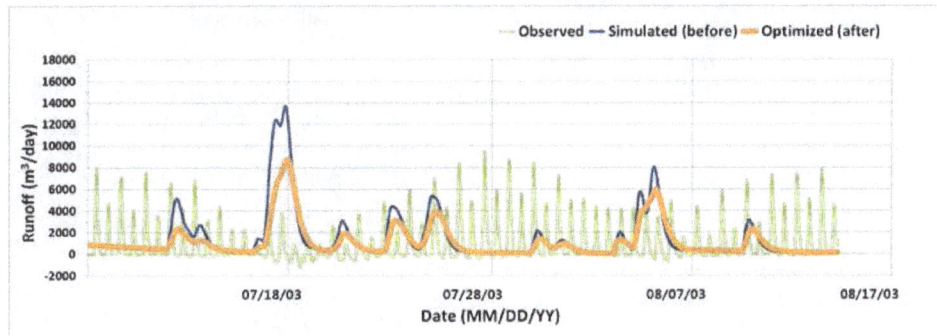

Figure 4. *Runoff at Kimpo Junction Before & After Parameter Optimization.*

4. Results and Discussion

The runoff obtained from optimized hydrological simulation is substituted into the water balance (Eq. (1)) in order to calculate the Land/Simulation-based TWSC. The effective precipitation of each basin is calculated by subtracting the loss from the total precipitation. The mean precipitation (P) over the entire region estimated from the Thiessen Polygon method is 32.89 cm/month.

$$P = \sum_i \frac{A_i P_i}{A_T} \qquad (2)$$

Where, A_i is the area of each basin, P_i is the effective precipitation of each basin, A_T is the total area. The depth of runoff (R) calculated by dividing the runoff by the total basin area (8,758.9 km^2) is 18.91 cm/month, and the depth of the evapotranspiration (E) is 8.79 cm/month. Now the Land/Simulation-based TWSC is obtained by substituting P, R and E values into Eq. (1): 5.19 cm/month.

Lee et al. [12] found that the GRACE-based TWSC of the study basin from July 15 to Aug 14, 2003 was 1.45 cm/month, while the Land/GLDAS-based TWSC obtained from the water balance using the GLDAS runoff data was 3.96 cm/month. Comparing these values with the Land/Simulation-based TWSC obtained above (5.19 cm/month), we find that the GRACE-based, and the Land/GLDAS-based approaches underestimate the TWSC during the wet season (Table 3).

It is noteworthy that while the GRACE- and the Land/GLDAS-based TWSCs are relatively straightforward to obtain, the degree of complexity in calculating the Land/Simulation-based TWSC is much higher. Therefore, if an appropriate correction factor is introduced to the GRACE-based TWSC (3.58 for Land/Simulation-based estimates and 2.73 for Land/GLDAS-based, in the case), TWSC can be calculated utilizing the gravitational data in a much simpler way than traditional methods. When correction factors corresponding to various weather patterns for the basin of interest become available, the usefulness of the GRACE-based approach can be greatly improved.

5. Conclusion

Reliable estimation of terrestrial water storage is important for the sustainable management of water resources. The GRACE-based approach using the gravitational field variation has proven its effectiveness in estimating TWS changes. The largeness in spatial and temporal scales of GRACE-based estimation, however, limits its applicability to smaller areas or time-varying weather conditions.

This work attempted to check the accuracy of three different methods to estimate TWSC for a relatively small basin during the rainy season. TWSC estimated from direct GRACE-approach was compared with the estimates from the water balance with two runoff values: One from GLDAS and the other from hydrologic simulation.

As a result, we found that the GRACE-, and the Land/GLDAS-based approaches underestimate the TWSC during the wet season, compared to the Land/Simulation approach. It suggests that we can save a lot of time, cost

Table 3. *Comparison of Three Different Approaches.*

	GRACE-based	Land/GLDAS-based	Land/Simulation-based
TWSC (cm/month)	1.45	3.96	5.19

and resources required for hydrologic analysis if we have appropriate correction factors to multiply since the GRACE-based TWSC is relatively easy and less costly to obtain. When enough information is accumulated for various basins with diverse weather conditions, GRACE can be a useful tool to estimate TWSC with high accuracy and effectiveness.

Acknowledgements

This research was sponsored by the NRF (National Research Foundation of Korea) under the contract number 2013-052502.

References

[1]   B. D. Tapley, S. Bettadpur, J. C. Ries, P. F. Thompson, and M. M. Watkins, "GRACE measurements of mass variability in the earth system," Science, 2004, vol. 305, pp. 503–505.

[2]   M. Rodell, J. S. Famiglietti, J. Chen, S. I. Seneviratne, P. Viterbo, and S. Holl, "Basin scale estimates of evapotranspiration using GRACE and other observations,"Geophys. Res. Lett., 2004, vol. 31, pp. L20504.

[3]   M. Rodell, J. Chen, H. Kato, J. S. Famiglietti, J. Nigro, and C. R. Wilson, "Estimating groundwater storage changes in the Mississippi River Basin (USA) using GRACE," Hydrogeol. J., 2007, vol. 15, pp. 159-166.

[4]   P. J. F. Yeh, S. C. Swenson, J. S. Famiglietti, and M. Rodell, "Remote sensing of groundwater storage changes in Illinois using the Gravity Recovery and Climate Experiment (GRACE)," Water Resour. Res., 2006, vol. 42, pp. W12203.

[5]   S. Swenson, J. Wahr, and P. C. D. Milly, "Estimated accuracies of regional water storage variations inferred from the Gravity Recovery and Climate Experiment (GRACE),"WaterResour. Res., 2003, vol. 39, pp. 1223.

[6]   J. Wahr, S. Swenson, I. Velicogna, "Accuracy of GRACE mass estimates,"Geophys. Res. Lett., 2006, vol. 33, pp. L06401.

[7]   J.W. Crowely, J.X. Mitrovica, R.C. Bailey, M.E. Tamisiea, and J.L. Davis, "Land water storage within the Congo Basin inferred from GRACE satellite gravity data,"Geophys. Res. Lett., 2006, vol. 33, pp. L19402.

[8]   M. Rodell, I. Velicogna, and J.S. Famiglietti, "Satellite-based estimates of groundwater depletion in India," Nature, 2009, vol. 460, pp. 999-1002.

[9]   M. Becker, W. LLovel, A. Cazenave, A. Guntner, and J. F. Cretaux, "Recent hydrological behavior of the East African great lakes region inferred from GRACE, satellite altimetry and rainfall observations," C. R. Geosci., 2010, vol. 342, pp. 223–233.

[10]   B. R. Sacanlon, L. Longuevergne, and D. Long, "Ground referencing GRACE satellite estimates of groundwater storage changes in the California Central Valley, USA,"WaterResour. Res., 2012, vol. 48, pp. W04520.

[11]   O. Lenk, "Satellite based estimates of terrestrial water storage variations in Turkey," J. Geodyn., 2012, vol. 67, pp. 106-110.

[12]   S.-I. Lee, J. S. Kim, and S. K. Lee, "Estimation of average terrestrial water storage changes in the Korean Peninsula using GRACE satellite gravity data," J. Korean Water Resour. Asso., 2010, vol. 45, pp. 805-814.

[13]   NASA GES DISC, 2012, Available online at: http://daac.gsfc.nasa.gov/.

[14]   Water Management Information System (WAMIS). 2012. Available online at: http://www.wamis.go.kr/.

[15]   T. H. Syed, J. S. Famiglietti, M. Rodell, J. Chen, and C. R. Wilson, "Analysis of terrestrial water storage changes from GRACE and GLDAS," Water Resour. Res., 2008, vol. 44, pp. W02433.

[16]   U.S. Army of Civil Engineers, HEC-HMS Manual, 2012.

[17]   National Geographic Information Institute (NGII), 2012, Available online at: http://www.ngii.go.kr/.

14

Assessment of organic pollutants of water samples in River Getsi and River Gwagwarwa in Kano State Nigeria

Y. Mohammed[1], A. Ekevwe[2]

[1]Department of Chemistry, Nigeria Defence Academy, Kaduna, Kaduna State
[2]Department of Chemistry, Federal College of Education (Technical) Bichi, Kano State

Email address:

sawaba83@gmail.com (Y. Mohammed), upambrose@yahoo.com (A. Ekevwe)

Abstract: Water samples of River Gwagwarwa and River Jakara were analyzed for organic pollutants. The organic parameters were determined using the standard methods of America Public Health Agency (APHA) and was extracted and analyzed using Gas chromatography-mass spectrometer (GC-MS). Thirteen different organic compounds were detected at different percentage values at the two sampling stations. The compounds fall within five classes of organic compounds, which include carboxylic acid, acid chloride, ester, aldehyde and acid anhydride. The distribution pattern of the organic pollutant at the two sampling stations depict the pattern; River Gwagwarwa > River Getsi .The study shows that organochloride was the predominant organic pollutant present in the samples

Keywords: Gas Chromatography, Mass Spectrometer, Carboxylic Acid, Acid Chloride, Ester, Aldehyde, Organochloride, Organic Pollutant

1. Introduction

Water is the most common liquid on our planet, vital to life form. The total water on earth is enormous, estimated to 1.5 x 1018 metric tons; this quantity is 300 times larger than the mass of the entire atmosphere [1]. Unfortunately, most of these are not accessible because they appear in ice-caps, oceans, in underground aquifer (ground water-bearing beds) and some are even in the air as moisture .Only a small fraction of water is on earth surface and directly accessible to man as rivers, streams and springs [2].

Water can sometimes be said to be pure but it can never entirely 100% pure. It inevitably carries traces of other substances – various organic compounds, particles, gases, minerals and ions which impart to its physical, chemical and bacteriological characteristics [3].

In cities of Nigeria, with particular reference to Kano state, a major industrial and commercial centre with a population of over 7,000,000 people according to the national census figure of 2006.The various component of the natural environment are often adversely affected by these human activities resulting in the devastation of components of the environment such as air, land and water [4].

The quality of water is continuously changing as a result of the reaction of water with contact media affected by anthropogenic influences, such as domestic or municipal waste [5]. The behavior of organic compounds is dependent upon their molecular structures, size and shape and the presence of functional groups that which are important determinants of toxicity, (Adeola, 2004). There are many different types of organic pollutants, examples are:- Hydrocarbons, Polu Aromatic Hydrocarbon's, Polycyclic Biphenyl's, Detergents,plastic, persistent organic pollutant,pesticide among others.

Many organic compounds, to a varying degree, resist photolytic, biological and chemical degradation. These are referred to as persistent organic pollutants (POPs). POPS are often halogenated and characterized by low water solubility and high lipid solubility, leading to their bioaccumulation in fatty tissues. They are also semi-volatile, enabling them to move long distances in the atmosphere before deposition occurs, (Ritter et al., 2007).

Although many different forms of POPs may exist, both natural and anthropogenic, many of these compounds have been or continue to be used in large quantities and, due to their environmental persistence, have the ability to bioaccumulate and biomagnify. Some of these compounds such as polycyclic biphenyls (PCBs), may persist in the environment for periods of years and many bioconcentrate by factors of up to 70,000 fold, (Ritter et al., 2007).

Pesticides (synthetic organic chemicals) are widely used in fruit and vegetable production because of their susceptibility to insects and diseases attack. Consequently, food safety is a major public concern worldwide as residues of pesticides could affect the ultimate consumers especially when these commodities are freshly consumed. Given the potential risk of pesticides and heavy metals for public health, the use of pesticides in fruit and vegetable production is subjected to constant monitoring.

An assessment report carried out by Ritter et al., (2007) on several organic pollutants concluded that a number of the organic substances assessed in the report have been implicated in a broad range of adverse human health environmental effects including impaired reproduction and endocrine dysfunction, immunosuppression and cancer. In many cases, the substances are considered as possible human carcinogens by the International Agency for Research on Cancer consumed [11,12,13&14]..

2. Study Site

River Gwagwarwa originate from Gwagwarwa quarter under Nassarawa local government area of Kano State. Gwagwarwa is a highly populated town in Kano. This is because of its semi-industrial nature and proximity to Sabon Gari (a densely residential and commercial settlement) in Kano. Therefore the River cut across domestic, industrial and agricultural areas which makes it to carry along pollutant due to the activities of the areas it pass through.

River Getsi originate from Bompai industrial areas, Bompai in Nassarawa local government area of Kano State.

Pollution loads in River Getsi are through industrial operations in Bompai and proximate localities.

3. Material and Method

Water samples were collected at various points along River Getsi and River Gwagwarwa in the morning and evening on each sampling day. 100 cm3 of water sample was collected at each designated point which is 20 metres to the next point[8,9]. 10 samples were collected in each sampling session which are composited to a total of 1 litre. The samples were labeled and taken to the laboratory for further analysis[8,9,10]. This procedure was repeated throughout the sampling. Appropriate quantities of the composite samples were measured and treated according to the standard methods of American Public Health Agency (APHA).

4. Procedure

$50cm^3$ of each composite water sample was measured and added into a cleaned 250 cm^3 separatory funnel. $50cm^3$ each of diethyl ether and trichloromethane were measured and added into the separatory funnel.. The resultant mixtures was vigorously shaken and gas released intermittently by controlling the lid. The mixture was allowed to stand on a retort stand for 5 minutes and the organic layer was collected in a cleaned glass sample bottle, labeled and kept for further GC-MS analysis [7].This process was repeated for all the composite samples.

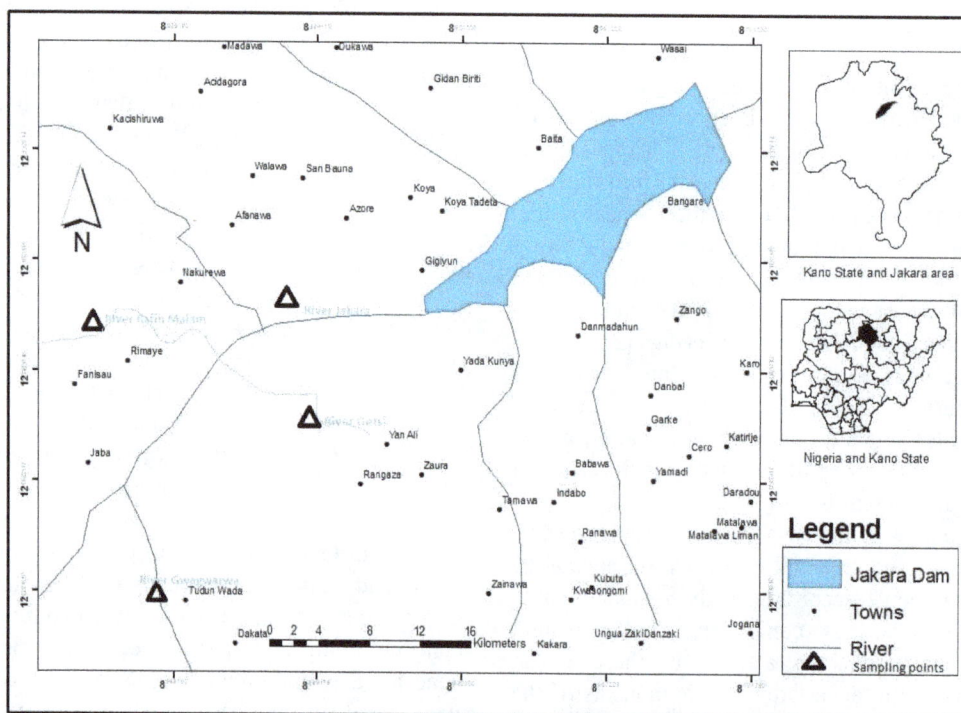

Figure 1. Map Showing River Getsi and River Gwagwarwa with other Rivers across River Jakara

Table 1. Average Percentage (%) Value of Organic Compound Detected at River Getsi Samples

S/No	Compound Detected	% Value
1	Dodecanoic Acid	1.98 ±0.47
2	Tetradecanoic Acid	2.39 ±0.48
3	Palmitic Acid	9.59 ±2.86
4	Methyl Octadecanoate	7.03 ±1.65
5	Oleic Acid	34.07 ±5.88
6	Docosanoic Anhydride	9.75 ±1.37
7	Octadecanoic acid 1,2,3 Propanetriyl Ester	31.93 ±8.13
8	9- hexadecanoic Acid	11.58 ±2.79
9	Octadecadienoyl Chloride	55.19 ±0.0
10	Hexadecanoic Acid 1 – {{{2-Aminoethylhodroxy Phosphinyl} Oxy} Methyl }– 1 , 2 Ethenediyl Ester	9.32 ±0.0

Table 2. Average Percentage (%) Value of Organic Compound Detected at River Gwagwarwa

S/No	Compound Detected	% Value
1	Dodeconoic Acid	1.93 ±0.51
2	Tetradecanoic Acid	2.39 ±0.40
3	Palmitic Acid	9.17 ±2.47
4	Methyl Octadecanoate	7.39 ±1.39
5	Oleic Acid	29.46 ±6.42
6	Docosanoic Anhydride	12.03 ±1.44
7	Octadecanoic acid 1,2,3 Propanetriyl Ester	41.05 ±8.66
8	9-Hexadecanoic Acid	24.81 ±0.0
9	1, 3 Octadecanal	12.94 ±0.0
10	Octadecadienoyl Chloride	66.66 ±0.0

5. Result and Discussion

The % values of the various organic compounds detected in the composite water sample collected from River Getsi are presented in table 1.0. Ten different organic compounds were detected at different % value. The compounds falls within four classes of organic compounds viz, carboxylic acid, acid chlorides, acid anhydride and esters. The distribution of the compounds depicts a pattern; carboxylic acid > esters > acid chlorides = acid anhydride. Highest percentage value of 55.19 was recorded for octadecadienoyl chloride and the least % value of 2.31 was recorded both for dodecanoic acid and tetradecanoic acid.

Exposures to Dodecanoic acid and tetradecanoic acid can cause mild irritation of the upper respiratory tract and mucous membrane at higher concentration which is in accordance with U.S Department of Health and Human Behaviour. While exposures to Octadecadienoyl chloride are very toxic and dangerous, it causes severe burns and eye damage. Human exposure present at level greater or equal to 0.1% is identified as probable, possible or confirmed human carcinogen by International Agency for Research on Cancer (IARC).

These classes of organic compound arises in the wastewater due to discharges of complex chemicals and solvent used in commercial, agricultural and industrial operations[15&16].

The % values of the various organic compounds detected in the composite water sample collected from River Gwagwarwa are presented in table 2.0. Ten different organic compounds were detected at different % value. The compounds falls within five classes of organic compounds viz, carboxylic acid, esters, acid anhydride, acid chlorides and aldehyde. The distribution of the compounds depicts a pattern; carboxylic acid > esters > acid chlorides = aldehyde = acid anhydride. Highest percentage value of 66.66 was recorded for octadecadienoyl chloride and the least % value of 1.76 was recorded for dodecanoic acid.

Exposures to Dodecanoic acid can cause mild irritation of the upper respiratory tract and mucous membrane at higher concentration which is in accordance with U.S Department of Health and Human Behaviour. While exposures to Octadecadienoyl chloride are very toxic and dangerous, it causes severe burns and eye damage. Human exposure present at level greater or equal to 0.1% is identified as probable, possible or confirmed human carcinogen by International Agency for Research on Cancer (IARC).

These classes of organic compound arise in the wastewater due to the discharges of complex chemicals and solvent used in commercial, agricultural and industrial operations[15&16].

References

[1] S.O Ajah and O. Osidayo (1981). Pollution studies on Nigerian Rivers: Water quality of some Nigeria Rivers, Environ Pollution, serv, B, 2: 87-95

[2] O. Dimitrovska, B. Markoski, B.A Toshevska, I. Mileveka and S. Gorin (2012). Surface water pollution of major rivers in the Republic Of Macedonia, Procedia Environ Sci, 14, 32-40

[3] .J Driver (1997); The geochemistry of natural waters: Surface and groundwater environments. 3rd ed. Upper Saddle Rivers, NJ: Prentice Hall.

[4] A.L. Vittoli, C. Trivisano, C. Gessa, M. Gherardi, A. Simoni and G. Vienello (2010). Quality of Municipal wastewater compared to surface waters of the river and artificial canal network in different areas of the eastern Po Valley (Italy). Water qual Expo Health, 2 (1), 1-13).

[5] O. Osidayo, P.D Adegbeuro and M.G Adewole (2011): The impact of industries on surface water quality of River Ona and River Alero in Oluyole industrial estate, Ibadan, Nigeria. African Journal of Biochemistry, 10 (4), 696-702.

[6] M.O Said (2008) Chemical analysis of water samples in Kano state. Ph.d Thesis, Bayero University, Kano. Nigeria. PP 125-128.

[7] G. Wyasu and Kure, O.A (2012): Determination of organic pollutants in hospital waste water and food samples within Ahmadu Bello University Teaching hospital (ABUTH) Shika, Zaira- Nigeria. Available online at www.pelagiaresearchlibrary.com

[8] APHA(1998); Standard Methods for the Examination of Water and Wastewater. America Public Health Association, 18th ed, Academic Press, Washington, D.C Pp. 200-240.

[9] APHA(2005); Standard Methods for the Examination of Water and Wastewater. America Public Health Association,19th ed, Academic Press, Washington, D.C Pp. 80-95.

[10] Burton, F.L Tchobanoglous, G. And Stensel, H.D (2003); Waste Water Engineering (Context Treatment, disposal and Reuse) Metcalf & Eddy Inc (4th Ed) McGraw-Hill book company New York).

[11] Damià, B. (2005) ; Emerging Organic Pollutants in Waste Waters and Sludge. Springer, Berlin.

[12] David T Allen and David R. Shunnart, (2000); Green Engineering – Environmentally Conscious Design of Chemical Processes, pp 201 – 207 prentice Hall

[13] Eldon D. Enger and Bradltey F. Smith (2010); Environmental Science (Study of Interrelationship) 12th Edition, McGraw-Hill Publishers, New York Pp 335 – 425.

[14] EPA (2007); United State Environmental Protection Agency, National Water Quality Inventory" Report to Congress for the 2002 Reporting Cycle-Profile Washington DC.

[15] European Commission (2006); Environmental fact sheet: reach a new chemical policy for EU commission, Luxembourg.

[16] Eichelberger, J W Belymer., T.D and Budde, W.L (1988); Determination of Organic compounds in Drinking Water by Liquid and Solid Extraction and Circularly column Gas Chromatography/Mass Spectrometry (Method 525 2, Revised 2.0) National Exposure Research Laboratory Office of Research and Development USEPA Cincinnati, Ohio 45268.

15

Surface water quality of Gorai river of Bangladesh

S. Z. K. M. Shamsad[1], Kazi Zahidul Islam[1], Muhammad Sher Mahmud[2] *,
A. Hakim[2]

[1]Department of Soil, Water and Environment, University of Dhaka, Dhaka, Bangladesh
[2]Department of Soil Science, University of Chittagong, Chittagong, Bangladesh

Email address:

mahmud240@yahoo.com (M. S. Mahmud)

Abstract: Some important characteristics of water quality of Gorai river system were evaluated for use in domestic, industrial, agriculture, recreation and aquaculture purposes. Twenty three water samples were collected both in the post monsoon period (November) and in the pre monsoon (May) period. The study revealed that most physical parameters and inorganic elements are not a serious problem for Gorai river system under decreased Ganges flow. A trend of organic and NO_3 pollution in some downstream areas of higher anthropogenic activities were observed. The water of Gorai river system is fairly rich with N, P and S probably due to urban run-off and livestock activities. No heavy metal toxicity was recorded. The Ca content was high in water samples representing Ganges calcareous floodplain.

Keywords: Water Quality, Gorai River, Pre Monsoon, Post Monsoon

1. Introduction

Bangladesh is the lowest riparian of three major river systems of South Asia, namely, the Ganges-Padma, the Brahmaputra-Jamuna and the Meghna-Barak [1]. Among them, the Padma is the major river of Bangladesh, and most important in the cases of international water resources sharing issues [1,2,3]. The Ganges originates from the Gangetri iceberg of the Himalaya and enters Bangladesh as Padma via West Bengal of India at Rajshahi district [4].

The river Gorai is the lone and largest perennial distributary's of the Padma. The Padma and Gorai are the major rivers of Kushtia region and are also supplying fresh water to the Southwest region of Bangladesh for hundreds of years. This fresh water flow is the key to the maintenance of an environmental, social and ecological balance in the region [2,4].

The surface water system of Bangladesh consists of the major river networks, world largest delta and the massive flood plains, which become inundated for a short period during the monsoon season and used for cultivation for the rest of the year to supply most of the agricultural crops [4]. Bangladesh lies across the delta of four major rivers. These rivers and their distributaries discharge about 5 million cubic feet of water per second into the Bay of Bengal at peak periods. The rivers contribute to the agriculture and general economy of the country by providing navigation, fish, water for irrigation and fresh alluvial sediment replenishing the soil [4,5].

Water is essential to plant and animal life; it is our best solvent, it carries of our water, and it modifies our climate [6]. Water is an indispensable component of the earth environment. Water is not only essential to life but it is the predominant inorganic constituents of living matter, forming in general nearly three quarters of the weight of the living all [7,8,9].

Water is not only important because it contributes to plant growth, but also because it is a transporting agent for dissolved materials, nutrients, chemicals and solids [5,10].

The availability of water supply adequate in terms of both quantity and quality is essential to human existence [7,8]. Water quality is influenced both by natural and anthropogenic intervention where the former includes the local climate, geology etc. and the latter covers the construction of dams and embankments, agricultural practices, indiscriminate disposal of industrial effluents etc. [8,11].

Water quality is and will continue to be a major economic and environmental issue. Water quality concerns have often been neglected because good quality water supplies have been plentiful and readily available. The situation is now changing in many areas of the world including Bangladesh [4,5,12]. Water quality study is necessary for its proper use. The study of water quality is of much importance in production of crops. This water quality depends on many parameters among which the most important is the presence of the nutrients, responsible for fertility. Water quality refers to the characteristics of water those will influence its suitability for a specific use, i.e., how well the quality meets the need of the user [13].

The main sources of fresh water in Bangladesh are the different surface water bodies including rivers, canals, lakes, ponds and beels. The Gorai, A major offshoot of the mighty Padma flows by Kushtia town. Due to loss in water flow in the Padma and carelessness and negligence, now-a-days the Gorai runs out of water for best periods of the year. Gorai is drying. Shrinking water flows and land grabbing has turned the Gorai river into a narrow canal [2]. The Gorai river is the main distributary of fresh water from the Padma/Ganges to the South-West region. The dry season flow of the Ganges has decreased and since 1988, there has been a resultant hastening of the natural decline of the Gorai River as it becomes totally cut-off from the Ganges during dry season [2]. However, the massive withdrawal of dry season Ganges outflow has already had a serious impact not only on water quality of the Padma/Ganges dependent areas but also on agriculture, fishery, forestry, industry and navigation over the last two decades [14] and the salinity is increasing in the river of coastal regions of Bangladesh and has increased at least 2 ppt over the last few years, which is a serious threat to overall environment [3]. So, a field research was conducted the dry season (November-May) to study the water quality of the Gorai river system of Kushtia region under decreased Ganges flow (DGF). The studied area lies in the Southwestern part of Bangladesh approximately between latitude 23°40′30″ and 24°89′ N and longitude 88°42′ and 89°21′30″. Being a deltaic part of the Gangetic deltaic plain, it is bounded in the North by Padma river, separating it from Natore and Pabna districts in the East by Pabna district; in the South by Meherpur, Chuadanga and Jhenaidah.

This research was aimed at making an environmental impact assessment of water quality deterioration caused by the decreased outflow of the river Ganges and evaluation of Gorai water quality for domestic, industrial, agriculture, recreation and aquaculture.

2. Materials and Methods

The study area lies in the Indian platform and Eocene Hinge zone passes through the Southeastern extremity of the study area. On the basis of a preliminary survey, twenty-three (23) water samples were collected from different locations at the Kushtia point of Gorai river. Sampling sites for water were selected as per sampling techniques [15] which represent the whole area of the Kushtia region of Gorai river. The high-density PVC bottles used for water sampling were thoroughly cleaned by rinsing with 8M HNO_3 followed by repeated washing with water sampled so as to avoid contamination [15]. The sampling bottles were kept air tight and labeled properly

for identification.

Each sample was acidified in the field for Fe determination. Aeration during sampling was avoided as far as possible. Each sample was composite of 10 sub-samples to minimize errors and heterogeneity [5]. Variable determinants such as temperature, electrical conductivity (EC), pH and dissolved oxygen (DO) of water samples were measured in the spot using thermometer, portable EC meter, pH meter and DO meter respectively [15,16]. Samples collected from the study area were carefully transported to the laboratory, preserved in a refrigerator and were immediately analyzed for finding intended physical and chemical parameters of water.

Analyses of different physical and chemical parameters of Gorai river were carried out in the laboratory. The temperature of water samples were measured by the mercury thermometer (0°-50°C range) immediately after collection by dipping the thermometer in sample for about one minute [9]. Total suspended solids (TSS) were measured gravimetrically [8]. The pH of water samples were determined directly by a pH meter taking 50ml of filtered water sample in a 100ml clean beaker [17]. The total hardness (Ht) as $CaCO_3$ was directly measured titrimetrically [8]. A rapid determination of total dissolved solids (TDS) of water samples were made simply by multiplying the measured electrical conductivity (EC) values (in µS/cm) by 0.64 [17]. The electrical conductivity (EC) of water samples were measured both in spot and in the laboratory directly by Electrical Conductivity meter (EC meter) [17]. Both NH_4-N and NO_3-N were determined by micro Kjeldahl's distillation method [17]. Sodium and potassium of the filtered water samples were directly determined by flame photometry at 589nm and 766nm of wavelength respectively [17] and calcium (Ca), magnesium (Mg), iron (Fe), manganese (Mn), zinc (Zn) and copper (Cu) concentration of water samples were measured directly by atomic absorption spectrophotometer [16]. The chloride content of the water samples were determined by Mohr volume method [17] and bicarbonate (HCO_3^-), carbonate (CO_3^{2-}) content of the water samples were determined volumetrically [17]. The sulphate (SO_4^{2-}) and phosphate (PO_4^{3-}) content were determined by spectrophotometer [17].

3. Result and Discussion

The physico-chemical characteristics of the water samples of the study area of the Kushtia point of Gorai river are presented in the tables 1 and 2 and the chemical constituents of water of the study area are presented in the tables 3, 4, 5 and 6.

The Average temperature of water samples of the study area was approximately 19.9°C and in the range of 19°C to 22°C (Table 1) in the post monsoon period (November) and was approximately 32.8°C (Range 32°C to 33.5°C) in the pre monsoon (May) period (Table 2). The temperature of water samples of both periods showed no extreme variations at the time of collection and found suitable for domestic and industrial uses and irrigation purposes [8,9,18,19].

Table 1. Physico-chemical properties of water of Gorai river system in the post-monsoon period under decreased Ganges flow

Sample no	Location	Temp. °C	TSS mg/l	pH	DO Mg/l	EC µS/cm	TDS mg/l	Ht (mg/l) as CaCO₃	%OM
1	Talbaria	19.5	1.5	7.79	7.2	246	157	126	0.000
2	Baradia	19.5	1.3	7.68	7.2	238	152	122	0.014
3	Shalda	20.0	1.4	7.87	7.6	254	163	134	0.014
4	Ghoraghat	20.5	1.4	7.55	7.5	258	165	128	0.014
5	Goirtia	19.0	1.6	7.22	6.4	264	169	132	0.022
6	Shawria	19.5	0.9	7.48	6.6	254	163	142	0.028
7	Raidanga	19.5	1.7	7.43	6.8	257	164	136	0.022
8	Kaiya	20.0	1.9	7.52	6.8	208	133	114	0.022
9	Rainipara	21.0	1.8	7.37	6.9	217	139	122	0.014
10	Kashimpur	22.0	1.6	7.39	6.4	233	149	144	0.022
11	Daspur	22.0	1.6	7.60	6.0	243	155	148	0.028
12	Shawta	19.5	2.3	7.65	6.2	237	152	132	0.030
13	Borunia	19.5	2.4	7.48	6.6	250	160	126	0.040
14	Varola	20.	2.3	7.61	6.6	235	150	128	0.028
15	Charpara	20.0	2.3	7.64	6.7	231	148	118	0.028
16	Borudia ghat	19.5	2.4	7.72	6.5	228	146	126	0.030
17	Kumarkhali ghat	19.0	2.6	7.65	6.1	241	154	138	0.040
18	Khayarchara	19.0	2.8	7.44	6.5	243	155	134	0.028
19	Pathorbari	19.5	2.7	7.65	6.3	250	160	140	0.048
20	Kamlapur	20.0	2.7	7.77	6.7	242	155	138	0.040
21	Jagolbar	20.5	2.9	7.63	6.8	251	161	142	0.048
22	Janipur	20.5	3.0	7.66	6.6	264	169	130	0.040
23	Muragacha	20.5	3.0	7.75	6.7	254	163	134	0.048
SD			0.8257	0.153	0.1546	0.4030	14.0348	9.0263	8.6019	0.0121
Range			19-22	0.9-3.0	7.22-7.87	6.0-7.6	208-264	133-169	114-148	0-0.048
Average			20	2.09	7.6	7.0	243	156	132	0.028

TSS=Total Suspended Sediments; DO=Dissolved Oxygen; EC=Electrical Conductivity; TDS=Total Dissolved Solids; Ht=Hardness

Table 2. Physico-chemical properties of water of Gorai river system in the pre-monsoon period under decreased Ganges flow

Sample no	Temp. °C	TSS mg/l	pH	DO Mg/l	EC µS/cm	TDS mg/l	Ht (mg/l) as CaCO₃	%OM
1	32	1.3	7.66	7.2	233	149	138	0.008
2	32	1.4	7.64	7.7	248	159	114	0
3	32	1.5	7.65	7.6	264	169	126	0.013
4	32	1.4	7.56	6.5	265	170	136	0.016
5	32.5	1.5	7.66	6.6	250	160	156	0.013
6	32.5	1.1	7.65	6.8	228	146	124	0.016
7	32.5	0.9	7.63	6.5	241	154	142	0.008
8	32.5	1.3	7.67	6.9	235	150	130	0.013
9	33	1.5	7.69	6.4	219	140	116	0.023
10	33	1.8	7.64	6.1	241	154	132	0.020
11	33	2.0	7.65	6.3	241	154	132	0.016
12	33	1.9	7.66	6.2	237	152	112	0.023
13	33	2.0	7.68	6.6	248	159	154	0.008
14	33.5	2.3	7.63	6.0	237	152	128	0.020
15	33.5	2.3	7.66	5.8	233	149	126	0.016
16	33.5	2.2	7.70	6.6	242	155	126	0.016
17	33.5	2.4	7.78	6.1	243	156	128	0.020
18	33	1.9	7.77	6.5	270	173	132	0.023
19	33	2.5	7.90	6.4	231	148	158	0.036
20	33	2.4	7.79	7.1	251	161	132	0.028
21	33	2.4	7.80	6.5	228	146	132	0.043
22	33	2.5	7.82	6.3	278	178	140	0.043
23	33	2.5	7.76	6.7	264	169	118	0.043
SD	0.4910	0.134	0.0795	0.4726	14.984	9.6653	12.313	0.0117
Range	32-33.5	0.9-2.5	7.56-7.9	5.8-7.7	219-278	140-178	112-158	0-0.043
Average	32.8	1.96	7.69	6.58	244.65	156.65	131.83	0.0202

TSS=Total Suspended Sediments; DO=Dissolved Oxygen; EC=Electrical Conductivity; TDS=Total Dissolved Solids; Ht=Hardness

The pH value of water in the study area ranged from 7.22 to 7.87 (Table 1) in the post monsoon period and 7.56 to 7.90 (Table 2) in the pre monsoon period, which are within the permissible limit for irrigated agriculture [19,20] and industrial and domestic use [8].

The hardness (Ht) in water of the study area of Gorai

river system ranged from 114 mg/l to 148 mg/l with an average value of 131.1 mg/l (Table 1) in the post monsoon period and 112 mg/l to 158 mg/l with an average value of 131.83 mg/l (Table 2) in the pre monsoon period. Significant changes in hardness due to seasonal variations were not observed in the Gorai river system as the natural processes by which water is made hard were not found to exist here [18]. According to hardness scale the water of the Kushtia point of Gorai river falls to the soft classes, which are suitable for most of the intended uses [8,18].

The electrical conductivity (EC) of water is an indicator of salinity hazard and gives the total salt concentration in water [21,22,23,24]. In the Gorai river system, the EC value of water at different locations varied from 208 to 264 µS/cm with an average value of 243.4 µS/cm (Table 1) in the post monsoon period and 214 to 278 µS/cm with an average value of 244.65 µS/cm (Table 2) in the pre monsoon period which are "excellent to good" for irrigation according to Wilcox (1955) irrigation water quality classification and surface water quality [18].

The TDS values of water of the study area ranged from 133 to 169 mg/l with an average value of 155.74 mg/l (Table 1) in the post monsoon period and 140 to 178 mg/l with an average value of 156.65 mg/l (Table 2) in the pre monsoon period. The TDS and EC values of the sampled

water shows moderate concentration of dissolved solids and non-saline water [21,25,26] and within permissible limit for utilization [19,20].

From table 1 it can be seen that the Dissolve Oxygen (DO) content of the water sample of Gorai river varied considerably ranging from 6.0 to 7.6 mg/l with an average value of 7.0 mg/l in the post monsoon period and 5.8 to 7.7 mg/l with an average value of 6.58 mg/l (Table 2) in the pre monsoon period. It was observed that (Table 1 and 2) DO values of water bodies under study were higher in post monsoon than in pre monsoon period. The ambient temperature of the study area was colder in post monsoon than in the pre monsoon period, which may have influence to dissolve more oxygen in the colder climate than the warmer ones [8].

Table 1 shows that the range of organic matter (OM) content of the water samples is trace to 0.048% with an average value of 0.028% in the post monsoon period. The OM content varied from trace to 0.043% with an average value of 0.0202% in the post monsoon period. It is very interesting to note that following the similar trend as in salinity the OM content increases as water moves downstream from Charpara to Muragacha with increasing settlement areas and urban runoff [8].

Table 3. Cationic composition (me/l) of water of Gorai river system in the post monsoon period under decreased Ganges flow

Sample no	Na	K	Ca	Mg	Fe	Mn	Zn	Cu	NH$_4$
1	0.61	0.070	1.8	0.51	0.023	0.0041	0.00056	Trace	0.031
2	0.63	0.071	1.8	0.51	0.025	0.0043	0.00062	Trace	0.0135
3	0.62	0.071	1.85	0.53	0.021	0.0039	0.00066	Trace	0.036
4	0.64	0.072	1.9	0.51	0.024	0.0051	0.0018	Trace	0.030
5	0.69	0.074	1.9	0.52	0.027	0.0048	0.0030	Trace	0.033
6	0.72	0.070	2	0.57	0.033	0.0052	0.0024	Trace	0.040
7	0.70	0.071	2	0.61	0.031	0.0055	0.0022	Trace	0.037
8	0.74	0.072	1.95	0.59	0.019	0.0054	0.00098	Trace	0.041
9	0.78	0.072	2	0.61	0.027	0.0055	0.00155	Trace	0.043
10	0.72	0.074	2.05	0.65	0.020	0.0053	0.0022	Trace	0.039
11	0.71	0.072	2	0.66	0.035	0.0049	0.0042	Trace	0.037
12	0.70	0.071	1.15	0.65	0.022	0.0053	0.00055	Trace	0.037
13	0.72	0.070	2	0.69	0.031	0.0054	0.0032	Trace	0.035
14	0.96	0.071	2.05	0.674	0.025	0.0059	0.0024	Trace	0.035
15	0.75	0.071	2.1	0.674	0.024	0.0051	0.0007	Trace	0.035
16	0.76	0.073	2.1	0.65	0.040	0.0053	0.0015	Trace	0.040
17	0.74	0.072	2.1	0.65	0.043	0.0042	0.0003	Trace	0.043
18	0.76	0.074	2.15	0.63	0.041	0.0047	0.0022	Trace	0.045
19	0.71	0.072	2.1	0.57	0.035	0.0051	0.00092	Trace	0.041
20	1.74	0.075	2.2	0.61	0.047	0.0055	0.00155	Trace	0.043
21	0.72	0.076	2.25	0.65	0.043	0.0053	0.0024	Trace	0.043
22	0.74	0.076	2.3	0.67	0.041	0.0060	0.0018	Trace	0.041
23	0.76	0.077	2.3	0.67	0.047	0.0056	0.0022	Trace	0.047
SD	0.0697	0.0020	0.2333	0.0606	0.0091	0.0007	0.00099		0.0069
Range	0.61-0.78	0.070-0.077	1.8-2.3	0.51-0.69	0.019-0.047	0.0039-0.0060	0.0003-0.0042		0.03-0.047
Average	0.71	0.0728	2.037	0.612	0.0323	0.0051	0.00173		0.0385

Ignoring the seasonal variation the average value of sodium (Na) and potassium (K) of the water samples of the study area was 0.71 me/l and 0.0728 me/l in the post monsoon period (Table 3) and 0792 me/l and 0.081 me/l in the pre monsoon period (Table 4). While the calcium (Ca) and magnesium (Mg) content of water samples ranged from 1.8 to 2.3 me/l and 0.51 to 0.69 me/l with an average value

of 2.037 me/l and 0.612 me/l respectively in the post monsoon period (Table 3). In the pre monsoon period the Ca and Mg content ranged between 2.0 to 2.6 me/l and 0.05 to 0.856 me/l with an average value of 2.276 me/l and 0.698 me/l respectively (Table 4). It is evident that the values of Na, K, Ca and Mg content of the water samples were within the recommended limits for irrigation, industrial, domestic and aesthetic purposes [19, 27].

Table 4. *Cationic composition (me/l) of water of Gorai river system in the pre monsoon period under decreased Ganges flow*

Sample no	Na	K	Ca	Mg	Fe	Mn	Zn	Cu	NH₄
1	0.70	0.077	2	0.58	0.028	0.0047	0.00064	Trace	0.033
2	0.72	0.077	2.1	0.59	0.033	0.0062	0.00066	Trace	0.041
3	0.70	0.077	2.1	0.61	0.022	0.0036	0.00070	Trace	0.049
4	0.74	0.079	2.15	0.53	0.040	0.0070	0.0024	Trace	0.033
5	0.76	0.081	2.25	0.55	0.037	0.0067	0.0032	Trace	0.041
6	0.76	0.077	2.15	0.674	0.063	0.0065	0.0030	Trace	0.049
7	0.74	0.077	2.05	0.69	0.040	0.0067	0.0042	Trace	0.041
8	0.804	0.077	2.7	0.724	0.018	0.0068	0.0024	Trace	0.049
9	1.28	0.077	2.4	0.724	0.034	0.0068	0.0018	Trace	0.049
10	0.0804	0.086	2.1	0.75	0.056	0.0063	0.00055	Trace	0.041
11	0.76	0.077	2	0.76	0.043	0.0064	0.00082	Trace	0.041
12	0.76	0.077	2.2	0.74	0.0314	0.0062	0.0018	Trace	0.041
13	0.72	0.077	2.15	0.79	0.035	0.0066	0.00098	Trace	0.041
14	0.76	0.077	2.15	0.724	0.00	0.0063	0.0024	Trace	0.041
15	0.87	0.079	2.25	0.78	0.027	0.0059	0.0226	Trace	0.041
16	0.804	0.079	2.3	0.75	0.047	0.0059	0.0030	Trace	0.066
17	0.804	0.084	2.3	0.69	0.047	0.0057	0.0018	Trace	0.049
18	0.805	0.082	2.35	0.708	0.050	0.0057	0.0022	Trace	0.049
19	0.72	0.077	2.4	0.67	0.038	0.0062	0.00098	Trace	0.033
20	0.804	0.090	2.5	0.69	0.062	0.0065	0.00046	Trace	0.049
21	0.83	0.097	2.55	0.724	0.041	0.0060	0.0018	Trace	0.049
22	0.76	0.097	2.6	0.76	0.044	0.0050	0.00155	Trace	0.049
23	0.804	0.084	2.6	0.856	0.089	0.0042	0.0022	Trace	0.058
SD	0.115	0.006	0.204	0.0806	0.0156	0.0087	0.0044		0.0078
Range	0.70-1.28	0.077-0.097	2.0-2.6	0.53-0.856	0.00-0.089	0.0036-0.0070	0.00055-0.0226		0.033-0.066
Average	0.79	0.080	2.28	0.70	0.028	0.0060	0.0028		0.045

The concentration of iron (Fe), manganese (Mn) and zinc (Zn) in the post monsoon period ranged from 0.019 to 0.047 me/l, 0.0039 to 0.006 me/l and 0.0003 to 0.0042 me/l with an average value of 0.0323 me/l, 0.0051 me/l and 0.00173 me/l respectively (Table 3); while in the pre monsoon period the concentration ranged from 0.018 to 0.089 me/l, 0.0036 to 0.007 me/l and 0.0005 to 0.0226 me/l averaging 0.028 me/l, 0.006 me/l and 0.0028 me/l respectively (Table 4). It is evident that all the values of Fe, Mn and Zn in the study area are within the recommended limits for irrigation, industrial, domestic and aesthetic purposes [19,28]. But the amount of copper (Cu) was found at a trace or not detectable in both the pre monsoon and post monsoon periods in the water of Gorai river system.

Table 5. *Anionic composition (me/l) of water of Gorai river system in the post monsoon period under decreased Ganges flow*

Sample no	CO₃	HCO₃	Cl	SO₄	NO₃	PO₄
1	0.2	2.41	0.35	0.049	0.037	0.012
2	0.25	2.47	0.35	0.049	0.040	0.0123
3	0.25	2.49	0.40	0.054	0.045	0.020
4	0.3	2.57	0.40	0.054	0.040	0.0088
5	0.3	2.63	0.35	0.057	0.034	0.0095
6	0.35	2.65	0.35	0.054	0.037	0.0081
7	0.3	2.71	0.30	0.054	0.034	0.0088
8	0.35	2.87	0.30	0.057	0.026	0.0088
9	0.35	3.10	0.25	0.057	0.022	0.0088
10	0.25	2.78	0.30	0.054	0.031	0.0095
11	0.25	2.98	0.30	0.065	0.031	0.0095
12	0.35	2.80	0.30	0.049	0.031	0.0081
13	0.35	2.72	0.30	0.003	0.034	0.0084
14	0.3	3.10	0.35	0.057	0.034	0.0088
15	0.25	2.80	0.40	0.057	0.0242	0.0081
16	0.3	2.98	0.40	0.054	0.031	0.0088
17	0.25	3.15	0.40	0.049	0.047	0.0095
18	0.3	3.06	0.45	0.054	0.037	0.010
19	0.35	3.15	0.35	0.054	0.041	0.012
20	0..5	3.245	0.35	0.049	0.034	0.0088
21	0.4	3.15	0.35	0.057	0.041	0.018
22	0.4	3.43	0.30	0.054	0.047	0.020
23	0.4	3.43	0.40	0.063	0.041	0.020
SD	0.06	0.30	0.049	0.012	0.0067	0.0041
Range	0.2-0.4	2.41-3.43	0.25-0.45	0.003-0.0524	0.022-0.047	0.0081-0.020
Average	0.32	2.90	0.35	0.052	0.036	0.011

When the ammonium (NH_4) content of water sample of the Gorai river system are in consideration in the post monsoon period, sample 4 (collected from Ghoraghat) shows the lowest and sample 23 (collected from Muragacha) shows the highest value of NH_4 content with the amount of 0.030 and 0.047 me/l

respectively (Table 3). In the pre monsoon period (Table 4) the range of NH_4 content in the water of Gorai river system ranged from 0.033 to 0.066 me/l with an average value of 0.045 me/l. The water sample (sample no 16) collected from Borudia represent the highest value of NH_4.

Table 6. Anionic composition (me/l) of water of Gorai river system in the pre monsoon period under decreased Ganges flow

Sample no	CO_3	HCO_3	Cl	SO_4	NO_3	PO_4
1	0.25	2.49	0.45	0.057	0.054	0.0123
2	0.25	2.59	0.45	0.060	0.052	0.0133
3	0.30	2.67	0.40	0.057	0.067	0.044
4	0.30	2.60	0.40	0.070	0.045	0.0095
5	0.40	2.80	0.35	0.070	0.037	0.012
6	0.50	2.80	0.45	0.076	0.040	0.0088
7	0.40	2.91	0.40	0.065	0.034	0.0095
8	0.50	3.245	0.40	0.065	0.022	0.0095
9	0.50	3.245	0.40	0.060	0.0242	0.0088
10	0.25	2.98	0.30	0.065	0.034	0.0095
11	0.30	3.245	0.40	0.070	0.034	0.010
12	0.50	3.245	0.35	0.049	0.034	0.0088
13	0.40	3.06	0.35	0.067	0.034	0.0088
14	0.25	3.245	0.40	0.063	0.042	0.0095
15	0.40	3.15	0.50	0.070	0.034	0.0088
16	0.30	3.245	0.50	0.065	0.036	0.0095
17	0.40	3.34	0.50	0.054	0.065	0.010
18	0.40	3.43	0.50	0.063	0.041	0.0095
19	0.30	3.43	0.35	0.057	0.054	0.014
20	0.40	3.51	0.40	0.063	0.032	0.0095
21	0.50	3.508	0.45	0.073	0.050	0.020
22	0.50	3.57	0.35	0.057	0.065	0.020
23	0.60	3.57	0.50	0.073	0.047	0.0114
SD	0.104	0.335	0.059	0.0068	0.0125	0.0076
Range	0.25-0.60	2.49-3.57	0.30-0.50	0.049-0.076	0.022-0.067	0.0088-0.044
Average	0.387	3.13	0.40	0.064	0.043	0.0125

Tables 5 and 6 show that the nitrate (NO_3) content of the water samples of Gorai river system of study area varied considerably ranging from 0.022 to 0.047 me/l and 0.022 to 0.067 me/l with an average value of 0.036 me/l and 0.043 me/l in the post monsoon period and in the pre monsoon period respectively. From the values there is an indication of NO_3 pollution in some areas of higher human and livestock population [8].

The average concentration of carbonate (CO_3), bicarbonate (HCO_3) and chloride of the samples were 0.317 me/l, 2.90 me/l and 0.348 me/l in post monsoon period; while in the pre monsoon period the values were 0.387 me/l, 3.13 me/l and 0.4 me/l respectively. The sulfate (SO_4) and phosphate (PO_4) concentration of the water samples of the study area ranged from 0.0003 to 0.63 me/l and 0.0081 to 0.02 me/l with an average value of 0.0529 me/l and 0.011 me/l in the post monsoon period; while in the pre monsoon period SO_4 and PO_4 concentration varied from 0.049 to 0.079 me/l and 0.0088 to 0.044 me/l averaging 0.064 me/l and 0.0125 me/l respectively.

4. Conclusion

The samples were analyzed for intended water quality parameters following internationally recognized and well established analytical techniques. From the present

investigation it was observed that there were no extreme variations of river water temperature and the ambient temperature. Dissolved Oxygen (DO) of the water of study area was higher in the post monsoon than in the pre monsoon period. The pH values are within the permissible limit and no significant changes were observed due to seasonal variation. Hardness fall "soft" classes and are suitable for most of the intended uses. Electrical conductivity (EC) of collected water samples is "excellent to good". The TDS and EC values of the sampled water show moderate concentration of dissolved solids and non-saline water. There was a trend of increasing TDS, EC and OM as the water moves downstream. It is evident that all the values of sodium (Na), potassium (K), calcium (Ca), magnesium (Mg), iron (Fe), manganese (Mn), zinc (Zn), copper (Cu), ammonium (NH_4), nitrate (NO_3), carbonate (CO_3) and bicarbonate (HCO_3) falls under the permissible limit and there were no toxicity problem. Water samples showed no extreme variations in the concentrations of cations and anions and it was true for both post and pre monsoon periods. No toxic concentrations were observed for the heavy metals. Thus, most of the inorganic elements are not a serious problem in terms of water contamination in Gorai River. Higher concentrations of nitrogen and phosphorus were recorded especially around the locations of higher human population and livestock activities. As

Gorai River system meandering through Ganges calcareous alluvial floodplain its water has high Ca content.

References

[1] BUP (Bangladesh Unnayan Parishad), Resources, Environment and Development in Bangladesh with particular References to the Ganges, Brahmaputra and Meghna Basin. Bangladesh Unnayan Parishad (BUP). Academic publishers, Dhaka- 1994, pp. 1-79.

[2] BWDB (Bangladesh Water Development Board), Environmental and Social Impact Assessment of Gorai River Restoration Project. Main Report. Environmental and GIS Support Project for Water Sector Planning. Ministry of Water Resources, GOB, Dhaka-2001, pp. 1-185.

[3] Ecofile, Periodical on Life and nature. Vol. 3&4. Unnayan Shamannay, Dhaka-2003, pp. 9-39

[4] A.A. Rahman, S. Huq, G.R. Conway, Environmental Aspect of Surface Water system of Bangladesh. The University Press Limited, Dhaka-2000, pp. 7-265.

[5] S.Z.K.M. Shamsad, Mohammad Saiful Islam, Muhammad Qumrul Hassan, Ground water quality and hydrochemistry of Kushtia district, Bangladesh. J. Asiat. Soc. Bangladesh-1999, Sci. 25 (1): 1- 11.

[6] R.L. Doanhue, R.W. Miller, J.C. Shickluna, Soils: An Introduction to Soils and Plant Growth. 5th ed., Prentice-Hall of India (pvt.) Ltd. New Delhi-1999, pp. 450-465.

[7] S.E. Manahan, Environmental Chemistry. CRL Press Inc. Boca Raton, USA-1994, pp.179-200

[8] H.S. Peavy, D.R. Rowe, G. Tchobanoglous, Environmental Engineering. McGraw Hill, New York-1985, pp. 14-56.

[9] P. K. Gupta, Methods in Environmental Analysis: water, Soil and Air. Agrobios (India), Jodhpur-2000, pp. 5-76.

[10] J.L. Hatfield, D.L. Karlen, Sustainable Agricultural Systems. Lewis Publishers. Boca Raton, Florida. USA-1994, pp. 21-46.

[11] S.O. Ryding, W. Rast, The Control of Eutrophication of Lake and Reservoirs. Man and the Biosphere Series Vol. 1. United Nations Educational, Scientific and Cultural Organization (UNESCO). Parthenon, Carnfoth, Lancashire-1989, pp. 3-314.

[12] M.S. Islam, M.Q. Hasan, S.Z.K.M. Shamsad, Quality of irrigation water in the Kushtia District of Bangladesh. *J. Biol. Sci.*-1998, 7 (2): 129-138.

[13] V.E. Hansen, O.W. Israelsen, G.E. Stringham, Irrigation: Principles and Practices. 4th ed., John Wiley and Sons, New York-1980, pp. 1-5.

[14] M.Q. Hassan, M.S. Islam, Hydrogeo-environmental Impact on Kushtia District, Bangladesh: A Study on Pre- and Post-Farakka Conditions. In: Proc of the Workshop on Groundwater and Environment, BGS-Goethe Institute, Dhaka-1997, pp. 84-93.

[15] G. R. Chhatwal, M. C. Mehra, M. Sataka, T. Katyal, M. Katyal, T. Nagahiro, Encyclopidia of environmental pollution and its control. Vol. II, water pollution. Anmol Publications. New Delhi-1992, pp. 70-254.

[16] A.L. Page, R.H. Miller, D.R. Keeney, Methods of Soil Analysis (ed.), Part 2. Am. Soc. Agron. Soil Sci. Am. Madison, Wis. USA-1982, pp. 159-446.

[17] M.L. Jackson, Soil Chemical Analysis, Prentice Hall, Inc. Englewood Cliffs, N.J. USA-1967, pp. 227-261.

[18] M.L. Davis, D.A. Cornwell, Introduction to Environmental Engineering. 3rd ed. McGraw Hill, Boston, USA-1998, pp. 284-289.

[19] DOE (Department of Environment), Bangladesh Gazette, No. DA-1; Department of Environment. Ministry of Environment and Forest-1997, pp. 1324-1327.

[20] UCCC, Guidelines for Interpretations of water Quality for Irrigation. Technical Bulletin, University of California Committee of Consultants, California, U.S.A-1974, pp. 20-28.

[21] A.M. Michael, Irrigation Theory and Practices. Vikash Publishing House Ltd., New Delhi-1992, p. 740.

[22] L.V. Wilcox, Classification and Use of Irrigation Waters. United States Department of Agriculture. Circ. 969, Wasington, D.C.- 1955, p. 19.

[23] N.C. Brady, R.R. Well, The Nature and Properties of Soils. 13th ed. Pearson Education, Inc. New Delhi, India-2002, pp. 261-269.

[24] R.D. Misra, M. Ahmed, Manual of Irrigation Agronomy, Oxford and IBH Publishing Co. Pvt. Ltd., New Delhi-1987, pp. 248-271.

[25] L.A. Richards, Diagnosis and Improvement of Saline and Alkali Soils, U.S. Department of Agriculture Handbook, Vol. 60, Washington, D.C.- 1954, p. 160.

[26] D.K. Todd, Ground Water Hydorlogy. 2nd ed., John Wiley and Sons Inc.New York-1980, pp. 10-138.

[27] R. S. Ayers, D. W. Westcot, Water Quality for Agriculture. Irrigation and Drainage. Paper No. 29. Food and Agriculture Organization of the United Nations. Rome-1985, pp. 1-117.

[28] BWPCB (Bangladesh Water Pollution Control Board), Bangladesh Drinking Water Standard. Bangladesh Water Pollution Control Board, GOB, Dhaka-1976.

16

Sustainable groundwater exploitation in Nigeria

David O. Omole

Department of Civil Engineering College of Science and Technology, Covenant University, Canaanland, Km 10 Idiroko Road, Ota, Nigeria

Email address:
david.omole@covenantuniversity.edu.ng, omojohnny@yahoo.com

Abstract: In this study, a critical review of the groundwater resources history and management in Nigeria was done. The aim was to identify reasons why groundwater is increasingly being exploited in recent times, and to explore ways through which the exploitation could be done sustainably. This was achieved through literature review. It was observed that an estimated 60% of Nigeria's population get drinking water from ground resources. This high statistic is mostly the resultant effect of infrastructural decay in the potable water supply sector of the country. In Nigeria, just 14 % of the country's population get regular water supply through piped sources while the remainder of the country's 162.5 million people draw their water supply mainly from surface and groundwater sources. Groundwater, in Nigeria, is accessed mainly in form of shallow (hand-dug) and deep (boreholes) wells. Well water withdrawal activities are mostly un-regulated and, therefore, the water sources are often subjected to avoidable abuses and pollutions. In particular, rapid urbanization, agricultural and industrial activities are major contributors of pollutions to groundwater sources. Also, unavailability of data such as geographical information on water quantity, hydrology, state of aquifer and withdrawal limits have contributed to the unsustainable use of groundwater in Nigeria. It was recommended that proper regulation of groundwater resources and its protection under the land use act of 1978 could be instrumental to its sustainable exploitation.

Keywords: Groundwater, Sustainable, Sub-Sahara Africa, Nigeria, MDG, Exploitation

1. Introduction

The unsustainable abstraction of groundwater has recently become a subject of global debate[1]. It is an issue that became prominent when global water consumption increased by nearly 1000% within a space of 50 years (1950-2000), mainly as a consequence of agricultural irrigation in[1]. According to OECD[2], agriculture is responsible for the use of 70% of all freshwater including groundwater. Factors which have made groundwater use quite attractive for agricultural is the relatively cheap cost of getting the water to the farm by sinking boreholes on location rather than piping or channelling the water over long distances[1]. Other contributing factors include cheap technology, breakdown of public utilities, the relative clean state of GW, and rapid and unplanned urbanization which makes new connection to public utilities nearly impossible[1]. Other causes are municipal and industrial supplies respectively. It is reported that more than half the world population obtain drinking water from groundwater sources[3-4]. GW has been instrumental to the partial success of millennium development goal 7c (MDG7c)

which aims at reducing by half the number of people in the world without access to clean water and improved sanitation. The target for water was reached in 2010, five years ahead of the deadline of 2015. Today, only an estimated 783 million people are yet to have access to clean drinking water, with over 2 billion people gaining access between 1990 and 2010[5].

Prominent among countries where the MDG on clean drinking water has not been reached are sub-Saharan African countries (Fig. 1). UNICEF/WHO[5] reported that four out of every ten persons without access to clean drinking water are in sub-Sahara Africa. In spite of this statistic, about 75% of the population within the sub-Sahara Africa rely on groundwater sources for clean drinking water[4, 6-8].

Nigeria is strategic among the sub-Sahara Africa countries mainly because of its current and projected population. It is the opinion of the international development committee of the House of Commons that the failure of Nigeria to meet the MDG is the failure of Africa. This is because one in every five African is a Nigerian[9]. Unfortunately, Nigeria receives the least support among

SSA countries from aid institution, thus contributing to the slow rate of achieving MDG7c[10]. With a growth rate of 3%, Nigeria's growth rate is one of the highest in the world. The country's population became more than tripled within 50 years (1960-2010). If this trend is repeated, it means Nigeria might have a population of 320 million people by 2060. The population factor directly impacts on demand for water, thus making it a major consideration for water planners.

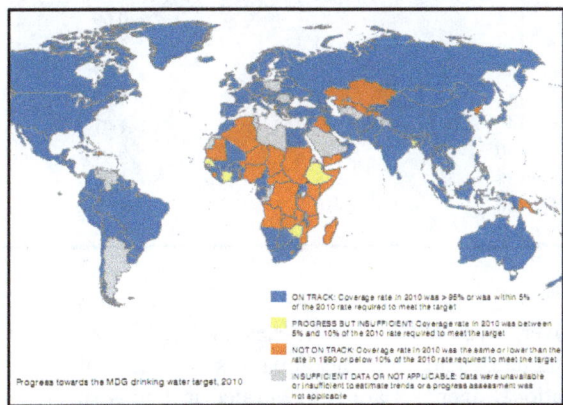

Fig 1. Countries off the track of achieving MDG 7c are shown in orange shade (Source:[5])

1.1. Chronicle of Water Supply in Nigeria

Groundwater development in Nigeria began as far back as 1917 when the Nigeria Geological survey (NGS) was established with part of its objectives being the determination of a geological map for the country and prospection for groundwater in the arid regions of the northern parts of the country. It was found that the country is comprised of two main types of groundwater formations namely the pore-type water in stratified rocks and fissure-type water found in crystalline rocks (Fig. 2). It is to the credit of NGS that the concrete lined hand dug well was introduced into Nigeria around 1928, a technique that is still widely used across the country. The NGS proceeded to upgrade the prospecting for water from hand dug wells to drilled wells in 1938. In 1963, the US Geological Survey partnered with the NGS in conducting an extensive survey of the Sokoto and Chad basins, both within the northern (arid) sections of the country[11].

In 1976, the Federal ministry of water resources (FMWR) and its subsidiaries of eleven river basin development authorities (RBDA) were created and merged with the water section of the NGS. The twelfth RBDA was created at a later date. The FMWR was charged with the responsibility of managing the water resources of the country, provision of water for irrigation and municipal supply, basic hydrological data collection, storage and analysis for national water planning purposes[13]. These are further complemented by State water agencies (SWA) in the thirty six states of the federation and the federal capital territory[10-11, 13].

In 1976, the Federal ministry of water resources (FMWR) and its subsidiaries of eleven river basin development authorities (RBDA) were created and merged with the water section of the NGS. The twelfth RBDA was created at a later date. The FMWR was charged with the responsibility of managing the water resources of the country, provision of water for irrigation and municipal supply, basic hydrological data collection, storage and analysis for national water planning purposes[13]. These are further complemented by State water agencies (SWA) in the thirty six states of the federation and the federal capital territory[10-11, 13].

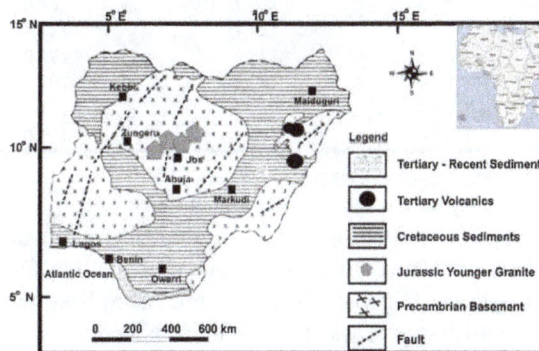

Figure 2. Geological Map of Nigeria (source:[12])

2. Challenges to Sustainable Groundwater Use in Nigeria

Groundwater withdrawal is in itself not wrong. Neither is groundwater abstraction by private individuals wrong. However, anthropogenic activities always have impact on the environment. This is more so when a large and increasing proportion of the population continue to impact on groundwater due to factors that can be regulated. Some of these factors are:

2.1. Insufficient Funding

The performance of FMWR, the RBDAs and SWAs and has been less than efficient due to insufficient and unsustainable public spending. A breakdown of the national budget reveals that 70% of the budget is expended on recurrent expenditures such as salaries, staff allowances, rudimentary maintenance of facilities and skeletal services. This leaves 30% of the budgetary allocation for capital expenditures[10, 14]. The Nigerian national budget for the period 2002 to 2011 is presented in Table 1.

From Table 1, it can be observed that allocation to water resources development in Nigeria has not risen consistently with population increase, thus suggesting that increasing water demands have not been factored into budgetary allocations. The table also shows that over the period of ten years, investment in provision of water amounts to an average of $0.30/capita/annum. In contrast, analysis shows that aid for water from donor countries towards the provision of water for MDG7c amounted to about

$1.1/capita/annum for the period 1996 - 2001[16]. Rationally, grants from external sources ought to be less than National allocation. Since this is in reverse, it can be deduced that the country may not have the capacity to sustainably meet its obligation in the provision of water for

its citizens. This deduction becomes relevant when compared to the fact that about 34% of the population obtain water for domestic use from surface water bodies and other unsafe sources[17].

Table 1. Brief summary of Nigeria's budget allocation for the past ten years (☐ billion)

Year	Total Expenditure (Billion Naira)	Allocation to Water Supply (Billion Naira)	allocation to Water supply (USD million)	Population (million)	Per capita/per annum investment (cents)	Percentage of Water Supply to Total Expenditure (100%)
2002	724.5	5.5	34.4	129	8	0.76
2003	921.2	6.4	40.0	133	9	0.69
2004	1125.1	18.5	115.6	136	30	1.64
2005	1478.6	26.2	163.8	139	35	1.77
2006	1586.8	29.7	185.6	143	39	1.87
2007	2116.1	22.7	141.9	146	29	1.07
2008	3107.8	28.0	175.0	150	35	0.90
2009	2776.9	47.7	298.1	154	58	1.72
2010	3266.2	37.5	234.4	158	46	1.15
2011	3542.0	20.4	125.0	162	23	0.58

Source:[14-15]; NB: NGN160 = USD 1

Aside from national budgetary allocation, international aid has been another very important source of funding for water projects in Nigeria. However, this also has not resolved the difficulty of water provision as Nigeria is still classified among water short nations[5]. USAID[10] reported that Nigeria receives the least sum in aids in the Sub-Sahara Africa. Gleick[16] also reported that more than half of global aid for the provision of water and sanitation was given to just ten countries while countries where about 60% of their population lacked access to clean drinking water received just 12% of the funds.

2.2. Hydrological Factors

Hydrologically, groundwater is inseparable from surface water. Therefore, variations in either one affect the entire hydrology system. Annual rainfall in Nigeria varies between 250 mm in the arid north and 4000 mm in the rainforests in the south[18]. Paradoxically, the economy of the arid north is more agriculture-based than the south. Therefore, the northern section of the country requires a lot of irrigation for the cultivation of produces such as cotton, wheat, onions, tomatoes, millet, sorghum, and groundnuts which is sold in all parts of the country. The northern part of the country also produces most of the beef consumed by the rest of the country. Therefore, the lack of national capacity to supply piped water coupled with low precipitation has led to heightened use of groundwater in northern Nigeria[10].

2.3. Infrastructure Overload

In southern Nigeria, the highest demand for water arise from municipal water supply, industries, and agriculture. Due to job opportunities and other factors, nearly 27% of Nigeria's population live in the eight coastal states of the country while the remaining 28 States and the Federal capital territory accommodate the rest of the population[19].

Moreover, over 60% of industries in Nigeria, all banks headquarters, and the headquarters of several other international agencies are concentrated in Lagos state[19], thus serving as an attraction for job seekers. This high population density and rural-urban migration in selected states of the country has led to rapid urbanization with attendant pressure on available infrastructure. As an example, the infrastructure in Lagos which was designed for one million people currently supports a population of more than 15 million people[9]. Thus, the absence of expansion programmes and/or maintenance of existing infrastructure has led to infrastructure overload, and in many cases, abandonment. This leaves a vast proportion of the population, industries and small scale enterprises fewer choices among which is groundwater exploitation.

2.4. Weak Institutions

Weak institutions and governance systems is counter development. The weakness of the institutions is linked to weak enforcement of the existing policies, laws and regulation. Poor management and incompetent managers also contribute significantly to the weak institutions The SWAs, for instance, are unable to meet up with the responsibility of providing water for people in their states partly because public water utilities are being operated without sound accounting principles[20]. Water is often provided free of charge and in many cases, unaccounted water is as high as 83%[20]. This inefficient and unsustainable water provision system has run aground many SWAs.

2.5. Data Management System

Without complete information on the hydro-geological systems in the country information, it becomes impossible to model the aquifer systems, identify fossil aquifers for special resource administration or to enforce restrictions on purposes for which a user might be licensed to withdraw groundwater as is being practiced in developed nations. It also becomes difficult to make informed decisions on policies and regulations that make for sustainable use of water from such formations. Fossil aquifers are also known as non-renewable underground water reservoirs because of their extremely slow rate of recharge[21]. It often takes several hundreds to thousands of years to have significant accumulation of water in them. These aquifers are often characterized with very clean water and are mostly found in arid regions[21-22]. An example of fossil aquifer in Nigeria (but which shares boundary with five other countries) is the Chad basin with reserves of 170,000 – 350,000 Mm3. It is being exploited at the rate of 250 Mm3/annum but is being recharged at a rate of less than 1 Mm3/annum[21-22]. At the current rate, it would be exhausted between 680-1400 years. This is based on the assumption that it is not hydraulically linked to an adjacent draining aquifer. This type of aquifer ought not to be left to the whims of anyone because of its value and depleting nature.

In terms of hydrological data collection, storage, and analysis, the RBDAs operate below capacity. Most of the RBDAs still operate analogue systems of data recording. In places where computers exist, epileptic electricity supply prohibits their use, thus leading to inefficient data collection and storage. With the expected volume of data to be managed, data analysis becomes tedious. This in turn leads to limited decision taking at national level on hydrological matters.

2.6. Resource Protection Through Regulations and Enforcement

In terms of resource protection, the National Environmental Standards and Regulations Enforcement Agency (NESREA) was established in 2007 to replace the Federal environmental protection agency (FEPA) which operated from 1988 to 2007[23]. The latter was scrapped due to reported operational lapses[23]. However, due to other perceived more environmentally sensitive issues such as oil spills and hazardous wastes management, NESREA has not paid adequate attention to surface and groundwater resources[23]. The first known environmental regulation on surface and groundwater was published only in 2011 with its objective being 'to restore, enhance and preserve the physical, chemical and biological integrity of the nation's surface and ground waters and to maintain existing water uses'[23-24]. This regulatory provision appears to be a step in the right direction if it can be given the necessary attention and drive for accomplishment.

2.7. Indiscriminate Groundwater Exploitation

Due to lack of capacity on the part of responsible institutions to meet with the ever increasing demands for water in Nigeria, there has been heightened and unrestrained exploitation of groundwater. Nearly anyone that can afford to sink a well has gone ahead to do so without recourse to expert advice on hydro-geological data, safe yield, technology, or excess draw down in water table[11]. This chaotic situation has given rise to affiliated problems such as the involvement of unqualified well drillers and capital loss. With frequent reports of failed wells, failed pumps[11], and groundwater contamination arising from drilling mud and septic tanks interferences, capital flight arising from the systemic failure would be enormous when quantified.

3. The Way Forward

In Nigeria, all land and mineral resources within it are held in trust for the people by the Governor of each State Government through the land use Act of 1978[25]. By extension, it could be deduced that all water resources ought to be held in trust also. However, this is not clearly defined. The act allows persons who wish to build on a piece of land the right of ownership for 99 years once a certificate of occupancy has been obtained. This may be renewed at expiration. However, if crude oil or gold is discovered on the same land, the State reserves the right of ownership of such resources and the right to revoke such certificate of occupancy in order to secure the resource in public trust. In the case of groundwater however, the State has been silent, probably because it recognizes its own incapacitation at providing water for people. Thus, the riparian system is very much in use. The occupiers of the land simply exploit the water at will without any form of permit, control, or penalty. It is therefore suggested that control of groundwater resources within each state should be tied to the 1978 land use act. That is, groundwater should be accorded the same protective status as precious resources such as gold and crude oil. Therefore, special permits should be attached to rights to own a well. However, in view of the high poverty rate in the country and the need to achieve the MDG on clean drinking water, licensing for the prospection of water should not attract additional charges. Rather, the licensing should impose additional responsibilities on prospective well owners in order to make them exploit the wells sustainably and with the full knowledge that the groundwater resource is not theirs to abuse at will. Responsibilities that could be attached to obtaining a license to exploit groundwater should include:

i.   Only qualified and certified operators could be allowed to drill wells.

ii.   Private drilling operators should be required to submit data to a central pool. This will help reduce costs of exploratory drilling and help furnish decision makers with information on sub-regional reconnaissance.

iii. Private drilling should also be fitted with remote sensing meters to monitor the withdrawal rates and to plan exit strategies.

Furthermore, in order to begin to take control of its water resources, Nigeria needs to upgrade its entire water resources data base management system. There should be collaborative effort between the FMWR, its subsidiaries, and its SWA counterparts in order to achieve this since there is always benefits in synergizing efforts and information sharing. Also, steps need to be taken to complete hydro-geologic mapping of the entire country in order to characterize and take inventory of all its water bearing formation. The FMWR and SWAs may be designed to be a 'one-stop shop' where detailed information on national water resources and prospecting licenses can be obtained. While the FMWR pre-occupies itself with the hydro-geological mapping, the SWAs should embark on sub-regional reconnaissance. This reconnaissance could be done by combining satellite imagery and aeromagnetic surveys with ground observations of hydro-geological features in order to obtain greater details on sub-surface conditions of groundwater resources in their immediate vicinity[22]. The SWAs may also need to be tasked with the duty of conducting regular exploratory drillings to determine individual aquifer properties such as depth, thickness, water quality, direction of flow, and other aquifer hydraulic properties. The cost of doing this may be defrayed by putting a system in place whereby those licensed to prospect for water are required by law to provide the coordinates of the proposed well, return core samples and provide other affiliated information. This information could be archived in a central data-base system for future analysis and decision taking on the host aquifer systems.

With respect to funding, budgetary allocations and implementation, the culture of transparency and accountability should be promoted. More results could be achieved with less monetary resources. The issue of fiscal accountability was mentioned by some aid agencies as a major concern in Nigeria[9-10, 26]. This may also be a reason why aid institutions have been committing less financial resources to the provision of water resources in Nigeria in spite of the strategic position of Nigeria in sub-Sahara Africa and the world. Seemingly, the Federal Government cannot solve the challenge of provision of potable water supply for most of its citizens because of competing national needs and debt burdens. It is also apparent that most of the population cannot help themselves because of the high rate of poverty. It is an established fact that the poor pay more per litre for water all over the world. This burden on the poor should be alleviated by allowing the aid from donor organizations to reach them. The opportunity afforded by aid institutions should therefore be encouraged and maximized by returning value for every dollar invested.

In addition, global attention is shifting from the provision of more infrastructures meant for the provision of more water to the efficient use of what is available. This has been described as the 'soft path' approach to water resources management[16, 27]. A situation where 83% of water provided by SWAs cannot be accounted for underscores the need to upgrade the agency's operational efficiency. If the surface water resources are well protected, managed and channelled, there would be less pressure on groundwater resources, thereby preserving them for future generations and/or contingent situations. The efficient use of water can be achieved by strengthening the existing water management institutions and giving more bite to regulations enforcement. Polluters should be made to pay and everyone should be made aware that groundwater is a common wealth that should be protected.

4. Conclusion

The current study reviewed the origin and the present status of groundwater exploration in Nigeria. Agriculture, municipal water supply and industrial/manufacturing concerns were identified as the largest consumers of groundwater. It found that currently, as much as 60% of Nigerians utilize groundwater for domestic purposes. The inordinate demand for groundwater in Nigeria has been exacerbated by inadequate and ill-maintained infrastructure, a short fall in public utility supply of water, improper management of available financial resources, inadequate funding of water and sanitation programmes, ineffective and ambiguous policies and regulations where it concerns groundwater resources, pollution of available surface water bodies, and availability of relatively cheap technology for groundwater exploitation. This has led to serious abuse of groundwater resources as evidenced by incursion of untrained personnel in the business of groundwater exploitation, high rate of failure of deep wells, indiscriminate well sinking resulting in excessive draw down and general lowering of the water table. It was recommended that all groundwater resource in Nigeria should be classified as a common wealth and accorded the same status as crude oil by insisting that licenses should be obtained before prospecting. In view of the high poverty rate among most Nigerians, it was further recommended that no additional charge should be attached to licensing beside from the cost of purchasing the land. However, additional responsibilities which would ensure that the resource is handled with care should be imposed on all prospective users. It was further suggested that efficient use of available water resources should be given more emphasis than the provision of more water.

References

[1] Shah T, Burke J, Villholth K (2007) Groundwater: a global assessment of scale and significance. In: Water for Food, Water for Life: A Comprehensive Assessment of Water Management in Agriculture. Earthscan and Colombo: International Water Management Institute, London.

[2] OECD. (2012). Development Co-operation Report 2012: Lessons in Linking Sustainability and Development, OECD Publishing. Available at: http://dx.doi.org/10.1787/dcr-2012-en. Accessed 14 December 2012.

[3] IAH (2010). IAH Commission on Groundwater and Climate Change. Available at http://www.iah.org/gwclimate/gw_cc.html. Accessed 14 December 2012

[4] IIED (2010). Groundwater, self-supply and poor urban dwellers: A review with case studies of Bangalore and Lusaka. Edited by Gronwall, J.T., Mulenga, ., and McGranahan, G for International Institute for Environment and Development, London. ISBN: 978-1-84369-770-1.

[5] UNICEF/WHO (2012). Progress on drinking water and sanitation: 2012 update. UNICEF and World Health Organization. ISBN: 978-92-806-4632-0.

[6] Ogunba, A. (2011). Nigeria's New Environmental Laws: What Implications for Groundwater Protection and Sustainability? Colloquium, South Africa.

[7] Omole D.O. (2010), Water Quality Modelling: Case study of the Impact of Abattoir Effluent on River Illo, Ota, Nigeria. LAP Lambert Academic Publishing GmbH & Co. KG, Saarbrücken, Germany. ISBN: 978-3-8433-7034-9.

[8] AWV. (2009). The Africa water vision for 2025: Equitable and sustainable use of water for socioeconomic development. UN Water/Africa. Available at http://www.icp-confluence-sadc.org/sites/default/files/African%20Water%20Vision%202025.pdf. Accessed 1 SEPT 2012.

[9] IDC (2009). DFID's Programme in Nigeria. International Development Committee. Eighth Report of Session 2008–09, vol. 1, House of Commons, London.

[10] USAID (2010). NIGERIA Water and Sanitation Profile. Available at: http://www.washplus.org/sites/default/files/nigeria.pdf Accessed 14 Dec. 2012

[11] Eduvie M.O. (2006). Borehole Failures and Groundwater Development in Nigeria. National Seminar on the Occasion of Water Africa Exhibition (Nigeria 2006), Lagos, Nigeria. Available at http://www.nwri.gov.ng/userfiles/file/Borehole_Failure_in_Nigeria.pdf Accessed 20 Dec. 2012.

[12] Adepelumi, A. A. and Fayemi, O. (2012). Joint application of ground penetrating radar and electrical resistivity measurements for characterization of subsurface stratigraphy in Southwestern Nigeria. *J. Geophys. Eng.* 9; pp 397-412.

[13] FMWR (2000). National Water Supply and Sanitation Policy. 1st edition, Fed Republic of Nigeria. Available at: http://www.nwri.gov.ng/userfiles/file/National_Water_Supply_and_Sanitation_Policy.pdf. Accessed 20 December 2012.

[14] CBN, 2012. Annual report for the year ended 31st of December. Available at: http://www.cenbank.org/Out/2012/publications/reports/rsd/arp-

2011/2011%20Annual%20Report_Complete%20Report.pdf Accessed 26 May 2013.

[15] WORLD BANK, 2012. World development indicators and global development finance. Available at: http://www.google.co.uk/publicdata/explore?ds=d5bncppjof8f9_&met_y=sp_pop_totl&idim=country:NGA&dl=en&hl=en&q=nigeria%20population Accessed: December 29, 2012.

[16] Gleick, P.H., Allen, L., Christian-Smith, J., Cohen, M.J., Cooley, H., Herberger, M., et. al. (2012). *The World's water volume 7: The Biennial Report on Freshwater resources.*

[17] Longe, E.O., Omole, D.O., Adewumi, I.K. and Ogbiye, A.S. (2010). *Water Resources Use, Abuse and Regulations in Nigeria.* Journal of Sustainable Development in Africa, Vol. 12 (2): 35-44.

[18] Adekunle, I. M., Adetunji, M. T., Gbadebo, A. M. and Banjoko, O. B. (2007). Assessment of Groundwater Quality in a Typical Rural Settlement in Southwest Nigeria. Int. J. Environ. Res. Public Health, 4(4), 307-318.

[19] Omole, D.O. and Isiorho, S.A. (2011). Waste Management and Water Quality Issues in Coastal States of Nigeria: The Ogun State Experience. Journal of Sustainable Development in Africa, 13(6):207-217.

[20] AfDB/OECD. African Economic Outlook: Nigeria (2007); Kauffmann, Celine and Perard, Edouard. New Partnership for Africa's Development (NEPAD)/OECD. Stocktaking of the water and sanitation sector and private sector involvement in selected African countries (2007).

[21] Margat, J., Foster, S. and Droubi, A. (2006). Concept and importance of non-renewable resources. In Foster, S. and Loucks, D.P. (Editors), 'Non renewable groundwater resources'. UNESCO Paris.

[22] Shaminder, P., Margat, J., Yurtsever, Y., and Wallin, B. (2006). Aquifer characterisation techniques. In Foster, S. and Loucks, D.P. (Editors), 'Non renewable groundwater resources'. UNESCO Paris.

[23] Ladan, M.T. (2012). Review of Nesrea Act 2007 and Regulations 2009-2011: A New Dawn in Environmental Compliance and Enforcement in Nigeria. Law, Environment and Development Journal 8(1): p. 116-140.

[24] FRN (2011). Environmental regulation on surface and groundwater, S.I. No. 22, Gazette No. 49, Vol. 98 of 24th May, 2011 of the Federal Republic of Nigeria, Abuja.

[25] FRN (1978). Land Use Act (1978) C.A.P. 2002. Federal Republic of Nigeria.

[26] DFID NIGERIA (2012). Operational Plan 2011-2015. Available at http://www.dfid.gov.uk/documents/publications1/op/nigeria-2011.pdf. Accessed 29 December 2012.

[27] Gleick, P.H. (2003). Global freshwater resources: Soft path solutions for the 21st century. *Science,* 302(1524).

17

Rating curve pattern in a compound meandering channel flow

Md. Abdullah Al Amin[1, *], Md. S. M. Khan[2], M. R. Kabir[3], Md. Ashraf-ul-Islam[4]

[1]Assistant Engineer, Bangladesh Water Development Board, Bangladesh
[2]Dept. of Water Resources Engineering, BUET, Dhaka, Bangladesh
[3]Dept. of Civil Engineering, UAP, Dhaka, Bangladesh
[4]Project Coordinator, RDRS, Dhaka, Bangladesh

Email address:

amin_01buet@yahoo.com(M. A. A. Amin), mostafakhan@wre.buet.ac.bd(M. S. M. Khan), mkabir@uap-bd.edu(M. R. Kabir),
ashraf.hiz@gmail.com(M. Ashraf-ul-Islam)

Abstract: Most of the river flow in the world can be characterized as compound meandering channel in which the discharge distributions are very complex. Engineers, planners and researchers are highly interested in predicting accurately as well as reliably the quantitative estimates of discharge in a compound meandering channel. A laboratory experiment has been performed in a compound meandering channel with symmetric cross-sections having floodplain width ratio of 1.00, 1.67, 2.33, 3.00 and depth ratio of 0.20, 0.30, 0.35, 0.40 using the large-scale open air facility in the Department of Water Resources Engineering, Bangladesh University of Engineering and Technology (BUET), Dhaka, Bangladesh. Point velocity data have been collected using an ADV (Acoustic Doppler Velocity meter) for different depth and width ratio at different locations of a compound meandering channel. Cross-sectional discharge is computed by area velocity method from the observation of velocity profile. The laboratory experiment shows that the rating curve pattern in a compound meandering channel follows the straight curve up to in bank flow and in the out bank flow it is in curvature nature.

Keywords: Meandering Channel, Discharge Distribution, Width Ratio, Depth Ratio, Flood plain, ADV

1. Introduction

Generally, natural rivers, streams and manmade surface drainage channels often overflow their banks during episodes of high flooding resulting in a huge potential damage to life and property as well as erosion and depositions of sediments. Many rivers have meandering compound channels possessing a main channel, which always carries flow in one or two floodplains, which only carry flow at above bankful stages. It has been established that a strong interaction between the faster moving main channel flow and slower moving floodplain flow takes place in a compound channel. This interaction results in a lateral transfer of a significant amount of longitudinal momentum which affects the discharge distribution in a channel flow. Discharge in a compound meandering channel is strongly governed by interaction between flow in the main channel and that in the floodplain and are different due to prevailing of different hydraulic conditions in the main channel and floodplain flow.

Stage-discharge curve known as rating curve in a compound meandering channel plays an important role in controlling floods, solving a variety of river hydraulics and engineering problems, designing stable channels, revetments and artificial waterways. There are limited reports concerning the stage-discharge relationship in a compound meandering channel. Most of the efforts of [1], [2], [3], [4], [5], [6], [7], [8], [9] and [10] were concentrated on the energy loss, different methods for stage-discharge assessment in the compound meandering sections. The present study is aimed at understanding the phenomenon of rating curve pattern in a compound meandering channel.

2. Methodology

The experimental study has been conducted in the open air facility of Water Resources Engineering Department, Bangladesh University of Engineering and Technology (BUET), Dhaka. The experimental setup is shown in the Fig.1 which consists of two parts, the permanent part and the temporary part. The permanent part is the experimental facility necessary for the storage and regulation of water circulating through the experimental reach. The temporary part is mainly brick walls which are used to vary the floodplain width for different setups. The experimental reach consists of a 670 cm long symmetric compound meandering channel, set at constant bed slope (S_o) of 0.001845 with fixed bed and banks and sinuosity ratio (S_r) of 1.20. Water is drawn by the centrifugal pump of discharge capacity 80 l/s from the storage reservoir then it discharges into the upstream reservoir and conveys water to the experimental reach through approach channel of 30 m in length and 3.1 m in width. To ensure a more smooth flow towards the approach channel guide vanes and tubes are placed between the upstream reservoir and the approach channel which are at right angle to each other. In order to prevent turbulence in the approach channel, PVC pipes (diffuser) are used. The water regulating function of the downstream end is provided by tail gate. The tail gate rotates around a horizontal axis. It is operated to maintain desired water level in the experimental reach. At the end of the experimental channel, water is allowed to flow freely so that backwater has no effect in the experimental reach. Behind the tail gate, the water falls into the stilling basin and passes through a transition flume which allows water for recirculation.

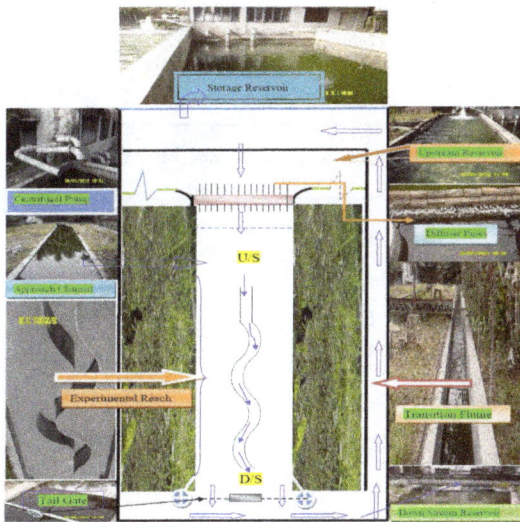

Figure 1. Schematic diagram of the laboratory experimental setup

Experiments were performed for four cases i.e. width ratio Wr = 1.00, 1.67, 2.33, 3.00 at four runs i.e. depth ratio Dr = 0.20, 0.30, 0.35, 0.40.

Case I: It represents no floodplain condition having width ratio Wr =1 and cross-sectional dimension of the Channel is 45.70 cm x 42 cm.

Case II: It indicates symmetric floodplain width 15.30 cm having width ratio Wr =1.67. The cross-sectional dimension of the main channel is 45.70 cm x 24.50 cm, left floodplain 15.30 cm x 18 cm and right floodplain 15.30 cm x 18 cm.

Case III: It indicates symmetric floodplain width 30.50 cm having width ratio Wr = 2.33. The cross-sectional dimension of the main channel is 45.70 cm x 24.50 cm, left floodplain 30.50 cm x 18 cm and right floodplain 30.50 cm x 18 cm.

Case IV: It indicates symmetric floodplain width 45.70 cm having width ratio Wr = 3.00. The cross-sectional dimension of the main channel is 45.70 cm x 24.50 cm, left floodplain 45.70 cm x 18 cm and right floodplain 45.70 cm x 18 cm.

Point velocities data have been collected by ADV (Acoustic Doppler Velocity meter) at different locations (upstream clockwise bend, upstream crossover, upstream anticlockwise bend etc) of a compound meandering channel. Each location is divided into 19 zones starting from left floodplain to right floodplain. The main channel is equally divided into nine zones (zone 1 to zone 9), the left floodplain is equally divided into 5 zones (zone 1 to zone 5) and right floodplain is divided into 5 zones (zone 1 to zone 5). The definition sketch of compound meandering channel is shown in Fig. 2 and the experimental run conditions are shown in the table 1.

Table 1. Experimental run conditions

Case	Run no.	Width Ratio (Wr)	Depth Ratio (Dr)	Location of the Reading
I	1	1.00	0.20	
	2		0.30	
	3		0.35	
	4		0.40	
II	5	1.67	0.20	Velocity reading at 0.1H, 0.2H, 0.4H, 0.6H, 0.8H from the water surface in the main channel and 0.1H', 0.2H', 0.4H', 0.6H', 0.8H' from the water surface in the flood plain.
	6		0.30	
	7		0.35	
	8		0.40	
III	9	2.33	0.20	
	10		0.30	
	11		0.35	
	12		0.40	
IV	13	3.00	0.20	
	24		0.30	
	15		0.35	
	16		0.40	

In each zone 3D point velocity readings are taken by ADV at five vertical points i.e. 0.1H, 0.2H, 0.4H, 0.6H, 0.8H for main channel and 0.1H', 0.2H', 0.4H', 0.6H', 0.8H' for floodplain. Sample of data collection by ADV is shown in the Fig. 3. In each vertical point 60 seconds point velocity readings are taken and average velocity of 60 seconds point velocity is used for plotting the velocity profile. Cross-sectional discharge is calculated by area velocity method from the observation of velocity profile. In this method a channel section is subdivided into a number of segments by a number of successive intervals. For all the cross-sections, discharge is calculated separately for the main channel, right and left flood plain; and then total discharge is obtained.

The discharge of the segment is calculated as follows

$$\Delta Qi = \Delta Ai.Ui \qquad (1)$$

Where ΔQi is the discharge in the ith segment, ΔAi is the cross-sectional area of the ith segment and Ui is the average velocity at the ith vertical.
The total discharge is computed as

$$Q = \Sigma \Delta Qi \qquad (2)$$

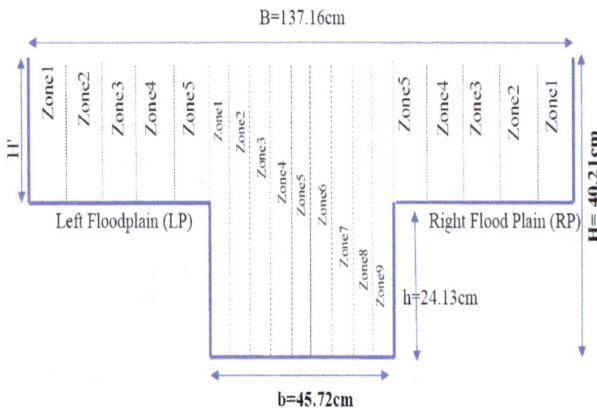

Figure 2. *Sketch of the compound meandering channel section*

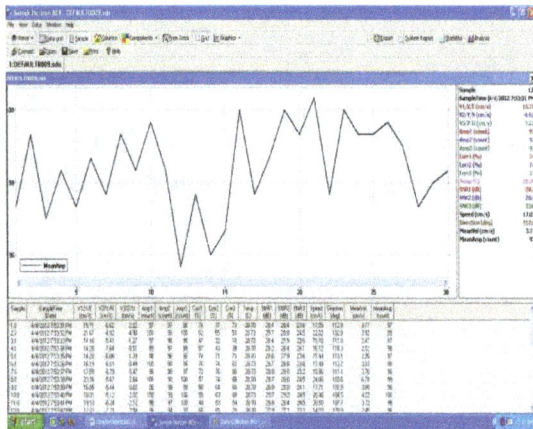

Figure 3. *Data collection by ADV*

3. Results and Discussions

Stage discharge curves with varying width ratios at upstream clockwise bend section, upstream cross-over section and upstream anticlockwise bend section are shown in the Fig. 4, Fig. 5 and Fig. 6. In all the cases, discharge increases with the increase of width ratio for the discharge just above the bank level of the main channel. For the discharges just above the bank level, the retarding influence of the flood plain takes its toll on the overall discharge of the channel. As the depth ratio increases, the retarding effect of the floodplain is counter balanced by the increase in greater flow area

provided by the floodplain. So for the higher water level (depth ratio), the discharge increases with the increase in depth of flow.

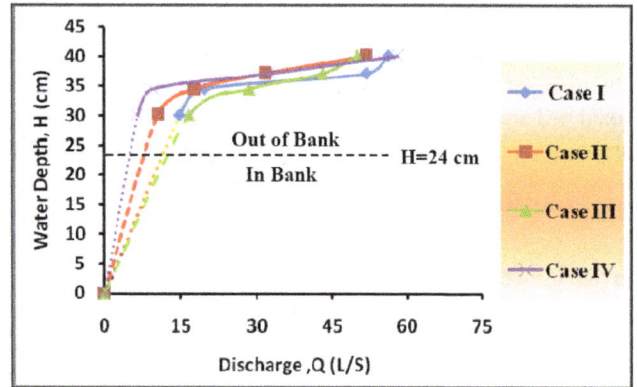

Figure 4. *Rating curve at upstream clockwise bend section*

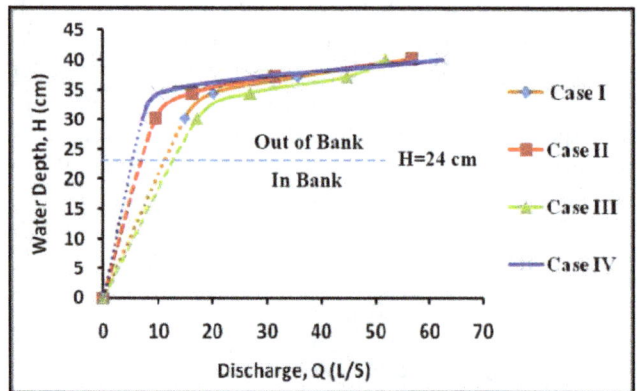

Figure 5. *Rating curve at upstream crossover section*

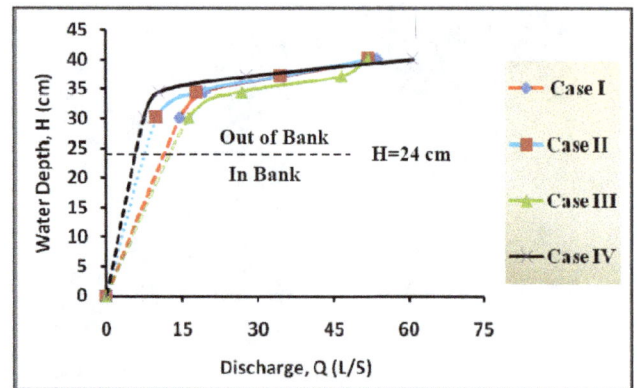

Figure 6. *Rating curve at upstream anti-clockwise bend section*

4. Conclusions

On the basis of present research concerning the rating curve pattern in a compound meandering channel with varying floodplain width the following conclusions are drawn: In a compound meandering channel, discharge increases with the increase of depth ratio and increasing rate of discharge is more at high water depth ratio. Because at low water depth ratio the slow moving flow in the floodplain interact with the fast moving main channel flow intensely and considerable momentum exchange takes place. But the intensity of interaction diminishes considerably with the increase of depth ratio. Rating curve pattern in a compound meandering channel follows the straight curve up to in bank flow and in the out bank flow it is in curvature nature.

Acknowledgements

First of all, the authors are very much thankful to the almighty, Allah for enabling them to complete their work successfully. The authors convey their sincere gratitude to head of the Department of Water Resources Engineering, Bangladesh University of Engineering and Technology, for giving all the facilities needed which helped the authors to reach at culmination of the work successfully.

The authors would like to thank the assistants of Physical Modelling Laboratory of BUET for their dexterous help to complete their laboratory experiment efficiently.

Finally, the authors would like to thank all of their well wishers and colleagues.

Nomenclature

The following symbols are used in this paper

ΔA_i = Cross-sectional area of the ith segment
B = top width of compound meandering channel
b = width of main channel
D_r = depth ratio (H-h)/h
H = total water depth
H' = depth of water above floodplain bed
h = height of the main channel
ΔQ_i = discharge in the ith segment
Q = total discharge
S_o = bed slope
S_r = sinuosity ratio
U_i = average velocity at the ith vertical.
W_r = width ratio [B/b]

References

[1] Al-Romaih, J. S. (1996). "Stage discharge assessment in meandering channels," PhD thesis, Univ. of Bradford, U.K.

[2] Groenhill, R. K., and Sellin, R. H. J. (1993). "Development of a simple method to predict discharge in compound meandering channels." *Proc. Institute Civil Engineers, Water, Merit and Energy*, 101, Water Board, (March), 37–44.

[3] Muto, Y. (1995). "Turbulent flow in two stage meander chanenls." PhD thesis, Univ. of Bradford, U.K.

[4] Patra, K.C., and Kar, S.K., Bhattacharya.A.K. (2004). "Flow and Velocity Distribution in Meandering Compound Channels." Journal of Hydraulic Engineering, ASCE, Vol. 130, No. 5. 398-411.

[5] Sellin, R. H. J., Ervine, D. A., and Willetts, B. B. (1993). "Behavior of meandering two stage channels." *Proc. of Institute Civil Engineers Water Maritime and Energy*, 101, (June), *Paper No. 10106*, 99–111.

[6] Shiono, K., and Muto, Y. (1993). "Secondary flow structure for inbank and overbank flows in trapezoidal meandering channels." *Proc., 5th Int. Symp. of Refined Flow Modl. and Turb. Measu.*, Paris (September), 645–652.

[7] Shiono, K., Muto, Y., Knight, D. W., and Hyde, A. F. L. (1999a). "Energy losses due to secondary flow and turbulence in meandering channels with over bank flow." *J. Hydraul. Res.*, 37(5), 641–664.

[8] Toebes, G. H., and Sooky, A. A. (1967). "Hydraulics of meandering rivers with flood plains." *J. Waterw. Harbors Div., Am. Soc. Civ. Eng.*, 93(2), 213–236.

[9] Wark, J. B., and James, C. S. (1994). "An application of new procedure for estimating discharge in meandering overbank flows to field data." *2nd Int. Conf. on River Flood Hydraulics*, March 22–25, Wiley, NewYork, 405–414.

[10] Willetts, B. B., and Hardwick, R. I. (1993). "Stage dependency for over bank flow in meandering channels." *Proc., Institute of Civil Engineers Water Maritime and Energy*, 101, 45–54.

18

Application of 2D morphological model to assess the response of Karnafuli River due to capital dredging

Sarfaraz Alam[*]**, M. Abdul Matin**

Department of Water Resources Engineering, Bangladesh University of Engineering and Technology, Dhaka, Bangladesh

Email address:

sarfaraz@wre.buet.ac.bd(S. Alam)

Abstract: The major sea port of Bangladesh is the Chittagong port located on the right bank of Karnafuli river of Bangladesh. This river port is considered as the lifeline of the economic activities of the country due to its increasing trade demand. Many port facilities have been planned to be implemented in future to meet this increasing demand. Due to manmade interventions the river flow becomes interrupted and thereby may cause the change in river morphology. Recently Chittagong Port Authority (CPA) has undertaken 3.5 Mm^3 Capital Dredging and 2.5 km long bank protection initiative from 3rd Karnafuli bridge to Sadarghat jetty. This study focuses mainly on the application of 2D mathematical model to assess the response of the river due to such development works on Karnafuli River. Necessary data have been collected from CPA and the model was set using the bathymetry of 2009. The river reach between Kalurghat and Khal no-18 has been selected for the study purpose. Time series discharge and water level data were used as boundary condition at upstream and downstream consecutively. Calibration and validation have been carried out with the recent water level data at Sadarghat and Khal no-10. After hydrodynamic calibration and validation the model was adjusted to match the change in bed level with the observed data. Different hydrodynamic and morphological assessments like variation of velocity, sediment transport and bed level changes have also been studied. It was further used to assess the effect of Capital dredging and Bank Protection near Sadarghat area. Model result shows slight increase in velocity and sediment transport due to project implementation. It also changes the rate of erosion/ deposition at some location of the selected reach. It is hoped that the results of the model simulation will be helpful to suggest the effect of possible future development work to be implemented on this river.

Keywords: Karnafuli River, Chittagong Port, Capital dredging, Morphodynamics, 2D Modeling

1. Introduction

Karnafuli river, which is the major river of Chittagong, originates from Lushai hills in Mizoram state and flows about 270 km south and southwest through the southeastern part of Bangladesh to reach the Bay of Bengal. During this course this meandering river passes Kaptai hydroelectric power plant, Halda-Karnafuli confluence and several bridges (Fig. 1).

The length of the river from Kaptai Dam to Halda-Karnafui confluence is about 45 km and from Halda-Karnafuli confluence to BN Academy is about 30 km. Karnafuli river is a tidal river having semi-diurnal characteristics. During flood period the flow travels long distance in the upstream direction of Halda river and very near to Kaptai Dam in upper Karnafuli river. This study mainly focuses on the lower part of Karnafuli river spanning from Kalurghat to Khal no-18. Lower Karnafuli river is the most important portion of the whole river due to the vast economic activities. Regular maintenance of this portion of river is necessary to keep it navigable for safe transportation of vessels. Recently capital dredging work at Sadarghat area is running to maintain sufficient draft. Thus the application of a 2D mathematical model to assess the effect of capital dredging at this location is highly envisaged.

A 2D model Delft3D has been applied to simulate the hydrodynamic and morphological processes of Karnafuli river. Delft3D consists of different modules such as Flow, MOR, Wave, WAQ. For this study the Flow module is used which is a multidimensional (2D or 3D) hydrodynamic and transport simulation program. This module is capable of calculating unsteady flow and transport phenomena resulting from tidal and meteorological forcing on curvilinear grid. It is also possible to take into consideration some other parameters such as temperature, salinity and different constituents and observe the 2D or 3D distribution of the results. To know and predict the navigability and erosion/deposition pattern it is important to know the hydrodynamic and morphological characteristics of the river which is reflected by its flow velocity, shear stress and sediment transport. Earlier in 1987, Department of Water Resources Engineering (WRE), Bangladesh University of Engineering and Technology (BUET) developed a mathematical model study to assess the 1D hydrodynamics of this river [5]. In 1990, Danish Hydraulic Institute (DHI)

and BUET applied a 2D model MIKE 21 at the entrance of Karnafuli river [7, 8]. Recently Chittagong Port Authority (CPA) and WRE of BUET had used a quasi 2D model to assess the effect of 3rd Karnafuly bridge on the navigability of this river within the port limit [1,2]. They also conducted a study to assess the effect of RCC jetty construction near Sadarghat area in 2010 [3]. The present research mainly deals with the hydro-morphological analysis using well known Delft3D morphological model.

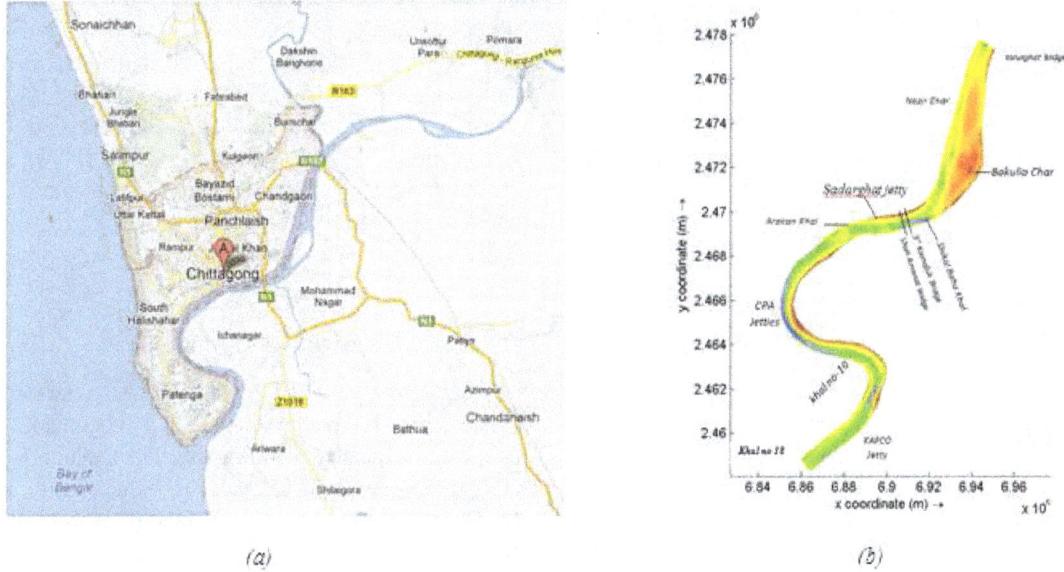

(a)   (b)

Figure 1. (a) Karnafuli river in Chittagong (source: Google) (b) Plan view of Karnafuli River

2. Model Description

Delft3D solves horizontal momentum, continuity and transport equation for hydrodynamic simulation. It solves Navier-Stokes equation for an incompressible fluid. The FLOW manual [6] of Delft3D extensively describes all the

$$\frac{\partial u}{\partial x} + \frac{\partial v}{\partial y} + \frac{\partial w}{\partial z} = 0 \qquad (1)$$

$$\frac{\partial v}{\partial t} + \frac{\partial vu}{\partial x} + \frac{\partial v^2}{\partial y} + \frac{\partial vw}{\partial z} + fu + \frac{1}{\rho_0}P_y - F_y - \frac{\partial}{\partial z}\left(v_V\frac{\partial v}{\partial z}\right) = 0 \quad (2)$$

$$\frac{\partial u}{\partial t} + \frac{\partial u^2}{\partial x} + \frac{\partial uv}{\partial y} + \frac{\partial uw}{\partial z} - fv + \frac{1}{\rho_0}P_x - F_x - \frac{\partial}{\partial z}\left(v_V\frac{\partial u}{\partial z}\right) = 0 \quad (3)$$

parameters and equations used in Delft3d-Flow module. Continuity (Eq. 1) and momentum equations (Eq.2-3) are shown which are to be solved by the model [10] for computation of velocities. Here u,v and w define the velocity components in three perpendicular directions and v_H, v_v are the horizontal and vertical eddy viscosity, f_u, f_v are coriolis paremeters which are neglected in this case. Horizontal pressure terms P_x and P_y for a certain depth z can be determined by (Boussinesq approximations) using Eq. (4 and 5) where ζ, ρ, ρ_0 and g denote water level elevation, density, reference density of water and gravitational

acceleration. The horizontal friction terms F_x and F_y (Reynold's stresses) are determined using the eddy viscosity concept. (Eq. 6 and 7).

$$\frac{1}{\rho_0}P_x = g\frac{\partial\zeta}{\partial x} + \frac{g}{\rho_0}\int_z^\zeta\left(\frac{\partial\rho}{\partial x} + \frac{\partial z'}{\partial x}\frac{\partial\rho}{\partial z'}\right)dz' \qquad (4)$$

$$\frac{1}{\rho_0}P_y = g\frac{\partial\zeta}{\partial y} + \frac{g}{\rho_0}\int_z^\zeta\left(\frac{\partial\rho}{\partial y} + \frac{\partial z'}{\partial y}\frac{\partial\rho}{\partial z'}\right)dz' \qquad (5)$$

$$F_x = \frac{\partial}{\partial x}\left(2v_H\frac{\partial u}{\partial x}\right) + \frac{\partial}{\partial y}\left(v_H\left(\frac{\partial u}{\partial y} + \frac{\partial v}{\partial x}\right)\right) \qquad (6)$$

$$F_y = \frac{\partial}{\partial y}\left(2v_H\frac{\partial v}{\partial y}\right) + \frac{\partial}{\partial x}\left(v_H\left(\frac{\partial u}{\partial y} + \frac{\partial v}{\partial x}\right)\right) \qquad (7)$$

The advection-diffusion equation [12] is also shown in Eq(8).

$$\frac{\partial[hc]}{\partial t} + \frac{\partial[hUc]}{\partial x} + \frac{\partial[hVc]}{\partial y} + \frac{\partial(\omega c)}{\partial\sigma} =$$
$$h\left[\frac{\partial}{\partial x}\left(D_H\frac{\partial c}{\partial x}\right) + \frac{\partial}{\partial y}\left(D_H\frac{\partial c}{\partial y}\right)\right] + \frac{1}{h}\frac{\partial}{\partial\sigma}\left[D_V\frac{\partial c}{\partial\sigma}\right] + hS \qquad (8)$$

Where S is indicating source and sink terms per unit area, D_H and D_V are horizontal and vertical diffusivity and ω is the vorticity. Mass concentration and scaled vertical co-ordinate are denoted by c and σ. The Suspended sediment reference concentration was estimated using van Rijn's formula [11]. Equation (9) shows the bed load transport rate formulation [6].

$$|S_b| = 0.006\rho_s w_s D_{50}^{(\ell)} M^{0.5} M_e^{0.7} \qquad (9)$$

Where S_b is bed load transport rate (kg/m/s) for median diameter D_{50}, sediment mobility number due to waves and currents M and excess sediment mobility number M_e. Fall velocity w_s was also computed using van Rijn equation. Bottom sediment change due to bed load transport is calculated using equation(10).

$$\Delta_{SED}^{(m,n)} = \frac{\Delta t f_{\text{MORFAC}}}{A^{(m,n)}} \left(\begin{array}{c} S_{b,uu}^{(m-1,n)}\Delta y^{(m-1,n)} - S_{b,uu}^{(m,n)}\Delta y^{(m,n)} + \\ S_{b,vv}^{(m,n-1)}\Delta x^{(m,n-1)} - S_{b,vv}^{(m,n)}\Delta x^{(m,n)} \end{array} \right) \qquad (10)$$

Here f_{MORFAC} is user defined morphological factor, $A^{(m,n)}$ is area of computational cell (m^2), $S_{b,uu}$ and $S_{b,vv}$ are computed bed sediment transport vector, $\Delta x^{(m,n)}$ and $\Delta y^{(m,n)}$ are cell widths in x and y direction and Δt is the time interval used for simulation.

3. Model Setup

3.1. Grid Generation

In order to continue the simulation a horizontal curvilinear grid is set up which covers the whole modeling area. Delft3D utility RGFGRID has been used to generate grid in Cartesian coordinate system (Fig. 2a). The grid extends from Kalurghat to Khal no 18 which is about 30 km. Grid was refined and finally 368 by 33 grids were taken. Local refinement was done near Sadarghat area in order to better approximate the results of imposed scenario at this location.

The model consists a total of 11744 cells having smallest dimension of about 50m by 25 m at Sadarghat area. Bathymetry data was collected from Chittagong Port

Authority (CPA) which was in ISLWL(Indian Spring Low Water Level). Delft3D QUICKIN was used to generate bathymetry for the model simulation.

3.2. Boundary Condition and Time Step

The model area spans between Kalurghat in the upstream direction and Khal no 18 in the downstream direction. Hourly discharge data was used as the upstream boundary condition and water level data was used as downstream boundary condition. Discharge data for upstream boundary was generated using one dimensional hydrodynamic model HEC-RAS of Halda-Karnafuli river network . Water level data was collected from Chittagong Port Authority (CPA).

Time step is another important consideration for stability and accuracy. Model was run for different type steps and spatially varying Courant number were calculated in order to check the stability. Finally one minute time step has been selected which was found to be sufficient enough for the accuracy and stability of the model.

3.3. Model Calibration and Verification

Water level was used as the calibration parameter for the Delft3D model developed. Data was collected from CPA for two stations located at Sadarghat and Khan no-10. Simulated and measured water levels for the month November (post-monsoon condition) and June (monsoon condition) at Khal no-10 were compared and adjusted which was found quite satisfactory (Fig. 3-4). For calibrating the model different values of roughness (Manning's n) have been tested to obtain adequate match with the observed field condition. The whole model area consists of varying roughness values which ranged between n= 0.023 to n=0.028.

After calibration, the model was validated at Sadarghat for the period 5th November to 16th November 2009 (Fig. 5). It also shows good agreement with the observed data.

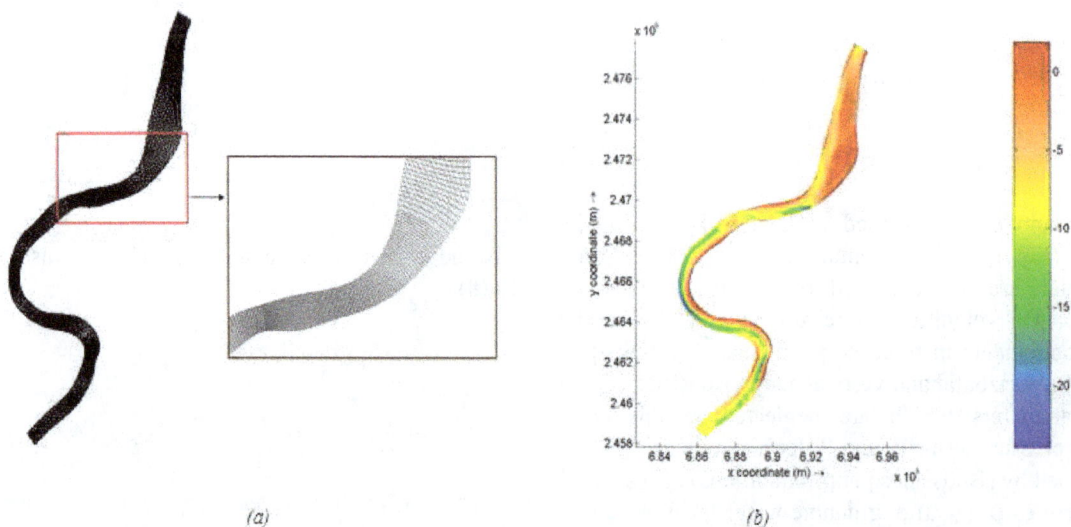

(a)   (b)

Figure 2. (a) Curvilinear grids with local refinement (b) Bathymetry of the model area

In regards to morphological calibration bed levels are computed by adjusting the coefficients and exponents of sediment transport predictor available in the module. However due to lack of measured sediment data the actual compliance of sediment transport capacity has not been achieved. Considerable efforts have been given to calibrate the morphological model by making comparison with the observed bed levels. Bed material sample has been collected and analyzed [4]. The median size of bed material was found to be range from 0.150 to 0.250 mm. The mean value 0.20 mm has been used as a representative size for morphological computation. Comparison of the computed bed level against the observed one at various section are shown in Fig. 6.

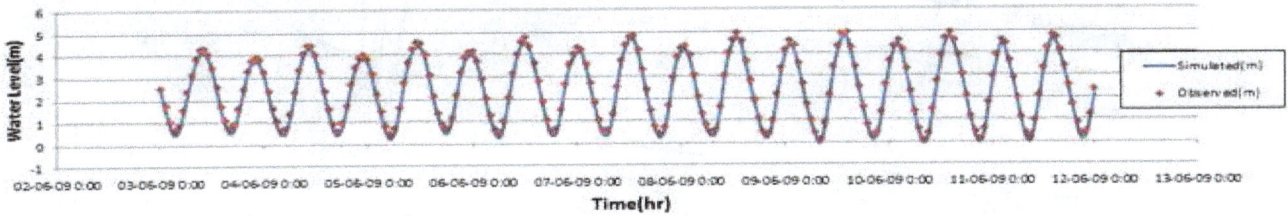

Figure 3. Water Level Calibration at Khal no-10 (3-12 June 2009)

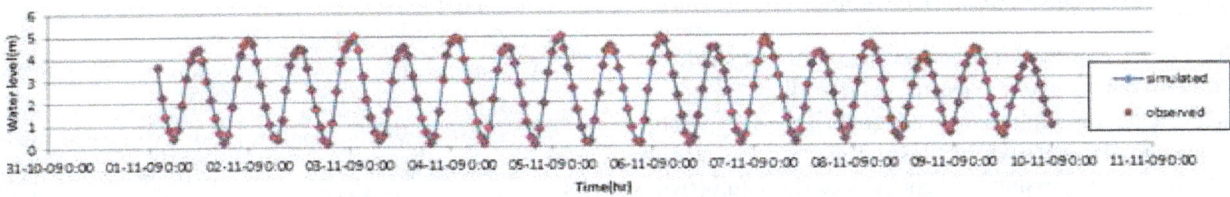

Figure 4. Water Level Calibration at Khal no-10 (1-10 November 2009)

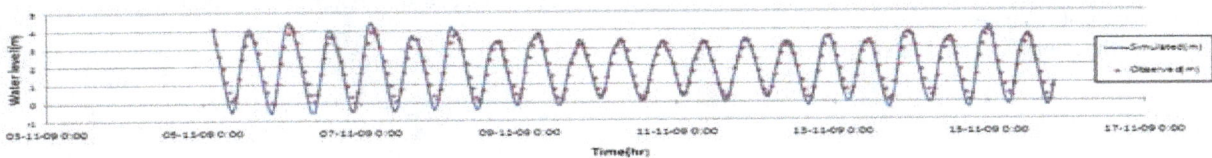

Figure 5. Water Level Validation at Sadarghat (5-16 November 2009)

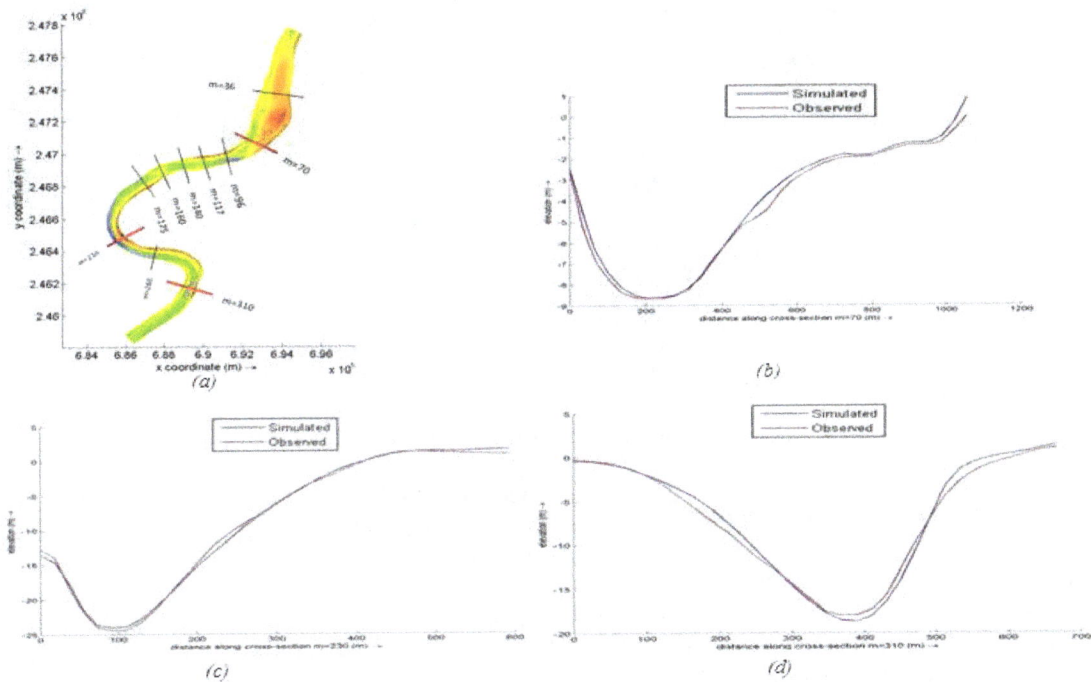

Figure 6. Bed level comparison between the simulated and observed one (a) Plan view of selected sections (red line) (b) Cross sections M= 70(c) M=230 and (d)M=310

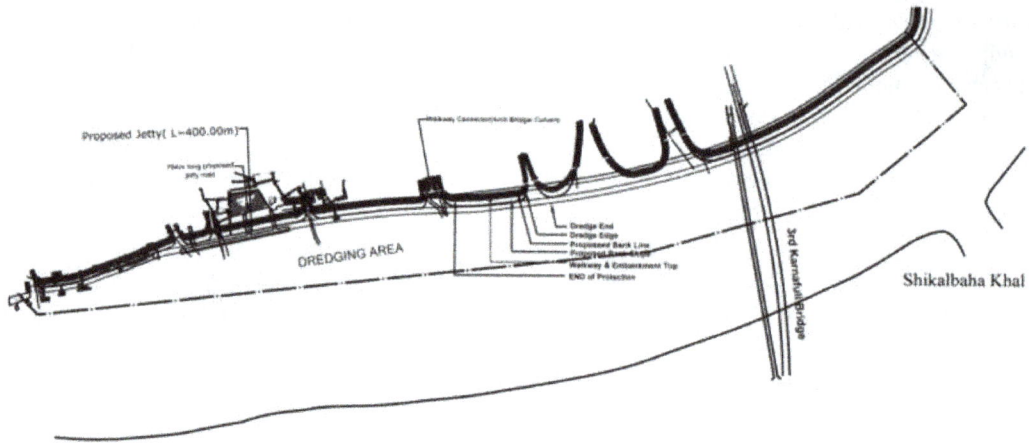

Figure 7. *Capital dredging project of Chittagong Port Authority, Bangladesh*

3.5. Capital Dredging and Bank Protection

3.5.1. Location of the Intervention (Capital Dredging)

The Port of Chittagong is located at the north of the Bay of Bengal on the Karnafuli River at a latitude 220 18'30" North and a longitude of 91048'30" East. The site is located in Chittagong encompassing the right bank of Karnafuli (river) and part of the river Karnafuli from 3rd Karnafuli bridge to Sadarghat area of the Port of Chittagong. Also the project area can be described below.

    a)   On the North: High Water Line / Existing High

Bank Line
b)   On the South: Dredge Limit
c)   On the East: 500 m upstream of 3rd Karnafuli bridge
d)   On the West: Approx. 100 m downstream of Sadarghat Jetty

The dredging site near Sadarghat area is shown in Fig. 7. The area extends from upstream of 3rd Karnafuli bridge to some distance downstream of the proposed jetty.

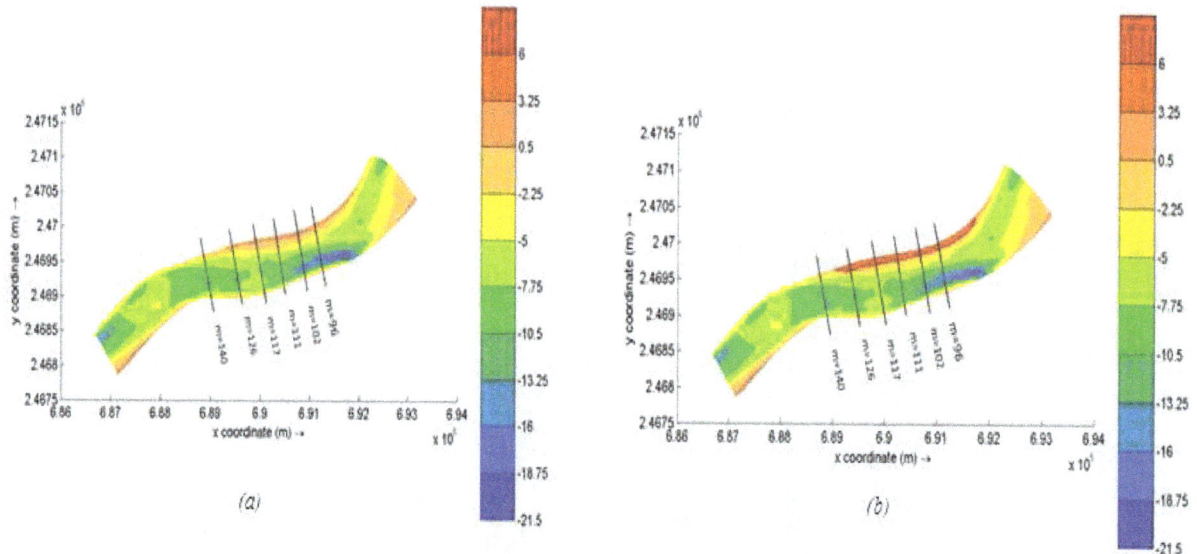

Figure 8. *Bathymetry (a) without and (b) with capital dredging and bank protection*

3.5.2. Application of Capital Dredging and Bank Protection

Capital dredging project was undertaken at Sadarghat to improve the navigability and flow condition. The developed two dimensional model was further used to assess the effect of capital dredging and bank protection project on the hydrodynamic and morphological properties of Karnafuli River at Sadarghat area. For this purpose the bathymetry of the river has been changed to match the planned section

according to the capital dredging project (Fig. 9). Changed cross section includes dredging up to -4 m ISLWL and river training works having the peak height 6 m ISLWL. Each cross section contains a series of steps, those are -4 m, 0 m, 6 m and lastly some area behind the bank for filling purpose having elevation of about 4 m ISLWL. Fig. 8 shows the changed cross sections at grid numbers M=96, 102, 111, 117 and 126 (see Fig. 8 for location). Grid numbers of some other features at Sadarghat are, Sikal Baha khal M=88 to 90; 3rd Karnafuli Bridge M=95.

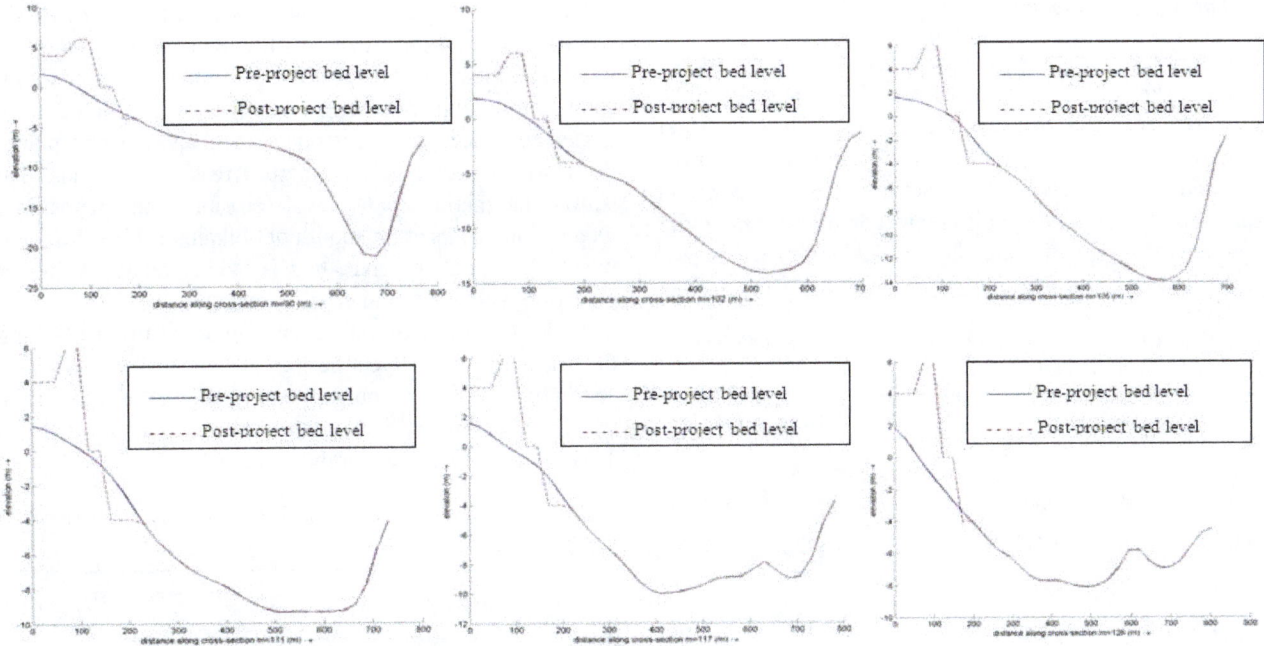

Figure 9. *Cross sections of pre- and post project situation*

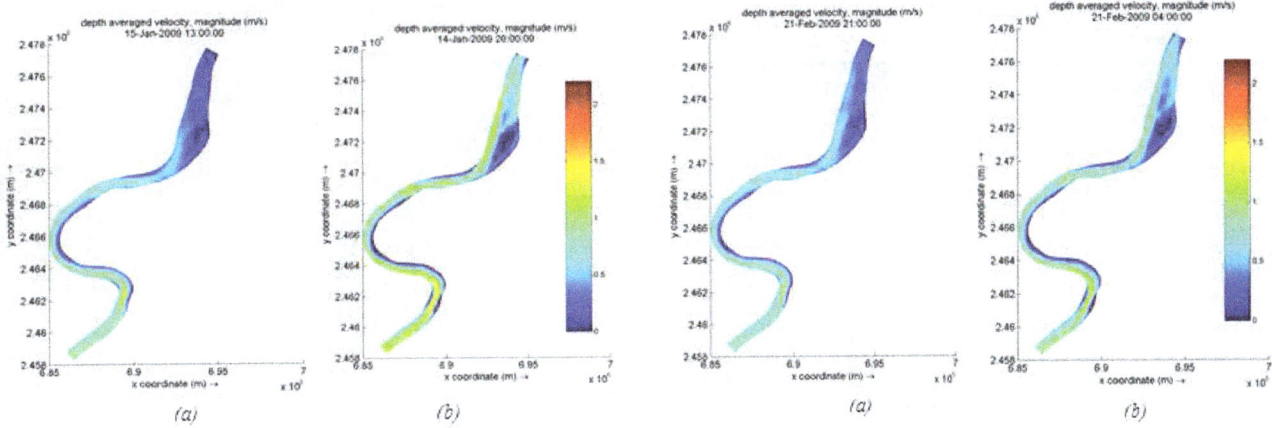

Figure 10. *Velocity variation (m/s) for spring tide lean period (a) flood (b)ebb*

Figure 12. *Velocity variation (m/s) for neap tide lean period (a) flood (b) ebb*

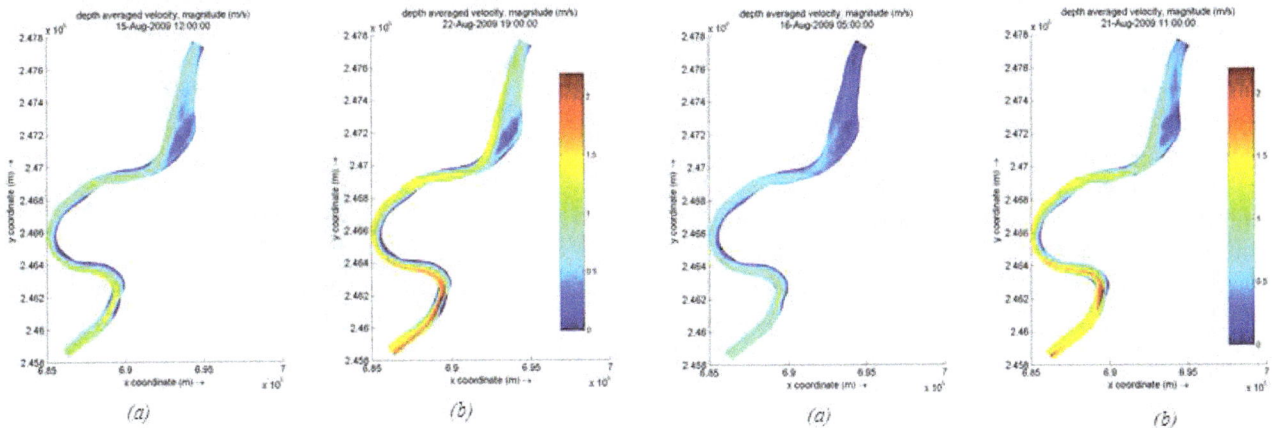

Figure 11. *Velocity variation (m/s) for spring tide monsoon period (a) flood (d) ebb*

Figure 13. *Velocity variation (m/s) for neap tide monsoon period (a) flood (b) ebb*

4. Results and Analysis

4.1. Variation in Velocity

The velocity variations at some salient points after model simulation are shown in Fig 10-13 as a visual evidence of the results. However Table 1 and 2 are also shown to interpret extent of such velocity variation under pre-project condition. It is seen that the velocity is higher along the thalweg and near the bends. Four different condition maximum depth-average velocities are shown (Table 1-2) where wet spring period shows maximum velocity of 1.323 m/s.

Table 1: Maximum longitudinal velocity (m/s) component during dry (January) season at different points (M,N)

Type	(96,20)	(101,20)	(105,20)	(111,20)	(140,20)
Spring					
Flood	0.389	0.451	0.527	0.571	0.555
Ebb	0.704	0.783	0.791	0.915	0.965
Neap					
Flood	0.380	0.390	0.460	0.510	0.521
Ebb	0.640	0.711	0.724	0.820	0.843

Table 2: Maximum longitudinal velocity (m/s) component during wet (August) season at different points (M,N)

Type	(96,20)	(101,20)	(105,20)	(111,20)	(140,20)
Spring					
Flood	0.78	0.965	1.013	1.06	1.13
Ebb	1.055	1.132	1.157	1.262	1.323
Neap					
Flood	0.150	0.1	0.14	0.18	0.17
Ebb	0.723	0.785	0.803	0.872	0.921

4.2. Variation in Sediment Transport

Similarly sediment transport characteristics of this river is influenced by the tidal effect. It was found that sediment transport rate is higher during wet season compared to the dry season which is due to the increased discharge during wet season. In all conditions the relative magnitude of sediment transport is almost the same. Some portion near Gupta bend is having higher transport rate. This causes higher erosion/deposition in this portion of the river. Very small portion near Arakan Khal is also having relatively high sediment transport. The narrow side channel on the right hand of Bakulia char is also having higher sediment transport, this leads to continuous erosion of the bed in this portion of the river. All other portions of the river are having relatively lower sediment transport with slight higher magnitude along the thalweg.

4.3. Erosion and Deposition Process

Sediment transport characteristics of any river are influenced by the tidal effect. Cumulative erosion/deposition throughout the river was calculated using the model Delft3D for one year simulation. Fig. 14-15 showing quarterly cumulative erosion/deposition . It is evident that change in river bathymetry is not significant for one year period and it shows similar type of change at any location. From the result it can been seen that erosion takes place at some places downstream of Kalurghat bridge. Areas around Sadarghat shows alternating pattern of erosion and deposition. Deposition occurs at the mouth of Shikalbaha khal. Just near the Arakan khal the river bed is having relatively higher elevation which is found eroding in nature.

Table 3 shows cumulative erosion/deposition (m) along the centerline for one year period. The rate also varies across each cross section. Along centerline maximum deposition was found 0.75m at 26 km and maximum erosion was found 1.7 m at 23 km from Kalurghat station.

Table 3. Cumulative erosion/deposition (m) throughout the river along centerline

Distance(km)	5	10	15	20	25
Erosion/Deposition(m)	-0.04	-0.76	-0.12	-0.25	+0.26

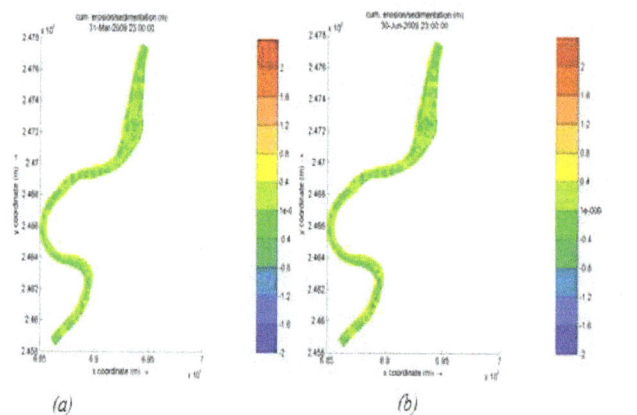

(a)   *(b)*

Figure 14. *Cumulative erosion/deposition at the end of (a) March (b) June*

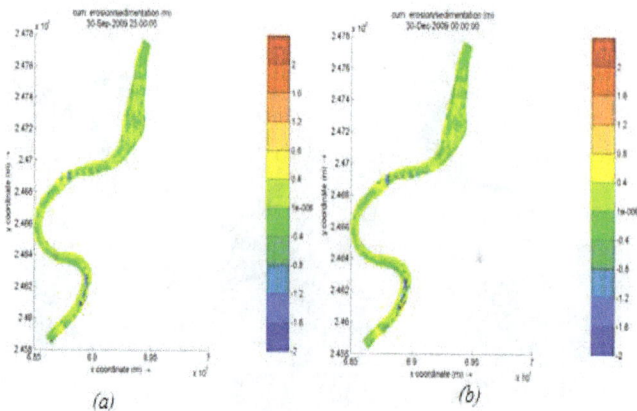

(a)   *(b)*

Figure 15. *Cumulative erosion/deposition at the end of (a) September (b) December*

Table 4: Model results showing maximum longitudinal (m component) velocity variation (m/s) during dry season (January) [Pre= Pre-Project, Post= Post-Project condition]

Type	(96,20)		(101,20)		(105,20)		(111,20)		(140,20)	
Spring	Pre.	Post.	Pre.	Post.	Pre.	Post.	Pre.	Post.	Pre.	Post.
Flood vel.	0.389	0.396	0.451	0.453	0.527	0.518	0.571	0.563	0.555	0.555
Ebb vel.	0.704	0.7084	0.783	0.785	0.798	0.791	0.915	0.905	0.965	0.988
Neap										
Flood vel.	0.380	0.379	0.390	0.391	0.460	0.452	0.510	0.493	0.521	0.531
Ebb vel.	0.640	0.645	0.711	0.715	0.724	0.721	0.820	0.812	0.843	0.860

Table 5: Model results showing maximum longitudinal (m component) velocity variation (m/s) during wet season (August)[Pre= Pre-Project, Post= Post-Project condition

Type	(96,20)		(101,20)		(105,20)		(111,20)		(140,20)	
Spring	Pre.	Post.	Pre.	Post.	Pre.	Post.	Pre.	Post.	Pre.	Post.
Flood vel.	0.78	0.79	0.965	0.963	1.013	1	1.06	1.045	1.13	1.14
Ebb vel.	1.055	1.072	1.132	1.140	1.157	1.157	1.262	1.263	1.323	1.347
Neap										
Flood vel.	0.150	0.155	0.1	0.1	0.14	0.14	0.18	0.18	0.17	0.175
Ebb vel.	0.723	0.727	0.785	0.789	0.803	0.801	0.872	0.874	0.921	0.937

4.4. Effect of Capital Dredging and Bank Protection

4.4.1. Effect on Velocity

Magnitude of maximum longitudinal velocity components along the river at different points of the selected reach for pre and post project condition have been shown in Table 4 and Table 5. The selected points cover both upstream and downstream portion of the Dredged area. It is evident that the dredged condition causes an increase in velocity at almost all the upstream and downstream locations near Sadarghat area (dry season and wet season).

4.4.2. Effect on Erosion/Deposition

Longitudinal variation of cumulative erosion/deposition within the study area (M=67-178) along three longitudinal sections (N=15, 20, 25) are shown in tabular format. Simulation result shows an increase in erosion or decrease in deposition rate in most of points. Some points showing slight increase in deposition. The overall magnitude of variation is not significant at any of the locations in the observed points.

Table 6: Model results showing cumulative erosion/ deposition (mm/day) along N=15,20,25 after one year period at different distances (km) between M=67 to178 [Pre=Pre-project condition, Post= Post-project condition]

Distance(km)	N=15		N=20		N=25	
	Pre.	Post.	Pre.	Post.	Pre.	Post.
0	0.183562	0.180822	0.048493	0.048493	0.006849	0.006849
1	0.410959	0.756164	0.821918	0.841096	0.027397	0.008219
2	0.013699	-0.0137	-0.05479	-0.06027	-0.9589	-0.96164
3	0.712329	0.684932	0.882192	0.90411	0.739726	0.767123
4	0.136986	0.136986	-1.17808	-1.23288	-0.38356	-0.41096
5	-0.21918	-0.21918	0.739726	0.739726	-0.31507	-0.30137
6	1.156164	1.156164	1.506849	1.452055	0.213699	0.205479

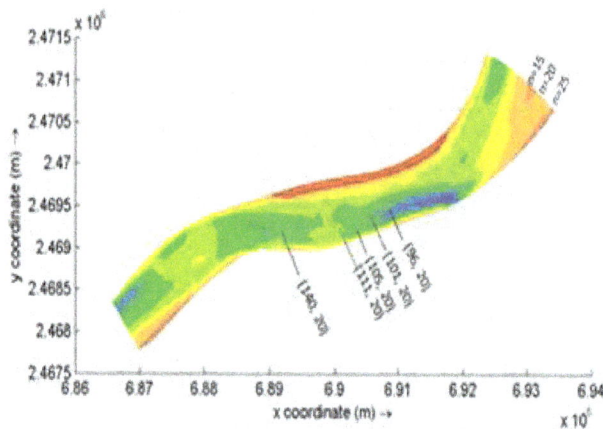

Figure 16. Location of different points at Sadarghat area

5. Conclusion

Morphological response of the river due to various development works need proper attention and quantification. The rate of erosion/deposition of this river is always required for estimating the bed level maintenance of the river reach under study. Therefore a two dimensional morphological model has been developed for the Karnafuli river to assess the erosion and deposition pattern due to dredging activities of the port. Considerable efforts have been made to calibrate and validate the model using the available data. Simulated velocity field, shear stress, sediment transport and erosion/deposition of the river reach under study have been generated from the model. In this paper some of typical results are given. Computed results show a variation of maximum depth-averaged velocity between 0.1 m/s and 1.323 m/s at the vicinity of the port jetty area (Sadarghat) under pre-project condition. However, post project velocity at the point downstream of the dredged area increased 3.33%. The increased velocity indicates a favorable condition for future maintenance dredging requirement. Annual cumulative erosion/ deposition along mid-line of the river was found to vary between -0.76 m and +0.26 m. Erosion/deposition rate of the same area varies between -69.96% (erosion) and +7.1304% (deposition) due to capital dredging. Overall the model results indicate that considerable improvement in the siltation behavior of the river in the vicinity of Sadarghat jetty area have been achieved under post project condition. As such no significant adverse effect on the river due to capital dredging has been observed from the model results.

Acknowledgements

Authors are gratefully acknowledging the cooperation rendered by Hydrographic division of Chittagong Port Authority (CPA) for providing the necessary data. Also gratitude renders to Deltares for making Delft3D open source which has been used in this study.

References

[1]   Alam, S. and Matin, M. A. (2012). "Application of Delft3d mathematical model in the river karnafuli for two dimensional simulation." ICACE 2012-Conference Proceeding,CUET, Chittagong. pp:7

[2]   Chittagong Port Authority (CPA 2010a), Capital Dredging and Bank Protection with Jetty Facilities; Report on "Mathematical Model Study on the Effect of 3rd Karnafuli Bridge on the Navigability of Karnafuli River within Port Limit." Dept. of WRE, BRTC, BUET, 2010

[3]   Chittagong Port Authority (CPA 2010b), "Mathematical Model study on the Effect of Korean (KEPZ) Project and Karnafuli River- Effect of Construction of a RCC Jetty on the left Bank of Karnafuli in the vicinity of the river." Dept. of WRE, BRTC, BUET, 2010

[4]   Chittagong Port Authority (CPA 2010c), Field Survey Report: "Capital dredging and bank protection with jetty facilities from Sadarghat jetty to 3rd Karnafuli bridge", July 2010, pp:32-39

[5]   Chittagong Port Authority (CPA 1987), "Mathematical Model Study of the River Karnafuli." Dept. of WRE and IFCDR, BRTC, BUET, 1987

[6]   Delft3D-Flow user manual (2011). "Simulation of multi-dimensional hydrodynamic flows and transport phenomena, including sediments." Version 3.15, revision 14499. pp:373,356

[7]   DHI, BUET (1990a), "Mathematical Model Study of Pussur-Sibsa River System and Karnafuli River Entrance", Interim Report, Volume 1 and 2

[8]   DHI, BUET(1990b), "Mathematical Model Study of Pussur-Sibsa River System and Karnafuli River Entrance", Final Report, ANNEX 1: Collection and Analysis of Data

[9]   L.C. van Rijn and D.J.R. Walstra(2003), "Modeling sand transport in Delft3D-Online", WL| Delft Hydraulics, Report, pp:15-17

[10] Schellingerhout, J (2012), "Modeling bio-physical interactions by tube building worm", Bachelor Assignment, Advanced Technology, University of Twente. Pp:17

[11] L.C. van Rijn (1984) "Sediment Transport, Part II; Suspended load transport", Journal of Hydraulic Engineering, vol.110, no.11; pp:1613-1641

[12] Wang, Li (2007), "Modelling of Cohesive Sediment Transport in the Maasmond Area", MSc Thesis (WSE-HI.07-18).

19

Ecological integrity of a peri-urban river system, Chiraura River in Zimbabwe

Beaven Utete[*]**, Rutendo Maria Kunhe**

Chinhoyi University of Technology Department of Wildlife and Safari Management P. Bag 7724, Chinhoyi

Email address:

beavenu@yahoo.co.uk(B. Utete), butete@cut.ac.zw(B. Utete)

Abstract: Ecological integrity of a peri-urban river system facing a plethora of anthropogenic pressures was assessed through multivariate analysis of physicochemical parameters correlated to the resident macroinvertebrate community. Monthly collection of macroinvertebrates and concurrent measurement of the physical and chemical parameters (dissolved oxygen concentration, percentage saturation of oxygen, pH, temperature, electrical conductivity and salinity) of water was done over a period of 5 months from November 2011- March 2012 in six sites across the Chiraura River. Macroinvertebrates were collected using the kick-net sampling technique, identified up to family level and enumerated at each site. Biodiversity indices were calculated for each site following the South African Scoring System version 5 (SASS5). A total of 1209 macroinvertebrates belonging to 49 families and11 orders were recorded in the Chiraura River. Most pollution sensitive taxa were found at sites 3 and 4 and the most pollution tolerant families were found at sites 1, 5 and 6. Sites4 and 5 of Chiraura River were the least polluted. Unsustainable anthropogenic activities, including industrial, domestic and urban agricultural activities affects water quality of Chiraura River. This is mainly through run-off and increased effluent to the river making routine water quality monitoring imperative.

Keywords: Lotic System, Macroinvertebrates, River Health, SASS 5, Water Quality, Biomonitoring

1. Introduction

There is severe pressure on the quantity and quality of freshwater resources in Zimbabwe due to organic and inorganic pollution coupled with over abstraction [1,2]. This is especially true in the tributaries of hypereutrophic Lake Chivero, the main source of water for Harare, the capital city of Zimbabwe. The increasing rate of water pollution and abstraction in Lake Chivero catchment is a result of population growth, industrialization, and greater demand for irrigation and livestock production [3]. Chiraura River, a subtributary of Lake Chivero and a main source of water for the animals in Mukuvisi Woodlands Nature Reserve Area, is increasingly becoming susceptible to anthropogenic pollution, capturing the attention of the park management because of the important biodiversity held within the park.

Assessment of water quality in aquatic systems in Zimbabwe has tended to be biased towards the analysis of physicochemical properties, with biological monitoringlargely neglected [4]. Physicochemical analyses provide, at best, a fragmented overview of the state of aquatic systems, as sporadic or periodic sampling cannot reflect fluxes of effluent discharge. In contrast, biological monitoring, premised on the fact that living organisms are the ultimate indicators of environmental quality or ecosystem ecological integrity [5], gives a time-integrated indication of the water quality components because of the capacity of reflecting conditions that are not present at the time of sampling and analysis. The unique composite picture of ecosystem conditions provided by biological monitoring can only be replicated by intensive and expensive chemical monitoring studies. In addition, if the aim is to maintain the diversity and health of biological communities, it is appropriate to monitor aquatic communities themselves rather than only abiotic factors. Biological monitoring is now recognized as one of the most valuable tools available to environmentalists [6, 7, 2].

Although macroinvertebrates are widely recognized as a biomonitoring tool, their application in river health assessments in Zimbabwe is of low precedence. To our knowledge, few available studies have only focused on some sections of rivers. These include the Mukuvisi River [4, 3], Nyaodza River in Kariba [8, 9] and Gwebi River [6].

Therefore this study seeks to contribute to the monitoring of peri-urban river health status in Zimbabwe. The objective of this study was to assess the ecological integrity of a peri-urban river system Chiraura River, facing a plethora of anthropogenic pressures correlating resident macroinvertebrates assemblages and abiotic factors.

2. Materials and Methods

2.1. Study Area and Design

The Chiraura River starts in the vlei behind the Honeydew farm near Greendale in the city of Harare. Upstream, it is fed by small rivulets which contribute substantially to the inflow. Burst sewer pipes, stream bank cultivation, urban run-off and informal motor vehicle industrial effluent in the catchment exposed the river to increasing levels of pollution. Headwater catchment area used to be characterised by wetlands, which are now being transformed into agricultural and settlement area. The transformation of the wetlands into agricultural and residential areas resulted in decreased flows, erosion and eutrophication in the Chiraura River. From this area, the river drains through Mukuvisi Woodlands Nature Reserve, a relatively undisturbed area. As it leaves the park, industrial effluent from the Msasa industry is discharged in the portion conjoined to the Mukuvisi River.

Six sites were selected along the Chiraura River based on the different land pattern use in the catchment. Site 1 (17.826930 °E; 31.1010 °S) was located in the headwater region characterized by intense agricultural activities at the horticultural Honey Dew Farm; site 2 (17.826936 °E; 31.1017 °S) covers the land adjacent to urban households where there is rampant dumping of solid waste and frequent burst sewer pipes. Site 3 (17.826940 °E; 31.1026 °S) was located in the wetland adjacent to the Mukuvisi Woodlands Nature Reserve area. Sites 4 (17.826942 °E; 31.1030 °S) and 5(17.826942 °E; 31.1032 °S) were located in the fenced section of the Mukuvisi Woodland Nature Reserve area. Site 6 (17.826942 °E; 31.1036 °S) was located at the confluence with the Mukuvisi River outside the Mukuvisi Woodland Nature Reserve area.

2.2. Water Quality Sampling and Analyses

At each site, dissolved oxygen (DO), Percentage saturation (% saturation), electrical conductivity (EC), temperature, Total Dissolved Solids (TDS), and pH were measured using potable HACH electronic meters.

2.3. Macroinvertebrates Sampling and Analysis

At each site, macroinvertebrates collection and stream health inference was done following the SAAS 5 protocol [10]. Macroinvertebrates were sampled by disturbing stones (kick sampling technique) with feet for a total of 5 minutes at each site while the net was held downstream to collect the disturbed organisms. Macroinvertebrate samples were separated from the mud and detritus, identified to

family level following studies by [11] and [12], sorted and counted before their release into the river.

2.4. Data Analysis

Analysis of variance (ANOVA) was used to compare means of physicochemical variables among sampling sites after testing for normality (Shapiro-Wilk test)and homogeneity of variance (Levene's test), and placed alongside local (Environmental Management Agency) and international (World Health Organisation) aquatic life threshold values. Pairwise comparison of the physicochemical values was done using the Tukeys 'test at 5 % significance level. Kruskal Wallis test was used to compare means of (SASS and ASPT scores) among sampling sites. Pearson's correlation was used to determine the relationship between the calculated SASS and ASPT scores and measured physical and chemical water quality data. Correspondence analysis (CA) was used to examine the direct effect of the physical and chemical (pollutant) characteristics of the water on variation in taxon composition among sampling sites as a means of identifying possible indicator taxa that were sensitive to water pollution. ANOVA, Kruskal Wallis, Pearson's correlation, Levene's test and Shapiro-Wilk test were performed using Palaeontological Statistics (PAST) software version 2.16 [13]. CA was performed using the programme CANOCO 5 [14]

2.5. Macroinvertebrate Diversityand Habitat Characterization

Diversity of macroinvertebrate families were calculated using biotic indices. The Shannon Weaver and the Pielou evenness index were calculated to assess the diversity and richness of macroinvertebrates in each river using PAST software version 2.16 [13].The ecological integrity of the river was assessed using the Habitat Assessment Matrix (HAM) strengthened by the Average Scores per Taxon at each site.

3. Results

3.1. Physical-Chemical Parameters

Physicochemical results measured in Chiraura River for the period (November 2011- March 2012) are summarized in Table 1 and a brief description of the highlights is done below:

The pH at most sites had values below 7 with the exception of site 4 and 5 with a mean pH of 7.1 and 7.3 respectively. There was a significant statistical difference (ANOVA; $p < 0.05$) in the pH values among sites. However, there was no significant (ANOVA; $p > 0.05$) temporal (monthly) difference in pH values at all sites in the Chiraura River. All sites in the Chiraura River had mean pHvalues below the EMA and WHO ranges (Table 1).

Conductivity differed significantly (ANOVA, $p < 0.05$) among sampled sites in the Chiraura River. Pairwise comparison revealed significant differences (Tukeys' test, p < 0.05) in conductivity between sites 1 (121.6 μS/cm^1) and 2 (128.4μS/cm^1) compared to sites 4 (41.6μS/cm^1), 5 (46.4μS/cm^1) and 6 (24.4μS/cm^1). Significant (ANOVA, p < 0.05) monthly variation was observed in conductivity values at all sites. Mean conductivity values fell within the acceptable EMA and WHO limits (Table 1).

There was a significant difference (ANOVA, p < 0.05) in Total Dissolved Solids among the sampled sites in Chiraura River. Pairwise comparison shows that there were significant differences (Tukeys' test, p < 0.05) in TDS values between sites: 4 vs.1; 4 vs.2; 4vs.3 and sites: 5 vs.1; 5vs. 2 and 5vs. 3. Significant (ANOVA; p <0.05) monthly/temporal differences in TDS values were observed across sites in the Chiraura River. Mean TDS values in the

Chiraura River were with the acceptable EMA and WHO values (Table 1). Dissolved oxygen levels did not differ significantly (ANOVA, p > 0.05) among sites sampled in Chiraura River during the study period. Sites1, 2, 3 and-5 had DO concentrations below acceptable EMA and WHO threshold limits (Table 1).% oxygen saturation: The six sites sampled had a significant difference (ANOVA, p < 0.05) in % oxygen saturation although there were no significant monthly variation at all sites. % oxygen saturation was significantly high at site 6 (107.9) in the Chiraura River. Significant temporal (monthly) variation (Anova,p < 0.05) was observed in temperature, for all the sampled sites. Mean temperature values recorded at all sites in the Chiraura River fell within the acceptable EMA and WHO threshold values (Table 1).

Table 1: Physico- chemical (Mean ± St Deviation.) variables measured at each sampling site in Chiraura River for the period (November 2011- March 2012). Note: Note: EMA = Environmental Management Agency. WHO = World Health Organisation.

Site	Temp°C	DO(mg/L)	TDS(mg/L)	pH(units)	% Saturation	Conductivity(µS/cm^1)
1	23.8	4.7	192.8	6.9	50.3	121.6
	±2.94	±1.11	±176.31	±0.21	±46.67	±22.143
2	23.7	4.9	201.4	4.2	50.5	128.4
	±3.00	±0.55	±185.21	±3.85	±46.49	±43.39
3	19.4	4.9	290.8	5.8	74.1	63.4
	±2.24	±0.33	±76.58	±0.2173	±2.66	±20.80
4	23.3	5.6	378.4	7.1	76.6	41.6
	±3.06	±0.68	±79.13	±0.22	±12.75	±20.56
5	25.9	4.3	373.2	7.3	64.1	46.4
	±1.26	±0.99	±94.16	±0.27	±15.06	±31.42
6	26.1	6.8	266.4	6.5	107.9	24.4
	±2.97	±3.36	±176.92	±0.77	±52.83	±15.94
EMA	35	5.0	1000	6-9	-	1000
WHO	27	5.0	1000	6.5-9	-	1000

3.2. Macroinvertebrate Diversity and Habitat Characterization

A total of 1209 macroinvertebrates belonging to 49 families and 11 orders were recorded in the Chiraura River. The highest number of taxa was recorded in the month of November (34) and the least number was in December (21 taxa) and January with 23 taxa. Shannon-Weaver diversity indices indicate that the month of November had the highest macroinvertebrate diversity (H = 3.353) with December having the least diversity (H =2.791). Pielou Evenness' index also show that the month of November had the highest evenness of macroinvertebrate assemblages (e^H/S =0.8409) and December had the least evenness (e^H/S =0.7763). The families: Aeshnidae, Elmidae, Belostomatidae, Athereceridae, Chironomidae, Oligochaetae, and Coenagrionadae were the most abundant taxa in the Chiraura River and were present at all sites sampled in this study. Some families that cannot tolerate

high levels of pollution such as the Tipulidae, Culicidae and Lestidae were found at sites 4 and 5 in the section where the river passes through the Mukuvisi Woodlands Nature Reserve.

There was a significant difference (Kruskal -ANOVA, p < 0.05) in the means of (SASS and ASPT scores) among sampling sites. No significant (Pearson test; p > 0.05) relationship was observed between the calculated SASS and ASPT scores and most of the measured physical and chemical water quality data during this study except for total dissolved solids (TDS). The ecological integrity of the river was assessed using the Habitat Assessment Matrix strengthened by the Average Scores per Taxon at each site. The HAM scores show that the three upper headwater sections (sites1, 2 and 3) of the river were in a poor condition whilst the lower sections (sites 4 and 6) were in fair condition only site 5 was in good condition. Of particular note is site 6 where the river confluences with the

Mukuvisi River that was also in fair- category (Table 3).

Table 2: *Habitat characterization of the Chiraura River for the period Nov 2011-Mar 2012.*

Site	ASPT	HAM score	Condition
1	2.48	65	poor
2	2.68	65	poor
3	2.46	60	poor
4	4.56	75	fair
5	4.84	85	good
6	4.06	70	fair

3.3. Influence of Environmental Variables on Macroinvertebrate Diversity

Canonical Correspondence Analysis show that the first two axes account for 62.3% of the variation in macroinvertebrate diversity and distribution in the Chiraura River. Percentage oxygen saturation had a positive association with Hydracarina, Corixidae, Ancylidae, Tipuliidae, and Hydranidae. Dissolved Oxygen (DO) had a strong influence on the diversity of the Helodidade, Gyrinidae, Nepidae and Hydrophilidae more prominently at site 6. Electrical conductivity had a significant relation with the Potamonautidae, Planobidae, Baetidae and Ephydridaedistribution and diversity particularly at sites 1, 2 and 3 in the Chiraura River (Figure 1). Temperature had a positive association with Notonectidae, Naucoridae, Belostomatidae, Culicidae, Lubellidae, Ceratopogonidae, Coenagrionadae, Dytiscidae and Oligochaeta. pH and Total Dissolved Solids (TDS) had a positive association with Elmidae, Pleidae, Dixidae, Chironomidae, Lymnaeidae, Lestidae, Gomphidae, Simullidae, Gerridae Vellidae and Pyralidae at site 4.

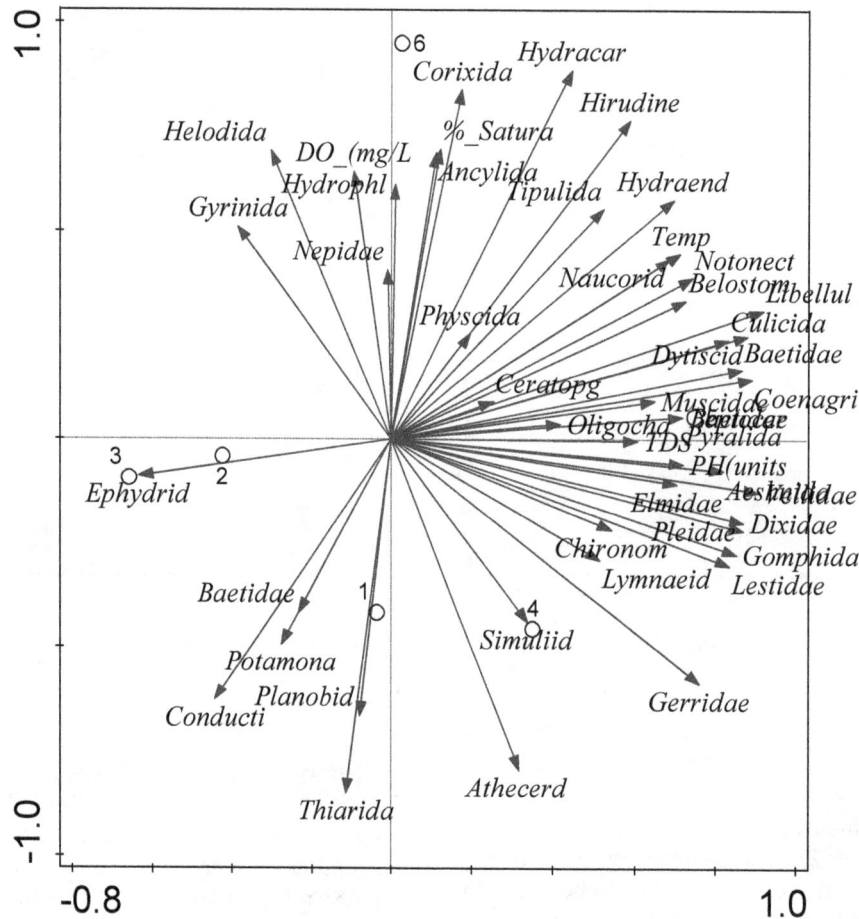

Figure 1.CA diagram showing relation between macroinvertebrates and environmental variables in Chiraura River

4. Discussion

Peri-urban rivers represent a pollution paradox as some sections are polluted and some sections have a functional self-purification capacity thus there is a tendency to exclude them in watershed management [7]. The peri-urban Chiraura River has some sites (sites 1, 2 and 3) which have poor water quality. These sites are located close to a horticultural farm and urban residential areas. The horticultural activities at these sites rely heavily on organic fertilizers, insecticides and pesticides which contribute to the pollution of the river. Sewage effluent discharges from adjacent urban households also contribute to the pollution of the river. Pollution tolerant families like the Chironomidae and Oligochaetae were abundant in sites 1, 2, 3 and 6 situated outside the Mukuvisi Woodlands Nature Reserve area. However, pollution sensitive taxa like the Trichoptera and Vellidae were observed at sites 4 and 5 which are located in the Mukuvisi Woodland nature reserve. The presence of tolerant families (e.g Potamonautidae, Planobidae, Thiaridae, Chironomidae in the more polluted sites (1, 2, 3 and 6) shows the usefulness of macroinvertebrates in indicating aquatic conditions [10]. However, absence/ presence of taxa is not only due to the abiotic conditions or anthropogenic impact but it might be just a natural phenomenon [15]. Unless baseline / reference conditions are established diversity indices can be misleading in indicating river conditions [5].

Using the HAM scores related to ASPT scores, sites 4 and 6 were in fair condition whilst site 5 was in good condition. This is because they are located in Mukuvisi Woodlands Nature Reserve area whose main objective is to preserve the pristine nature of the environment such that human and other environmental perturbations are low. There are no direct point sources of pollution. The good water quality in this section of the river appears suitable to support aquatic life because the highest macroinvertebrate taxa abundance (34 families) and diversity (H = 4.3) was recorded in site 5 in the month of November. Results of this research reveal that there are some sections of the Chiraura River which are in fairly good condition and some sections of the river are polluted. This can be highly attributed to human activities such as agriculture, waste disposal and industrial effluent by surrounding industries as well as the environmental preservation by the Mukuvisi Woodlands Nature Reserve.

Water quality is an urgent problem in most African cities and Harare is no exception, such that there is need for urgent measures to reduce pollution flows into Chiraura River. These include proper treatment and diversion of sewage effluent the river, protection of wetlands in the Chiraura river section so that no human activities such as agriculture and waste disposal takes place as these activities immensely contribute to the degradation of the river. The long-term approach to curbing the negative impacts of heavy pollution on the health of the Chiraura River is the systematic removal of the major nutrients causing pollution [16].Ways to achieve this goal include the use of a sediment scooper to physically remove sediments from the bottom of the river channel [17], including the nutrients contained within the sediments and river system, [2]. Proper construction, design, monitoring and maintenance of the effectiveness of riparian wetlands along polluted rivers such as the Chiraura River have the potential to curb eutrophication and in this study we suggest the use of the duckweed (*Lemna spp*) that has had success elsewhere to improve the self-purification capacity of rivers [16; 18]. Of concern is that the effects of poor water quality are very costly as they do not only affect human beings but also wildlife in Mukuvisi Woodlands Nature reserve where polluted water flows through it and is the main source of water for the animals.

Acknowledgement

Our special thanks to Dr Willie Nduku and Mr G. Nhokwara of WEZ and Mukuvisi Woodlands Association Nature Reserve& Environment Centre respectively for allowing us to conduct this research in the Mukuvisi Woodlands Association Nature Reserve and Environment Centre, Dr Tamuka Nhiwatiwa for help with the field equipment and Professor Taurai Bere for his incisive comments and critique on the script during its drafting. Mr Tatenda Dalu, Edwin Tambara and Edwin Zingwe for their help in the field and Mrs Elizabeth Munyoro for the assistance with laboratory work.

References

[1] Nhapi, I. 2004. Options for wastewater management in Harare, Zimbabwe. Dphil Thesis. .Wagenigen University, The Netherlands.

[2] Ndebele RM. 2012. Biological monitoring and pollution assessment of the Mukuvisi River, Harare, Zimbabwe. *Lake& Reservoirs: Research Management.* 17: 73-80.

[3] Magadza CHD.2003. Lake Chivero, A management case study. *Lakes& Reservoirs: Research Management.* 8: 69-81.

[4] Phiri C. 2000. An assessment of the health of two rivers within Harare, Zimbabwe, on the basis of macroinvertebrate community structure and selected physicochemical variables. *African Journal of Aquatic Science.* 25: 134–141.

[5] Dallas HF. 1997. A preliminary evaluation of aspects of SASS (South African Scoring System) for the rapid bioassessment of water quality in rivers, with particular reference to the incorporation of SASS in a national bio monitoring programme. *Southern African Journal of Aquatic Science.* 23 (1):79-94.

[6] Moyo NAG and Phiri C. 2002. The degradation of an urban stream in Harare, Zimbabwe. *African Journal of Ecology.* 40: 401-404.

[7] Nhapi I, Siebel MA and Gijzen HJ. 2009. A proposal for managing wastewater in Harare, Zimbabwe. *Water and Environment Journal.* 20(2): 101–108.

[8] Chakona A, Phiri C and Day JA. 2009. Potential for Trichoptera communities as biological indicators of morphological degradation in riverine systems. *Hydrobiologia.* 621: 155 167.

[9] Chakona A, Phiri C, Chinamaringa A and Muller N. 2009. Changes in biota along a dry land river in north-western Zimbabwe: declines and improvements in river health related to land use. *Aquatic Ecology.* 4: 1095–1106.

[10] Dickens C and Graham M. 2001. South African Scoring System (SASS) Version 5, Rapid Assessment Method for Rivers. Umgeni Water, Pietermaritzburg, South Africa.

[11] Thirion C, Mox E and WoestR.1995. Biological monitoring of streams using SSS4 – A User Manual. Department Of Water Affairs and Forestry Institute for Water Quality. Studies. South Africa. Report NOOOO/OO/REQ/1195.p 46.

[12] Gerber A and Gabriel MJM. 2002. Aquatic invertebrates of South African rivers: Field guide. Resource Quality Services, Department of Water Affairs and Forestry, Pretoria, South Africa 150 pp.

[13] Hammer O, David A, Harper T and Ryan PD. 2001. Paleontological Statistics Software Package for Education and Data Analysis. Paleontological Museum, University of Oslo, Sars Gate1, 0562 Oslo, Norway.

[14] Te Braak CJF and Šmilauer P. 2012. Conoco reference manual and users' guide: Software for ordination (version5.0) Macro computer Power Ithaca New York. U.S.A.

[15] Chakona A. 2005. The Macroinvertebrate Communities of Two Upland Streams in Eastern Zimbabwe with Reference to the Impact of Forestry. Msc Thesis. University of Zimbabwe.

[16] Machena C. 1997.The pollution and self-purification of Mukuvisi River. In: Lake Chivero, a Polluted Lake (Eds.N. A. G. Moyo). 75–91pp. University of Zimbabwe Publishers, Harare, Zimbabwe.

[17] Cherry DS, Currie RJ, Soucek DJ, Latimer HA and Trent GC. 2001. An integrative assessment of a watershed impacted by abandoned mined land discharges. *Environmental Pollution.* 111: 377-388.

[18] Kjeldsen TR, Lundorf A and Rosbjerg D. 1999.Barrier to unstainable water resources management, A case study *Hydro science Journal.* 44:529-539.

20

Fuzzy set approach–A tool to cluster Holy samples of groundwater quality parameters at Rameswaram, South India

V. Sivasankar[1, *], M. Kameswari[2], T. A. M. Msagati[3], M. Venkatapathy[4], M. Senthil Kumar[5]

[1]Department of Chemistry, Thiagarajar College of Engineering (Autonomous), Madurai – 625 015, Tamil Nadu, India
[2]Department of Mathematics, Thiagarajar College of Engineering (Autonomous), Madurai – 625 015, Tamil Nadu, India
[3]Department of Applied Chemistry, University of Johannesburg, Doornfontein Campus, P. O. Box 17011, Johannesburg, South Africa
[4]Department of Chemistry, A.A Government Arts College, Musiri – 621 221, Tamil Nadu, India
[5]Department of Civil Engineering, Sethu Institute of Technology, Virudhunagar – 626 115, Tamil Nadu, India

Email address:
vsivasankar@tce.edu(V. Sivasankar)

Abstract: A fuzzy set theoretic approach has been developed to study the potable nature of the holy groundwater samples in summer and winter by clustering method using equivalence relation. The physico-chemical parameters *viz.*, pH, Salinity, TDS, CH, MH, TH, Chloride and Fluoride are considered as attributes to develop the clusters. Based on the WHO recommendations, the linguistic approach has been developed for the water quality parameters of 22 holy groundwater samples in this study. Normalized eucilidean distance chosen for this study, measures the deviation of the determined quality parameters for any two holy groundwater samples. In the present paper, the seasonal changes in the quality of the water samples among the clusters at various rational alpha cuts are derived. The fluctuation in the water quality parameters was apparent such that the clusters contract from summer to winter with an exception of one sample with remarkable quality called Sethumadhava.

Keywords: Fuzzy Set, Potable Nature, Groundwater, Fuzzy Cluster, Alpha Cuts

1. Introduction

The physico-chemical quality of drinking water becomes as important as its availability. The water quality parameters with desirable, acceptable and not acceptable values, recommended by World Health Organisation (WHO) create awareness among the public, which enroute them towards the removal techniques [1, 2]. From the WHO guideline values, the groundwater samples are categorized for quality with respect to each water quality parameter. This may lead to a decision but with ambiguity, because it is derived out of only one parameter. So far, numerous research works have been carried out in the determination of water quality using fuzzy synthetic evaluation [3 – 5], fuzzy process capability [6], fuzzy clustering and pattern recognition method [7, 8], fuzzy logic approach [9], fuzzy logical rule [10],fuzzy simulink model [11], fuzzy logic drastic vulnerability map [12], fuzzy GNDCI-CNR method

[13] and fuzzy set theory [14]. In the present work, a new focus has been attempted using fuzzy equivalence relation to arrive at non-overlapping clusters of 22 groundwater samples by considering various agreement levels (alpha cuts).

2. Study Objectives and Methods

The objective of the present study is to obtain non-overlapping clusters of twenty two holy groundwater samples (Table 1) of Rameswaram temple in the summer and winter seasons based on the water quality parameters *viz.,* pH, Salinity, Total Dissolved Solids (TDS), Calcium Hardness(CH), Magnesium Hardness(MH), Total Hardness(TH), Chloride(Cl) and Fluoride(F).

2.1. Study Area

Rameswaram is located around an intersection of the

9°28'North Latitude and 79°3'East Longitude with an average elevation of 10 meters above the MSL, covering an area of 61.8 sq.kms and bearing a population of about 38,000, as on September 2007. This Indian Island having connection with main land assumes a shape of conch, is a Taluk with 1 Firka, 2 Revenue villages and 31 Hamlets. Climate prevails with a minimum temperature of 25°C in winter and a maximum of 36°C in summer. The average rainfall is 813mm [15, 16].

Table 1. Name of the holy groundwater samples

Sample No.	Name of the holy groundwater
1	Mahalakshmi
2	Savithri
3	Gayathri
4	Saraswathi
5	Sangu
6	Sarkarai
7	Sethumadhava
8	Nala
9	Neela
10	Kavaya
11	Kavacha
12	Kandhamadhana
13	Bramahathi
14	Ganga
15	Yamuna
16	Gaya
17	Sarwa
18	Siva
19	Sathyamrudham
20	Surya
21	Chandra
22	Kodi

2.2. Clustering of Groundwater Samples

Let S_1, S_2 ......S_{22} were the twenty two groundwater samples of Rameswaram temple are considered for clustering based on the criteria C_1, C_2.........C_8 *viz.*, pH, Salinity, TDS, CH, MH, TH, Cl and F.

Linguistic terms such as Excellent, Fairly Excellent, Good, Fairly Good and Poor were assigned to the chosen water samples with respect to the recommendations of the World Health Organisation [17].

2.3. Membership Functions and Fuzzy Relations

Membership function (µ) is a critical measure which represents numerically the degrees of elements belonging to a set. Distance measure is a term that describes a difference between fuzzy sets and can be considered as a dual concept of similarity measure.

The linguistic terms (Table 2) were converted into fuzzy numbers (membership functions) using probability technique. Using the fuzzy numbers, the Normalised euclidean distance (eqn.1)

$$\sqrt{\frac{1}{n}\sum_1^n \left[(\mu_A(x_i)-\mu_B(x_i))^2 +(\nu_A(x_i)-\nu_B(x_i))^2\right]} \quad (1)$$

was used to obtain similarity measures, which was found

by subtracting the distance measure from 1 using MATLAB (version 7). The obtained similarity measure possesses tolerance relation (R) between the undertaken groundwater samples (Tables 3 and 4). A fuzzy relation (R) is said to be fuzzy tolerance relation if R is reflexive [if $\mu_R(x,x)=1$, for every x $\in$ X] and symmetric [if $\mu_R[x,y]=\mu_R(y,x)$ for every x,y $\in$ X].

A fuzzy relation, R is said to be fuzzy equivalence relation R_E, if R is fuzzy tolerance relation and transitive closure [if μ_R satisfies $\mu_R(x,z)\geq$ min$\{\mu_R(x,y),$ $\mu_R(y,z)\}$ for every x,y,z $\in$ X.

An equivalence relation (R_E) in Tables 5 and 6 was determined from the computed tolerance relation by the following algorithm using Visual C^{++} on windows platform.

Step 1: R' =R o (R∪R)

Step 2: If R' ≠R, make R=R' and go to step 1.

Step 3: Stop: R' = RE

In the above o is the max-main composition of fuzzy relations and ∪ is the standard fuzzy union. By the consideration of reasonable alpha cuts ($\{x/\mu(x) > \alpha$, for some x $\in$X$\}$), the groundwater samples were clustered in the non-overlapping nature.

3. Results and Discussion

For grouping the 22 groundwater samples with respect to nine water quality parameters according to the WHO recommendations, the present work was initiated with a consideration of two suitable agreement levels (alpha cut), $\alpha = 0.85$ and 0.90 from four reasonable agreement levels, from $\alpha = 0.80$, 0.85, 0.9 and 0.95. The remaining two alpha cuts ($\alpha = 0.8$ and 0.95) generated a minimum number (two) of clusters, which does not reveal any significant impact in grouping the groundwater samples in accordance with the chosen water quality parameters. The non-overlapping clusters (grouped water samples) for the above two suitable alpha cuts in summer and winter seasons are depicted in Fig.1.

From the Fig 1, it can be accounted that the samples 7, 8, 11 and 20 & 21 remain as separate clusters both at 85% and 90% Agreement Level (AL). Thus the above said samples did not have any quality parametric changes at both the agreement levels. The samples 4 and 10 from a similar cluster at 85% AL were found to stay as independent clusters at 90% AL. Similarly, the samples 14, 15, 16 and 17 of same cluster at 85% AL got separated into an individual cluster at 90% AL. From these results, it is observed that, even at 0.05% difference in AL reveal some changes in the water quality characteristics.

The seasonal influence can also be witnessed from the merging of different clusters at 90% AL. Obviously, in summer, the samples 14, 15, 16 and 17 of a single cluster and the samples 20 & 21 of another single cluster, got merged with the cluster containing the samples 1-6, 8 and 17-21 in winter. Also it was observed that the samples 10 and 12 from two different clusters in summer grouped into a particular cluster in winter.

In summer, at 0.9 AL, there were nine clusters which get converged into five clusters in winter. Similarly, six clusters in summer at 0.85 AL get grouped and formed three separate clusters in winter.

A cluster with the samples 1,2,3,4,5,6,9,13,18,,19 & 22 in summer breaks into a separate cluster with the samples 9 & 22 and another cluster with the sample designated as 13 in winter. The above cluster separation from summer to winter indicates the change in the water quality parameters with respect to dilution influenced by seasons.

The samples 7 and 11, identified as separate clusters in summer were found to retain the same identity in winter at both agreement levels (0.9 and 0.85). This shows that the water quality parameters of both the samples falls within the standards fixed in this study as per WHO even after the seasonal influence.

In winter, the number of clusters at 0.9 AL and 0.85 AL were computed to be 6 and 3 respectively and this highlights the reduction of clusters indicates the role of dilution which causes recharging of groundwater and setting the water quality parameters with respect to the fixed standards as per WHO. In winter, at 0.9 and 0.85, there were two separate clusters formed with 7 and 10 & 12 as samples. The remaining samples have formed as a separate cluster by indicating their suitability within a particular water quality standard. The agreement level of 0.85 in both summer and winter was found with 6 and 3 clusters respectively.

In this case a separate cluster having the sample no.7 in both summer and winter was observed. Samples 12 and 10 from two separate clusters in summer were found to form a particular cluster in winter. This reflects that these two samples possess the quality characteristics within a particular limit, fixed in the study in accordance with WHO recommendations.

The identity of a single membered cluster containing a groundwater sample No. 7 (Sethumadhava) at both the agreement levels in two different seasons, ascertains the distinct quality of the water sample. Evidently, the present observation was in agreement with our earlier work [18].

4. Conclusions

Based on the above discussions, the following conclusions are drawn.

1. Among the four agreement levels (alpha cuts), the suitability was found for those at $\alpha = 0.85$ and 0.9, where appreciable number of clusters were generated.

2. From each appreciable cluster containing a group of water samples, the similarity of ground water quality among the samples with respect to the chosen parameters, was identified.

3. A cluster consisting of only one groundwater sample i.e. sample no.7 (Sethumadhava) at the suitable agreement levels in both summer and winter, was found to exhibit a distinct water quality by itself.

4. The contraction of clusters from summer to winter signifies the fluctuations in the water quality parameters at both the agreement levels.

Table 2. *Membership values assigned for the water quality parameters*

Parameters	Membership values (Linguistic forms)				
	1.0 (Excellent)	0.8 (Fairly Excellent)	0.6 (Good)	0.4 (Fairly Good)	0.2 (Poor)
pH	6.5 – 7.5	7.5 – 8.0	8.0 – 8.5	8.5 – 9.0	>9.0
TDS	<500	500 – 650	650 – 800	800 – 1000	>1000
CH	<75	75 – 100	100 – 150	150 – 200	>200
MH	<30	30 – 75	75 – 115	115 – 150	>150
TH	<100	100 – 250	250 – 350	350 – 500	>500
F	<0.5	0.50 – 0.75	0.75 – 1.0	1.0 – 1.5	>1.5
SAL	<200	200 – 350	350 – 500	500 – 600	>600
Cl	<250	250 - 350	350 - 450	450 - 600	>600

TDS-Total Dissolved Solids; CH-Calcium Hardness; MH-Magnesium Hardness; TH-Total Hardness; F-Fluoride; SAL-Salinity; Cl-Chloride

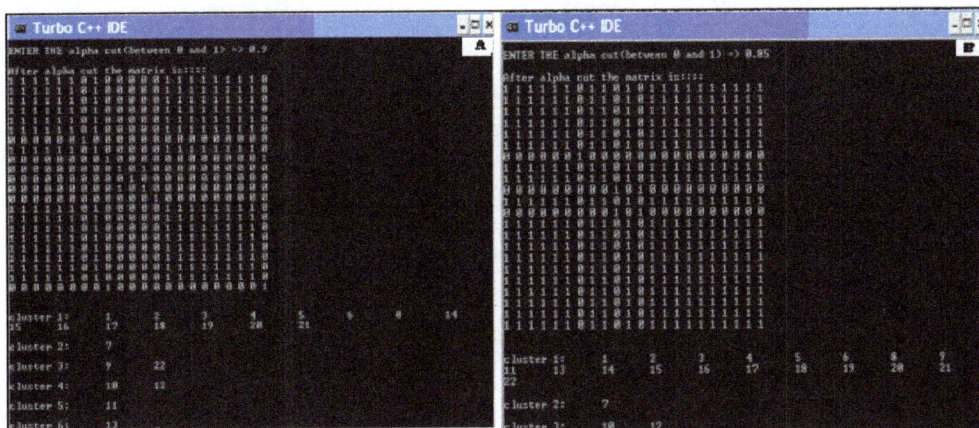

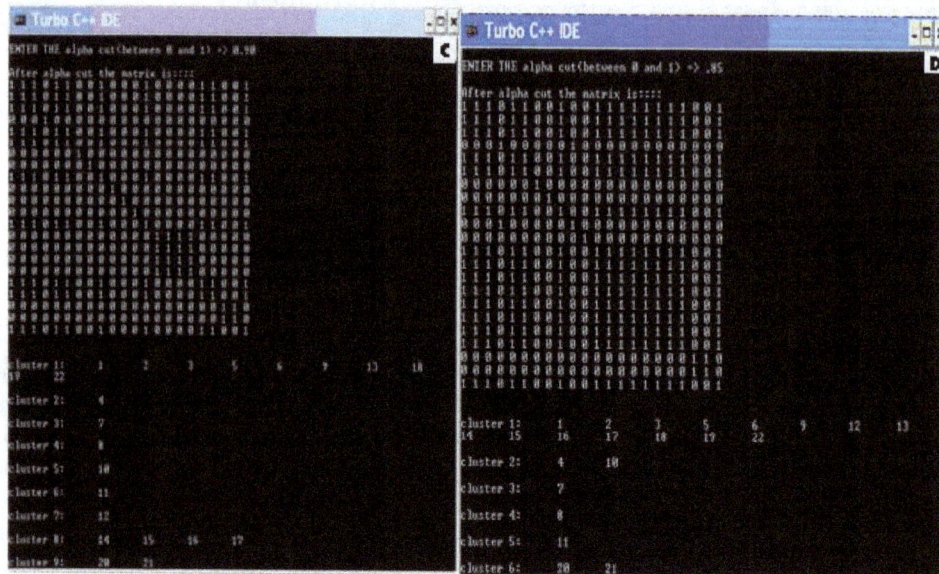

Figure 1. *Clustered (non-overlapping) groundwater samples by MAT LAB program for A) summer, α = 0.9   B) summer, α = 0.85   C) winter, α = 0.9   D) winter, α = 0.85*

Table 3. *Tolerance matrix in Summer*

1.000	0.900	0.900	0.735	0.842	0.859	0.576	0.735	0.827	0.765	0.842	0.745	0.827	0.800	0.827	0.859	0.788	0.929	0.878	0.827	0.827	0.859
0.900	**1.000**	0.900	0.735	0.842	0.900	0.588	0.700	0.859	0.788	0.842	0.745	0.827	0.735	0.776	0.776	0.708	0.929	0.929	0.800	0.800	0.900
0.900	0.900	**1.000**	0.800	0.929	0.900	0.613	0.735	0.900	0.813	0.813	0.788	0.859	0.827	0.859	0.827	0.765	0.929	0.929	0.827	0.827	0.900
0.735	0.735	0.800	**1.000**	0.842	0.776	0.613	0.735	0.827	0.878	0.765	0.745	0.800	0.776	0.755	0.700	0.661	0.745	0.788	0.776	0.776	0.735
0.842	0.842	0.929	0.842	**1.000**	0.878	0.633	0.765	0.929	0.827	0.776	0.827	0.878	0.842	0.842	0.788	0.735	0.859	0.900	0.788	0.788	0.878
0.859	0.900	0.900	0.776	0.878	**1.000**	0.613	0.735	0.900	0.813	0.813	0.788	0.859	0.755	0.776	0.755	0.692	0.878	0.929	0.776	0.776	0.900
0.576	0.588	0.613	0.613	0.633	0.613	**1.000**	0.588	0.639	0.661	0.558	0.676	0.684	0.576	0.564	0.531	0.495	0.582	0.619	0.542	0.542	0.613
0.735	0.700	0.735	0.735	0.765	0.735	0.588	**1.000**	0.776	0.726	0.708	0.813	0.827	0.735	0.668	0.668	0.619	0.692	0.745	0.639	0.639	0.735
0.827	0.859	0.900	0.827	0.929	0.900	0.639	0.776	**1.000**	0.842	0.788	0.842	0.900	0.776	0.776	0.735	0.676	0.842	0.929	0.755	0.755	0.900
0.765	0.788	0.813	0.878	0.827	0.813	0.661	0.726	0.842	**1.000**	0.827	0.755	0.842	0.726	0.726	0.692	0.639	0.776	0.827	0.788	0.788	0.765
0.842	0.842	0.813	0.765	0.776	0.813	0.558	0.708	0.788	0.827	**1.000**	0.684	0.788	0.708	0.726	0.745	0.684	0.827	0.827	0.842	0.842	0.765
0.745	0.745	0.788	0.745	0.827	0.788	0.676	0.813	0.842	0.755	0.684	**1.000**	0.878	0.745	0.708	0.676	0.626	0.735	0.800	0.646	0.646	0.813
0.827	0.827	0.859	0.800	0.878	0.859	0.684	0.827	0.900	0.842	0.788	0.878	**1.000**	0.776	0.755	0.735	0.676	0.813	0.878	0.735	0.735	0.859
0.800	0.735	0.827	0.776	0.842	0.755	0.576	0.735	0.776	0.726	0.708	0.745	0.776	**1.000**	0.900	0.859	0.842	0.788	0.765	0.776	0.776	0.755
0.827	0.776	0.859	0.755	0.842	0.776	0.564	0.668	0.776	0.726	0.726	0.708	0.755	0.900	**1.000**	0.900	0.878	0.842	0.788	0.827	0.827	0.776
0.859	0.776	0.827	0.700	0.788	0.755	0.531	0.668	0.735	0.692	0.745	0.676	0.735	0.859	0.900	**1.000**	0.929	0.842	0.765	0.827	0.827	0.755
0.788	0.708	0.765	0.661	0.735	0.692	0.495	0.619	0.676	0.639	0.684	0.626	0.676	0.842	0.878	0.929	**1.000**	0.776	0.700	0.788	0.788	0.692
0.929	0.929	0.929	0.745	0.859	0.878	0.582	0.692	0.842	0.776	0.827	0.735	0.813	0.788	0.842	0.842	0.776	**1.000**	0.900	0.842	0.842	0.878
0.878	0.929	0.929	0.788	0.900	0.929	0.619	0.745	0.929	0.827	0.827	0.800	0.878	0.765	0.788	0.765	0.700	0.900	**1.000**	0.788	0.788	0.929
0.827	0.800	0.827	0.776	0.788	0.776	0.542	0.639	0.755	0.788	0.842	0.646	0.735	0.776	0.827	0.827	0.788	0.842	0.788	**1.000**	1.000	0.735
0.827	0.800	0.827	0.776	0.788	0.776	0.542	0.639	0.755	0.788	0.842	0.646	0.735	0.776	0.827	0.827	0.788	0.842	0.788	1.000	**1.000**	0.735
0.859	0.900	0.900	0.735	0.878	0.900	0.613	0.735	0.900	0.765	0.765	0.813	0.859	0.755	0.776	0.755	0.692	0.878	0.929	0.735	0.735	**1.000**

Table 4 *Tolerance matrix in winter*

1.000	0.813	0.765	0.900	0.842	0.842	0.421	0.859	0.776	0.606	0.726	0.619	0.765	0.929	0.842	0.900	0.900	0.900	0.800	0.929	0.900	0.776
0.813	1.000	0.859	0.878	0.900	0.900	0.490	0.878	0.813	0.684	0.859	0.668	0.776	0.776	0.900	0.813	0.726	0.765	0.929	0.827	0.813	0.813
0.765	0.859	1.000	0.842	0.900	0.827	0.500	0.813	0.788	0.668	0.859	0.639	0.700	0.735	0.859	0.788	0.676	0.708	0.878	0.776	0.788	0.788
0.900	0.878	0.842	1.000	0.929	0.878	0.448	0.900	0.800	0.633	0.788	0.633	0.765	0.842	0.878	0.859	0.800	0.859	0.859	0.929	0.900	0.800
0.842	0.900	0.900	0.929	1.000	0.859	0.461	0.878	0.788	0.639	0.827	0.626	0.735	0.800	0.900	0.842	0.745	0.788	0.878	0.859	0.842	0.788
0.842	0.900	0.827	0.878	0.859	1.000	0.510	0.878	0.878	0.717	0.827	0.717	0.827	0.800	0.859	0.813	0.765	0.813	0.929	0.859	0.878	0.878
0.421	0.490	0.500	0.448	0.461	0.510	1.000	0.476	0.570	0.684	0.564	0.700	0.531	0.408	0.452	0.430	0.380	0.396	0.526	0.426	0.457	0.570
0.859	0.878	0.813	0.900	0.878	0.878	0.476	1.000	0.827	0.676	0.813	0.676	0.813	0.813	0.842	0.827	0.776	0.827	0.859	0.878	0.859	0.827
0.776	0.813	0.788	0.800	0.788	0.878	0.570	0.827	1.000	0.813	0.842	0.813	0.878	0.745	0.765	0.755	0.717	0.755	0.859	0.788	0.827	1.000
0.606	0.684	0.668	0.633	0.639	0.717	0.684	0.676	0.813	1.000	0.776	0.900	0.776	0.588	0.626	0.606	0.558	0.582	0.726	0.613	0.646	0.813
0.726	0.859	0.859	0.788	0.827	0.827	0.564	0.813	0.842	0.776	1.000	0.735	0.776	0.700	0.800	0.745	0.646	0.676	0.878	0.735	0.745	0.842
0.619	0.668	0.639	0.633	0.626	0.717	0.700	0.676	0.813	0.900	0.735	1.000	0.800	0.600	0.613	0.606	0.582	0.606	0.708	0.626	0.661	0.813
0.765	0.776	0.700	0.765	0.735	0.827	0.531	0.813	0.878	0.776	0.776	0.800	1.000	0.735	0.717	0.726	0.726	0.765	0.788	0.776	0.788	0.878
0.929	0.776	0.735	0.842	0.800	0.800	0.408	0.813	0.745	0.588	0.700	0.600	0.735	1.000	0.827	0.929	0.929	0.842	0.765	0.859	0.842	0.745
0.842	0.900	0.859	0.878	0.900	0.859	0.452	0.842	0.765	0.626	0.800	0.613	0.717	0.827	1.000	0.878	0.765	0.765	0.878	0.827	0.813	0.765
0.900	0.813	0.788	0.859	0.842	0.813	0.430	0.827	0.755	0.606	0.745	0.606	0.726	0.929	0.878	1.000	0.859	0.800	0.800	0.842	0.827	0.755
0.900	0.726	0.676	0.800	0.745	0.765	0.380	0.776	0.717	0.558	0.646	0.582	0.726	0.929	0.765	0.859	1.000	0.859	0.717	0.842	0.827	0.717
0.900	0.765	0.708	0.859	0.788	0.813	0.396	0.827	0.755	0.582	0.676	0.606	0.765	0.842	0.765	0.800	0.859	1.000	0.755	0.929	0.900	0.755
0.800	0.929	0.878	0.859	0.878	0.929	0.526	0.859	0.859	0.726	0.878	0.708	0.788	0.765	0.878	0.800	0.717	0.755	1.000	0.813	0.827	0.859
0.929	0.827	0.776	0.929	0.859	0.859	0.426	0.878	0.788	0.613	0.735	0.626	0.776	0.859	0.827	0.842	0.842	0.929	0.813	1.000	0.929	0.788
0.900	0.813	0.788	0.900	0.842	0.878	0.457	0.859	0.827	0.646	0.745	0.661	0.788	0.842	0.813	0.827	0.827	0.900	0.827	0.929	1.000	0.827
0.776	0.813	0.788	0.800	0.788	0.878	0.570	0.827	1.000	0.813	0.842	0.813	0.878	0.745	0.765	0.755	0.717	0.755	0.859	0.788	0.827	1.000

Table 5. *Equivalance matrix in Summer*

1.000	0.975	0.975	0.934	0.960	0.965	0.894	0.934	0.957	0.941	0.960	0.936	0.957	0.950	0.957	0.965	0.947	0.982	0.969	0.957	0.957	0.965
0.975	1.000	0.975	0.934	0.960	0.975	0.897	0.925	0.965	0.947	0.960	0.936	0.957	0.934	0.944	0.944	0.927	0.982	0.982	0.950	0.950	0.975
0.975	0.975	1.000	0.950	0.982	0.975	0.903	0.934	0.975	0.953	0.953	0.947	0.965	0.957	0.965	0.957	0.941	0.982	0.982	0.957	0.957	0.975
0.934	0.934	0.950	1.000	0.960	0.944	0.903	0.934	0.957	0.969	0.941	0.936	0.950	0.944	0.939	0.925	0.915	0.936	0.947	0.944	0.944	0.934
0.960	0.960	0.982	0.960	1.000	0.969	0.908	0.941	0.982	0.957	0.944	0.957	0.969	0.960	0.960	0.947	0.934	0.965	0.975	0.947	0.947	0.969
0.965	0.975	0.975	0.944	0.969	1.000	0.903	0.934	0.975	0.953	0.953	0.947	0.965	0.939	0.944	0.939	0.923	0.969	0.982	0.944	0.944	0.975
0.894	0.897	0.903	0.903	0.908	0.903	1.000	0.897	0.910	0.915	0.890	0.919	0.921	0.894	0.891	0.883	0.874	0.895	0.905	0.885	0.885	0.903
0.934	0.925	0.934	0.934	0.941	0.934	0.897	1.000	0.944	0.932	0.927	0.953	0.957	0.934	0.917	0.917	0.905	0.923	0.936	0.910	0.910	0.934
0.957	0.965	0.975	0.957	0.982	0.975	0.910	0.944	1.000	0.960	0.947	0.960	0.975	0.944	0.944	0.934	0.919	0.960	0.982	0.939	0.939	0.975
0.941	0.947	0.953	0.969	0.957	0.953	0.915	0.932	0.960	1.000	0.957	0.939	0.960	0.932	0.932	0.923	0.910	0.944	0.957	0.947	0.947	0.941
0.960	0.960	0.953	0.941	0.944	0.953	0.890	0.927	0.947	0.957	1.000	0.921	0.947	0.927	0.932	0.936	0.921	0.957	0.957	0.960	0.960	0.941
0.936	0.936	0.947	0.936	0.957	0.947	0.919	0.953	0.960	0.939	0.921	1.000	0.969	0.936	0.927	0.919	0.906	0.934	0.950	0.912	0.912	0.953
0.957	0.957	0.965	0.950	0.969	0.965	0.921	0.957	0.975	0.960	0.947	0.969	1.000	0.944	0.939	0.934	0.919	0.953	0.969	0.934	0.934	0.965
0.950	0.934	0.957	0.944	0.960	0.939	0.894	0.934	0.944	0.932	0.927	0.936	0.944	1.000	0.975	0.965	0.960	0.947	0.941	0.944	0.944	0.939
0.957	0.944	0.965	0.939	0.960	0.944	0.891	0.917	0.944	0.932	0.932	0.927	0.939	0.975	1.000	0.975	0.969	0.960	0.947	0.957	0.957	0.944
0.965	0.944	0.957	0.925	0.947	0.939	0.883	0.917	0.934	0.923	0.936	0.919	0.934	0.965	0.975	1.000	0.982	0.960	0.941	0.957	0.957	0.939
0.947	0.927	0.941	0.915	0.934	0.923	0.874	0.905	0.919	0.910	0.921	0.906	0.919	0.960	0.969	0.982	1.000	0.944	0.925	0.947	0.947	0.923
0.982	0.982	0.982	0.936	0.965	0.969	0.895	0.923	0.960	0.944	0.957	0.934	0.953	0.947	0.960	0.960	0.944	1.000	0.975	0.960	0.960	0.969
0.969	0.982	0.982	0.947	0.975	0.982	0.905	0.936	0.982	0.957	0.957	0.950	0.969	0.941	0.947	0.941	0.925	0.975	1.000	0.947	0.947	0.982
0.957	0.950	0.957	0.944	0.947	0.944	0.885	0.910	0.939	0.947	0.960	0.912	0.934	0.944	0.957	0.957	0.947	0.960	0.947	1.000	1.000	0.934
0.957	0.950	0.957	0.944	0.947	0.944	0.885	0.910	0.939	0.947	0.960	0.912	0.934	0.944	0.957	0.957	0.947	0.960	0.947	1.000	1.000	0.934
0.965	0.975	0.975	0.934	0.969	0.975	0.903	0.934	0.975	0.941	0.941	0.953	0.965	0.939	0.944	0.939	0.923	0.969	0.982	0.934	0.934	1.000

Table 6 Equivalance matrix in Winter

1.000	0.953	0.941	0.975	0.960	0.960	0.855	0.965	0.944	0.902	0.932	0.905	0.941	0.982	0.960	0.975	0.975	0.975	0.950	0.982	0.975	0.944
0.953	**1.000**	0.965	0.969	0.975	0.975	0.873	0.969	0.953	0.921	0.965	0.917	0.944	0.944	0.975	0.953	0.932	0.941	0.982	0.957	0.953	0.953
0.941	0.965	**1.000**	0.960	0.975	0.957	0.875	0.953	0.947	0.917	0.965	0.910	0.925	0.934	0.965	0.947	0.919	0.927	0.969	0.944	0.947	0.947
0.975	0.969	0.960	**1.000**	0.982	0.969	0.862	0.975	0.950	0.908	0.947	0.908	0.941	0.960	0.969	0.965	0.950	0.965	0.965	0.982	0.975	0.950
0.960	0.975	0.975	0.982	**1.000**	0.965	0.865	0.969	0.947	0.910	0.957	0.906	0.934	0.950	0.975	0.960	0.936	0.947	0.969	0.965	0.960	0.947
0.960	0.975	0.957	0.969	0.965	**1.000**	0.878	0.969	0.969	0.929	0.957	0.929	0.957	0.950	0.965	0.953	0.941	0.953	0.982	0.965	0.969	0.969
0.855	0.873	0.875	0.862	0.865	0.878	**1.000**	0.869	0.892	0.921	0.891	0.925	0.883	0.852	0.863	0.857	0.845	0.849	0.881	0.856	0.864	0.892
0.965	0.969	0.953	0.975	0.969	0.969	0.869	**1.000**	0.957	0.919	0.953	0.919	0.953	0.953	0.960	0.957	0.944	0.957	0.965	0.969	0.965	0.957
0.944	0.953	0.947	0.950	0.947	0.969	0.892	0.957	**1.000**	0.953	0.960	0.953	0.969	0.936	0.941	0.939	0.929	0.939	0.965	0.947	0.957	1.000
0.902	0.921	0.917	0.908	0.910	0.929	0.921	0.919	0.953	**1.000**	0.944	0.975	0.944	0.897	0.906	0.902	0.890	0.895	0.932	0.903	0.912	0.953
0.932	0.965	0.965	0.947	0.957	0.957	0.891	0.953	0.960	0.944	**1.000**	0.934	0.944	0.925	0.950	0.936	0.912	0.919	0.969	0.934	0.936	0.960
0.905	0.917	0.910	0.908	0.906	0.929	0.925	0.919	0.953	0.975	0.934	**1.000**	0.950	0.900	0.903	0.902	0.895	0.902	0.927	0.906	0.915	0.953
0.941	0.944	0.925	0.941	0.934	0.957	0.883	0.953	0.969	0.944	0.944	0.950	**1.000**	0.934	0.929	0.932	0.932	0.941	0.947	0.944	0.947	0.969
0.982	0.944	0.934	0.960	0.950	0.950	0.852	0.953	0.936	0.897	0.925	0.900	0.934	**1.000**	0.957	0.982	0.982	0.960	0.941	0.965	0.960	0.936
0.960	0.975	0.965	0.969	0.975	0.965	0.863	0.960	0.941	0.906	0.950	0.903	0.929	0.957	**1.000**	0.969	0.941	0.941	0.969	0.957	0.953	0.941
0.975	0.953	0.947	0.965	0.960	0.953	0.857	0.957	0.939	0.902	0.936	0.902	0.932	0.982	0.969	**1.000**	0.965	0.950	0.950	0.960	0.957	0.939
0.975	0.932	0.919	0.950	0.936	0.941	0.845	0.944	0.929	0.890	0.912	0.895	0.932	0.982	0.941	0.965	**1.000**	0.965	0.929	0.960	0.957	0.929
0.975	0.941	0.927	0.965	0.947	0.953	0.849	0.957	0.939	0.895	0.919	0.902	0.941	0.960	0.941	0.950	0.965	**1.000**	0.939	0.982	0.975	0.939
0.950	0.982	0.969	0.965	0.969	0.982	0.881	0.965	0.965	0.932	0.969	0.927	0.947	0.941	0.969	0.950	0.929	0.939	**1.000**	0.953	0.957	0.965
0.982	0.957	0.944	0.982	0.965	0.965	0.856	0.969	0.947	0.903	0.934	0.906	0.944	0.965	0.957	0.960	0.960	0.982	0.953	**1.000**	0.982	0.947
0.975	0.953	0.947	0.975	0.960	0.969	0.864	0.965	0.957	0.912	0.936	0.915	0.947	0.960	0.953	0.957	0.957	0.975	0.957	0.982	**1.000**	0.957
0.944	0.953	0.947	0.950	0.947	0.969	0.892	0.957	1.000	0.953	0.960	0.953	0.969	0.936	0.941	0.939	0.929	0.939	0.965	0.947	0.957	**1.000**

Acknowledgement

The authors thank the Principal and management of Thiagarajar College of Engineering (Autonomous), Madurai – 625 015, Tamil Nadu, India.

References

[1]   Biswas K, Bandhoyapadhyay D, Ghosh UC, Adsorption kinetics of fluoride on iron(III)-zirconium(IV) hybrid oxide, Adsorption, 13, pp. 83 – 94, 2009.

[2]   Karthikeyan M, Satheeshkumar KK, Elango KP, Defluoridation of water via doping of polyanilines, J Hazar. Mater. 163, pp. 1026-1032, 2009.

[3]   Lu RS, Lo SL, Hu JY, Analysis of reservoir water quality using fuzzy synthetic evaluation, Stoch. Environ. Res. Risk Assess. 13, pp. 327-336, 1999.

[4]   Jianhua W, Xianguo L, Jinghan, T, Ming J, Fuzzy Synthetic Evaluation of Water Quality of Naoli River Using Parameter Correlation Analysis, Chin. Geogra. Sci., 8(4), pp. 361-368, 2008.

[5]   Singh B, Dahiya S, Jain S, Use of fuzzy synthetic evaluation for Assessment of groundwater quality for drinking usage: a case study of Southern Haryana, India, Environ. Geol. 54, pp. 249 – 255, 2009.

[6]   Kahraman C, Kaya I, Fuzzy process capability indices for quality control of irrigation water, Stoch. Environ. Res. Risk Assess., 23, pp. 451- 462, 2008.

[7]   Chuntao R, Changyou L, Keli J, Sheng Z, Weiping L, Youling C, Water quality assessment for Ulansuhai Lake using fuzzy clustering and pattern recognition, Chinese Journal of Oceanology and Limnology, 26(3), pp. 339-344, 2008.

[8]   Yuan-yuan M, Xue-gang A, Lian-Sheng W, Fuzzy pattern recognition method for assessing ground vulnerability to pollution in the Zhangji   area, Journal of Zhejiang University Science A, ISSN 1009- 3095(Print); ISSN 1862-1775(Online), 2006.

[9]   Muhammetoglu A, Yardimci A, A fuzzy logic approach to assess groundwater pollution levels below agricultural fields, Environ. Monit. Assess. 118, pp. 337-354, 2006.

[10] Sei Z, Fuzzy groundwater classification rule derivation from quality maps, Water quality, Exposure and Health, 1, pp.114 – 122, 2009.

[11] Kumar NV, Mathew S, Swaminathan G, Multifunctional fuzzy approach for the assessment of groundwater quality, J Wat. Res. Prot., 2, pp.597 – 608, 2010.

[12] Rezaei F, Safavi HR, Ahmadi A, Groundwater vulnerability assessment using fuzzy logic: a case study in the zayandehrood aquifers, Iran, Environ Manage., 51, pp.267 – 277, 2013.

[13] Caniani D, Lioi DS, Mancini IM, Masi S, Sdao F, Fuzzy logic model development for groundwater pollution risk assessment, European Water, 35, pp.13 – 22, 2011.

[14] Dahiya S, Singh B, Gaur S, Garg VK, Kushwaha HS,

Analysis of groundwater quality using fuzzy synthetic evaluation, J Hazard. Mater., 147, pp.938 – 946, 2007.

[15] http://www.ramnad.tn.nic.in/SH2006.htm.

[16] http://en.wikipedia.org/wiki/Rameswaram

[17] World Health Organisation, Guidelines for drinking water quality,vol.2, Health criteria and other supporting information, 2nd ed., Geneva, 2006.

[18] Sivasankar V, Ramachandramoorthy T, An Investigation on the pollution status of holy aquifers of Rameswaram, Tamil Nadu, India, Environ. Monit. Assess., 156, pp. 307-315, 2009.

21

Gonadal state and condition factor of *Oreochromis niloticus* (Linnaeus 1758)in a hypereutrophic lake

Beaven Utete[*], **Edmore Happison Chikova**

Chinhoyi University of Technology, Department of Wildlife and Safari Management P. Bag 7724, Chinhoyi

Email address:

beavenu@yahoo.co.uk(B. Utete), butete@cut.ac.zw(B. Utete), mkaiyo@gmail.com (B. Utete)

Abstract: Condition factor, sexual maturity and length-weight relationship of 1132 *Oreochromis niloticus*inhabiting the hypereutrophic Lake Chivero in Zimbabwe were studied. Fish samples were collected through gill netting and from selected commercialfisheries around the lake from February - November 2011. Comparative analysis of age and size at maturity revealed that the *O. niloticus* reached maturity at 30±2.5 mm males and 50±3.6 mm females in total length. There was a significant difference in sexual maturity in both sexes with the inactive (I) stage more frequent. Both sexes of fish indicated a negative allometric increase in weight.August and November were determined as the most intensive and sensitive breeding periods for *O. niloticus* in Lake Chivero.Fishing activities in the lake need to be regulated in tandem with the breeding cycles of the *O. niloticus*.

Keywords: Allometric Growth,Maturity Stage, *Oreochromis Niloticus*, Pollution

1. Introduction

Fish is an important resource in the economic development of many societies. Besides being a cheap source of readily available protein, it contains more essential nutrients required by the body [1]. It is important to monitor the growth and development of fish to optimise yields [2]. In tilapia farming (both wild and domestic) length and weight measurements in tandem with gonadal maturity and age data are vital sources of information on their lifespan, age at maturity, mortality and development [3;4].

Fish condition factor (K) is based on the hypothesis that heavier fish of a particular length are in a better physiological condition [5].And the fact that fish condition factor is strongly influenced by both biotic and abiotic environmental conditions mean it can serve as a proxy to assess the status of the aquatic ecosystem in which the fish live and for further purposes of new stocking [6].In this short communication we assess the fish condition factor (K) and evaluate the sexual maturity of the *Oreochromis niloticus* in a hypereutrophic subtropical lake (Lake Chivero) in Zimbabwe. Primarily this research was prompted by the fact that*Oreochromis niloticus*forms the backbone of commercial fisheries in Lake Chivero which is highly productive all year round such that there is commercial fishing and angling for it all year round. This is despite the fact there are some sensitive stages where gonadal development and breeding is intensive in *O. niloticus* which might need a reduction in the fishing effort [7].

2. Materials and Methods

The study was conducted in Lake Chivero, Harare, Zimbabwe at an altitude of 1368.59 metres above sea level and extending from 17'59' south to30'59' east.The lake was built by the impoundment of Hunyani River in 1952 and has two inflowing rivers Mukuvisi and Marimba for the purpose of supplying water to the city of Salisbury (Harare) [8]. These small rivers provide the lake with approximately 160000 cubic metres of treated and untreated sewage effluent almost daily and this effluent is the main cause of eutrophication in the lake [9]. Gill net sampling was done from February –November 2011 in the lake at sites used by commercial fisheries. Some fish samples were obtained from commercial fisheries. A total fish population of 1132 (659 males and 473 females) were collected and assessed for total length (TL) and standard length (SL) to the nearest mm, and weight to the nearest 0.5 g. Fish were dissected to determine sex and gonadal maturity stages following procedures by[10]. The maturity stages were categorized as: I = Inactive; IA = Inactive-Active; A = Active; R = Ripe; RR = Ripe –Ready; S = Spent.

Temperature of the water was measured in situ using Eutech (Model CT10-1) electronic meter at each sampling site where fish were caught. Water temperature was measured as it has been shown in other studies to affect gonadal maturation [11].

2.1. Data Analysis

The fisheries data obtained were sorted and analyzed using SysStat 12 for Windows version 12.02.00 [12]. Age and maturity at reproductive age was correlated to the fish condition and temperature using the Pearson correlation test (significance $p < 0.05$). ANOVA was used to assess the difference in male and female tilapia condition factor and the gonadal states. T test was used to test for differences in standard and total lengths between male and female fish.Fish condition factor was calculated using the formulae:

Fish condition factor = $K = \dfrac{W \times 100}{SL^3}$

Where K = condition factor, W= weight of fish (g), and SL =standard length (cm) [13].

The length – weight relationships were estimated from the allometric formula:

$W = aL^b$

And this expression can be transformed logarithmically;

$\log W = \log a + b \log L.$

Where: W = weight, L = standard length, a= constant and b=exponent of the arithmetic form of the weight –length relationship and the slope of the regression line in the logarithmic form.

When b = 3 the fish grows isometrically, If b > 3, the fish shows positive allometric growth and If b < 3, then the fish shows negative allometric growth.

3. Results

3.1. Fish Condition Factor and Length Weight - Relationships

Fish condition factor was not significantly different (ANOVA; $p > 0.05$; F=0.6626) between and within the sexes in Lake Chivero. The only months where the condition factor was significantly different (ANOVA; $p < 0.05$;F=0.4532)was in August and September (Figure 1). There was no significant correlation (Pearson test; $p > 0.05$) between condition factor and gonadal maturity in both sexes of tilapia fish sampled in Lake Chivero. However, a strong relationship (r=0.78) existed between temperature and the inactive (I) stage of female fish. There was no significant correlation (Pearson test; $p > 0.05$) between temperature and the gonadal stages in both sexes during the entire sampling period.

Table 1.Relative percentage (%) abundance of different sexual maturity stages in male tilapia fish in Lake Chivero.

Month	Inactive	Inactive- Active	Active	Ripe	Ripe-Ready	SPENT
February	0.56	0.33	0.12	-	-	-
March	0.8	0.16	0.04	-	-	-
April	0.81	0.16	0.03	-	-	-
May	0.58	0.31	0.12	-	-	-
June	0.77	0.17	0.07	-	-	-
July	0.63	0.16	0.21	-	-	-
August	0.71	0.29	-	-	-	-
September	0.82	0.08	0.09	0.01	-	-
October	0.84	0.13	-	0.02	-	0.01
November	0.64	0.19	0.15	0.02	-	-
Overall	0.72	0.19	0.08	0.01	-	0.001

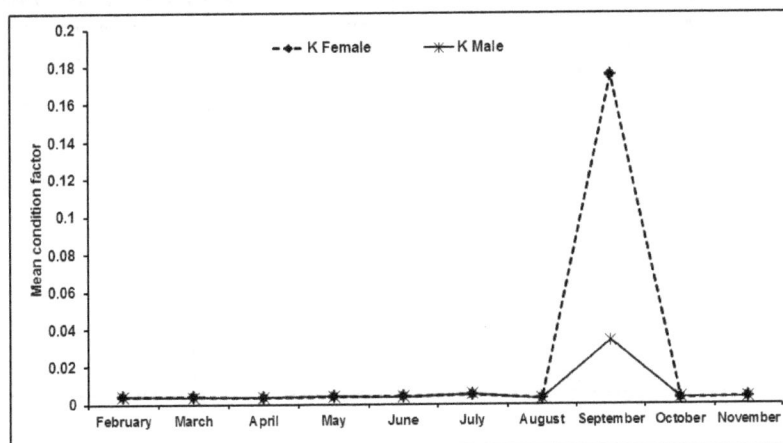

Figure 1.Condition factor of male and female O. niloticus in Lake Chivero for the period February -November 2011.

Standard lengths of fish ranged from 30 to 226 mm (males), 30 to 225mm (females) and total mean length between males and females did not differ significantly (student`s $t$=1.64, p>0.05). The total weight-standard length relationship was separately evaluated for all individuals. The BW-SL regressions did not differ significantly between sexes (student`s t=-0,077, p > 0.05). For both sexes of tilapia sampled in this study weight increased allometrically with size (males: student`s t=5.84, p<0.05 and females: student`s t=4.76, p<0.05). The slope values (b) observed for males and females (b < 3) indicated a negative allometric growth (Figures 2a & b).

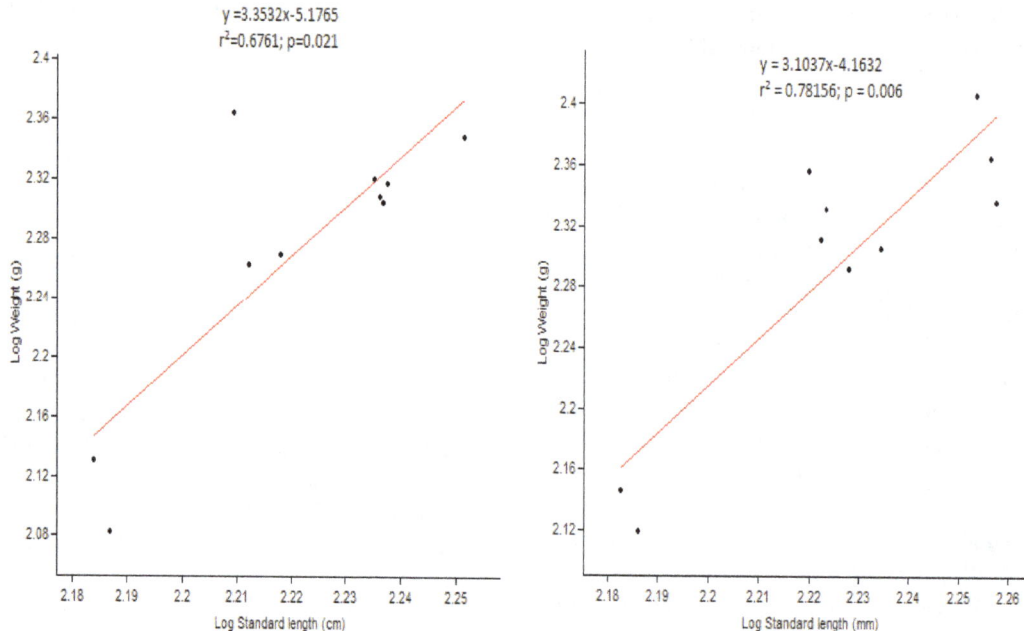

Figure 2a & b. Length -Weight relationships in (a) male and (b) female Oreochromis niloticus fish sampled in Lake Chivero for the period February – November (2011).

3.2. Gonadal Development andSexual Maturity inO.Niloticus in Lake Chivero

There were more males at gonadal maturation stages throughout the year than females. High percentages of mature active, ripe males were found in the months of April (78.8%) and September (81.3%) (Table1).Similarly females had a high percentage of mature active ripe stage between the months of February and September. Gonadal development in the female *O.niloticus* fish indicates dominating "inactive" (I) gonadal stage percentages in February, April, May, June, July and November (Table 2).

Inactive to active gonadal states were significantly high (ANOVA, p < 0.05) in March, August and September in the females. Active (A) percentages were also significantly high (p < 0.05) in October in the females. Inactive to active" (IA) ovaries were lower in February, 5.7%, and high from April to July, 29%. Active (A) figures were very low (10%). Ripe individuals were caught only in March and November recording 4.2% and 3.4% respectively. Ripe and running maturity stages in female fish were recorded in July (8.3%), September (2.6%), and October (4.9%) respectively (Table 2).

Table 2. Relative percentage (%) abundance of different sexual maturity stages in female tilapia fish in Lake Chivero.

Month	Inactive	Inactive - Active	Active	Ripe	Ripe-Ready	SPENT
February	0.78	0.11	0.11	-	-	-
March	0.35	0.48	0.13	0.04	-	-
April	0.56	0.29	0.15	-	-	-
May	0.72	0.11	0.17	-	-	-
June	0.67	0.29	0.05	-	-	-
July	0.48	0.30	0.13	-	0.09	-
August	0.51	0.49	-	-	-	-
September	0.62	0.08	0.25	-	0.05	-
October	0.26	0.23	0.24	0.23	0.05	-
November	0.68	0.21	0.08	0.03	-	-
Overall	0.53	0.25	0.14	0.05	0.02	-

3.3. Physical Parameter - Temperature

Water temperature was high in March and lowered in May and June. Constant temperatures were recorded in the summer stratification period of August –November (Figure 3). There was a significant difference (ANOVA; $P < 0.05$) in surface water temperature in Lake Chivero over the months of sampling.

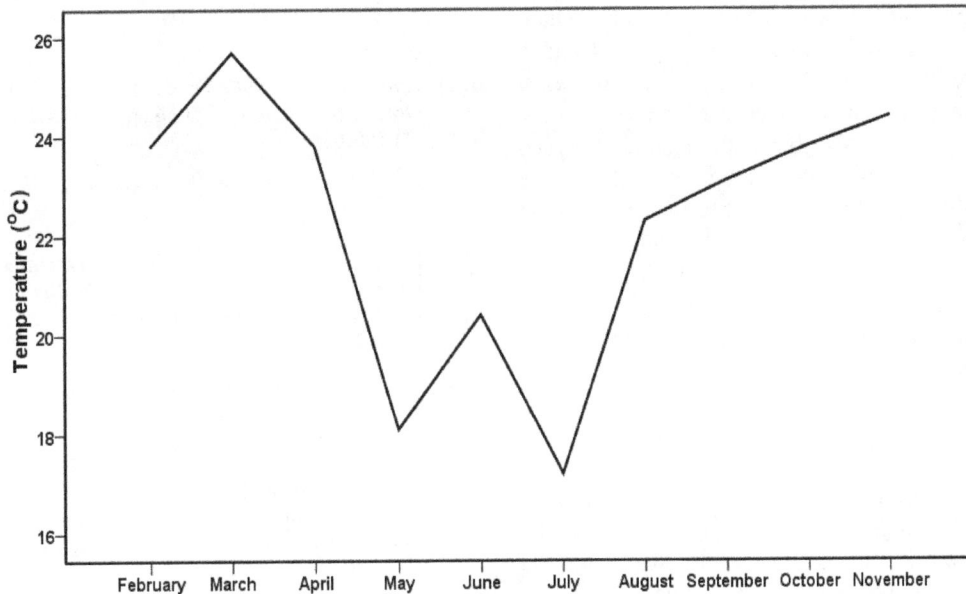

Figure 3. Water temperature of Lake Chivero for the period Feb –November 2011.

4. Discussion

Fish condition factor values shows an increase in the months of August, September and October, this attribute may be caused by the sexual maturation and spawning activities of larger fish in relation to improved feeding condition during the rainy season where phytoplankton production is high [14]. Nutrient inflow into Lake Chivero also increases with the onset of the rain season due to the increased surface run off in the catchment [15]. Moreover, large fish tend to have an improved condition factor; this was also discovered by [16] in Lake Timsah, Egypt, where larger fish had higher condition factor values as compared to small ones. Studies elsewhere [17;7] reveal that the weight –length relationship and fish condition factor depends on the fish species, the abiotic environment and fishing intensity in the water body.

For this study O. niloticus had a negative allometric growth which according to [6] implies that either large fish have changed their body shape to become more elongated or small fish were in better nutritional condition at the time of sampling. This good nutritional status can be correlated to the perennial presence of the phytoplankton as a result of pollution in Lake Chivero [14]. Studies elsewhere [18] show that length - weight relationship parameters of fish species depends on a series of factors such as season, gonad maturity, diet, health and environmental parameters. In this study the effect of an ever constant source of food seems to be significant in maintaining the good condition factor of the O niloticus. Most of the fish examined exhibited lower metabolism and gonad activity during the cool-dry season, which could have contributed to the decline in fish condition factor in June-August. Fish condition sometimes reflects food availability and growth within the weeks prior to sampling, but it is variable and dynamic. Individual fish within the same sample vary considerably, and the average condition of each population varies seasonally and yearly[19]. However, our study shows that there was no significant difference in the condition factor between sexes. This may be due to the excess food that exists in the lake as pollution of Lake Chivero is not seasonal [20; 15].

Results show that there were more males at gonadal maturation stages throughout the year than females. High percentages of mature active, ripe males were found in the months of April (78.8%) and September (81.3%). Similarly females had a high percentage of mature active ripe stage between the months of February and September. It appears the O. niloticus in Lake Chivero had a high breeding intensity between the months of April –September. What this implies is that fishing activities need to be minimal in this period. Winter fish kills in Lake Chivero occur from end of April and up to the end of July [8]. This means that there is a twofold pressure on the tilapia: one of reproduction and the other one a physiological constraint placed on the metabolism of the fish due to lowered oxygen levels and elevated ammonia concentration at lake turnover. As the temperature begins to increase from Augustoxygen levels decrease in the lake which also affects maturation and metabolism [21].

Sex and gonadal development are important variables in

some species, especially cichlids. The sex ratio of *O.niloticus* in Lake Chivero indicates an imbalance among males and females. For the total fish of 1132 caught 659 were males (58.2%) and 473 (41.8%) were females. Other studies have found that males tend to dominate in the cichlids population due to a number of reasons which include ability to survive postnatal mortality [21] and a better growth rate [22].Of note is that *O.niloticus* species in Lake Chivero breeds throughout the year but the breeding intensity was not equally distributed throughout the year. As such the lakemanagement should consider these breeding fluctuations possibly by establishing non-fishing zones during the most sensitive stages. Although pollution is detrimental in most aquatic systems, in Lake Chivero it appears that the fish condition factor is not significantly different in *O. niloticus* throughout the year possibly due to an incessant supply of nutrients.

Acknowledgements

The help of staff at the Lake Chivero Fisheries Research Station with field equipment and manpower is hereby appreciated as well as the guidance by Michael Tiki and Newman Songore in collecting the fishery data.

References

[1]   Sikoki, F.D and A.J.T. Otobotekere.1999. The Land People of Bayesa State Central Niger Delta.

[2]   Haimovici, M. and Velasco, G. 2000. Length- weight relationship of marine fishes from .Southern Brazil. .The ICLARM Quarterly 23(1):14-16.

[3]   Diaz, L. S, Roa, A, Garcia, C. B, Acero, A, and Navas, G. 2000. Length –weight relationships of demersal fishes from the upper continental slopes of Colombia. The ICLARM Quarterly 23(3):23-25.

[4]   Fafioye, O.O, and Oluajo, O. A. 2005. Length –weight relationships of five fish species in Epe .lagoon, .Nigeria. *African Journal Biotechnology* 4(7):749-751.

[5]   Kumolu-Johnson, C.A. and Ndimele, P.E. 2010. Length-weight relationships and condition factors of twenty-one fish species in Ologe Lagoon, Lagos, Nigeria.*Asian Journal of Agricultural Science*, 2: 174–179.

[6]   Shakir, H.A., Mirza, M.R., Khan, A.M. and Abid, M. 2008.Weightlength and condition factor relationship of Sperata Sarwari (Singhari), from Mangla Lake, Pakistan. *Journal of Animal and Plant Science* 18:158-161.

[7]   Hirpo, L. A. 2012. Breeding Season and Condition Factor of *Oreochromis niloticus* (Pisces: .Cichlidae) in Leke Babagaya. *International Journal of Agricultural Sciences*, 2(3):116-.120.

[8]   Marshall, B. 2011 .Fishes of Zimbabwe and their biology .Smithiana Monograph 3. The South .African Institute for Aquatic Biodiversity Grahamstown.

[9]   Moyo, N. A. G. 1997. Causes of Massive Fish Deaths in Lake Chivero. In N.A.G. Moyo (Ed.), *Lake Chivero: A Polluted Lake*. University of Zimbabwe Publications, Harare: 98-105.

[10] Holden, M. J and Raitt, D. F.S.1974. Manual of fisheries science.2. Methods of resource investigation and their application. *FAO* Fisheries Technical Paper, 115, Rev, 1:211pp.

[11] Kotos, A. A. 1998. Food, size, and Condition factor of *Oreochromis Niloticus* in Niger River.Nigeria. *Boil. Trop* 44 (3):566-658.

[12] Systat 2007. Mystat: A student version of Systat 32-bit UNICODE English. Version 12.02.00.

[13] Bagenal, T. B and Tesch, W. F. 1978. *Age and growth in methods of assessment of fish .production in freshwaters*. Ed Bagenal. T. Oxford Blackwell Scientific Publication pp .101-136.

[14] Tendaupenyu, P. 2012.Nutrient limitation of phytoplankton in five impoundmentson the Manyame .River, Zimbabwe.*Water SA* Vol 38 (1).

[15] Mtetwa, S. 1997.Effluent and Waste Water Standards in Zimbabwe. In N. A. G. Moyo (Ed.), *Lake Chivero: A Polluted Lake*. University of Zimbabwe Publications, Harare: 124-134.

[16] Mahomoud, W. F., Amin, A. M.M., Elboray, K. F., Ramadhan, A. M., and El-Halfwy, M. M. M.K. O. 20Reproductive biology and some observation on the age, growth and management of Tilapia zilli.(Gerv, 1848), from Lake Timsah, Egypt.*International Journal*.3: 15-25.

[17] Cherif, M. Zarrad, R. Gharbi, R and Hechmi J. 2008. Length –weight relationship for 11 fish species from the Gulf of Tunis. *Pan American Journal of Aquatic Sciences*, 4: 67-72.

[18] Froese, R. 2006. Cube law, condition factor and weight length relationships: history, metaanalysis and recommendations. *Journal of Marine Fisheries* 23: 45-52.

[19] Ayoade, A.A and Ikulala, A.A.O.O. 2007. Length weight relationship, condition facto and stomach contents of Hemichromis bimaculatus, Sarotherodonmelanotheron and Chromidotilapia.guentheri (Perciformes: Cichlidae) in Eleiyele Lake, South-western Nigeria. *Revista Biologia. Tropica*, 55: 969–977. [PubMed], [Web of Science ®]

[20] Zaranyika, M. F. 1997. Sources and Levels of Pollution Along Mukuvisi River. A Review. In N.A.G. Moyo (Ed.), *Lake Chivero: A Polluted Lake*. University of Zimbabwe Publications, Harare: 35-43.

[21] Fryer, G. TD.Iles. 1972. The Fishes of the Great Lakes of Africa. Their Biology and Evolution Oliver and Boyd, Edinburgh, Scotland.

[22] Mortuza, M.G. and Rahman, T. 2006. Length-weight relationship, condition factor and sex-ratio of freshwater fish, Rhinomugilcorsula (Hamilton) (Mugiliformes: Mugilidae) from Rajshahi Bangladesh. *Journal of Bio-science*, 14: 139–141.

[23] Utete, B, and Dzikiti, B. 2013. Comparative Study of Maize Bran and Chicken Manure as Fish Feed Supplement: Effects on Growth Rate of Oreochromis Niloticus in Pond Culture Systems. *International Journal of Aquaculture* 3(6):23-29.

22

The long empty canyon: A study of the old/new legal problems of the Nile basin

Shams Al Din Al Hajjaji

American Univeristy in Cairo/ Law Department, Egyptian Public Prosecution Bureau, Egypt

Email address:

salhajjaji@aucegypt.edu

Abstract: The Nile River Basin witnesses a long history of tension and negotiation among riparian states. There are two legal frameworks govern the Nile Basin. Firstly, the private legal framework reflected in legal history on the Nile. The most legal active period among Nile Basin states was the period between 1890th and 1930th. The legal solutions to the Nile Basin problems came to an end with the end of the colonization in Africa, especially the Nile riparian states. During this period, the tension among liberal states took a different shape. Harmon and Nyerere doctrine were introduced among the riparian states. This led to the refutation of most of the private legal framework from most of the independent states. Thus, riparian states started to explore new legal ground to regulate their relationship. On the other hand, the public legal framework represented in the work of the International Law Association, which started with Helsinki rules in 1966, and the 1997 UN Convention. Many scholars argue that the legal solution is the best one for the Nile question, based on the previous frameworks. However, this note argued that the international legal framework governing the international rivers generally and the Nile specifically cannot offer a solution to the disputes over the water of the Nile. This note discusses both the legal frameworks of the Nile on one hand. On the other hand, it highlights the points of indeterminacy of both frameworks to solve the Nile dilemma. It argues that the solutions of the present and future disputes through legal tools are not enough. This note goes beyond the most proposed recommendation to form a comprehensive treaty as the solution to the riparian problems. It asserts that the law is not a tool to end the states tension, rather than it is a tool to persevere good faith and prevent future dispute. A main role of the extra legal solutions must be played. It based its argument on substantive and formulate dilemma in the previous frameworks.

Keywords: Nile Basin, Egypt, Ethiopia, Sovereignty, Cooperation, Equitable and Reasonable Utilization, No Harm

1. Introduction

The relationship between Egypt and surrounding states is becoming more strained by the day. The demand of lower riparian states is increasing in regards to their share of the Nile water. These rising demands have caused political clashes among the riparian states, especially Egypt and Sudan from the upper riparian, and Ethiopia, Tanzania, and Kenya from the lower riparian states. This tension reached its peak when President Sadat declared that Egypt would be ready to go to war against Ethiopia, if it harms Egypt's interests in Nile water.[1] To maintain the status quo, many

scholars proposed a legal solution as one of the strong propositions in this case.[2] However, none of them offered an answer of why or how these states will enter in a new treaty regarding the Nile issues, especially given that some of them persisted on their acquired right to and share of the Nile water. I argue here that the Nile legal frameworks as they are interpreted cannot help the Nile Basin states to enter in a legal agreement unless under the existence of

---

[1] *TesfayeTafesse, The Hydropolitical Assessment of the Nile Question: An Ethiopian Perspective, 26 WATER INT'L 1, 2001, 4, See also, JuttaBrunnee and Stephen Troope, The Changing Nile Basin Regime: Does Law Matter?, 43 HARV. INT'L L.REV. 105, 106. [Hereinafter Does Law Matter] See aslo Sadat to Ethiopia: Leave Nile alone or it's war, The Gazette Montréal ,*

Saturday June 7, 1980,

http://news.google.com/newspapers?nid=1946&dat=19800607&id=IYkx AAAAIBAJ&sjid=caQFAAAAIBAJ&pg=1030,2287901 Last visit 24/12/2011

[2] *TakeleSobokaBulto, Between Ambivalence and Necessity: Occlusions on the Path Towards a Basin Wide Treaty in the Nile Basin, 20 COLO. J. INT'L ENVTL. L. &POL'Y 291 2008-2009, 318.[hereinafter Ambivalence and Necessity] See also, Christina M. Carroll, Past and Future Legal Framework of the Nile River Basin, 12 GEO. INT'LENVTL. L. REV. 269, 199-2000, 282.*

other factors, whether economic or political.

This paper is concerned with a certain legal occurrence, where there is a need to reach a new legal agreement in response to the existing dispute related to the current legal issues. The argument will be limited to the question of how this dispute affects the formulation of a new law. In other words, the thesis tackles the transition period between the old and new legal systems. The main question here and what I am trying to spread in my thesis is: "Is the existing legal framework fit to be a base for the new legal order of the Nile?" The answer is no, as I presented the Nile Basin states' argument, which they maintained – each from its own perspective- that such an argument is the suitable one. First, I present the Egyptian Legal argument, which is based on the historical and acquired rights of Egypt. Second, I tackle the issue of the Nyerere Doctrine, which leads me to discuss the issue of state succession and conflict between state continuity and state autonomy. Thirdly, I argue that the conflict between Sovereignty and Cooperation in the international water law is inevitable.

2. Overview of the Nile Legal Issues

2.1. Legal Framework of the Nile Basin State

2.1.1. Introduction

The controversial positions of states and scholars' position can be summarized in three main points. First, Egypt and Sudan accept the Nile conventions and consider them as acquired rights.[3] Secondly, Ethiopia, Kenya, and Tanzania refuse both acquired and historical rights, and they consider them to be a colonial conspiracy against the lower riparian states,[4] Thirdly, Congo, Uganda, and Rwanda accept the conventions, albeit after long negotiations with Egypt and Sudan, in order to take personal advantages.[5] Finally, Eritrea is an observer to previous states, and did not have any inclination to join

either pole.[6]

2.1.2. 1902 Treaty between Ethiopia and the United Kingdom

In 1902, the King of Great Britain Edward VII and the Ethiopian Emperor Menelik II signed a treaty regarding "the delimitation of the Frontier between Ethiopia and Sudan,"[7] which was part of the Egyptian territory while Egypt was under the British protection. The treaty was drafted both in English and Amharic. It consisted of five articles. While the first two are related to the determination of the boundaries between the two states, the last two articles deal with the future cooperation between the two empires. For the River Nile, article three was the only article dealing with the Nile Water.

2.1.3. 1925 Exchange Note between Italy and the United Kingdom

Between 1919 and 1925, both the British and the Italian governments exchanged notes on building the railroad from Eritrea to the Italian Somaliland. The exchange confirmed the right of both Egypt and Sudan to their share of the Nile water. In return for the exchange, Great Britain asked for the Italian government's recognition of such rights to ensure the execution of the railroad project. Italy has planned to build a railroad that will pass through Ethiopia, and the vicinity of Addis Ababa. The note was to ask the British colony its support to mediate between the Ethiopian government and the Italian Colonist. On the other hand, the British government asserted in the note that building the railroad is attached with the declaration of the Italian colony with the "prior hydraulic rights" of both Egypt and Sudan.[8]

2.1.4. 1929 Exchange Note between Egypt and the United Kingdom

OkothOwrio, a Kenyan scholar, argued that the 1929 note exchange was to "guarantee and facilitate an increase in the volume of water reaching Egypt."[9] However, the rights that this agreement guaranteed to Egypt were also maintained in the previous agreement. Besides, the 1929 Exchange note between Egypt and the United Kingdom ensured the continuity of the assigned share of water that

[3] Valerie Knobelsdorf, Note: The Nile Water Agreements: Imposition and Impacts of a Transboundary Legal System, 44 COLUM J. TRANSNATL. L. 634. 635

[4] See, Christina M. Carroll, Supra note 2 at 139, DerejeZelekeMekonnen, The Nile Basin Cooperative Framework Agreement Negotiations and the Adoption of a Water Security Paradigm: Flight Into Obscurity or a Logical Cul-de-sac? 21EUR. J. INT'L.2, (2010), [hereinafter The Nile Basin Cooperative Framework Agreement Negotiations], see also, DerejeZelekeMekonnen, Between the Scylla of water security and Charybdis of Benefit Sharing: The Nile Basin Cooperative Framework Agreement- Failed or Just teetering on the Brink?,GO. J. INT'L. 3 (2011), {hereinafter Benefit Sharing], see, TakeleSobokaBulto, Between Ambivalence and Necessity: Occlusions on the Path Toward A Basin – Wide Treaty in the Nile Basin, 20 COLO. J. INT'L ENVIRL. L. &POL'Y 291, (2008-2009) [hereinafter Between Ambivalence and Necessity], JuttaBrunnee. AZIZA MANSUR FAHMI, WATER MANAGEMENT IN THE NILE BASIN: OPPORTUNITIES AND CONSTRAINTS, http://www.isgi.cnr.it/stat/pubblicazioni/sustainable/133.pdf last visit 11/10/2011.

[5] Aaron Schwachach, The United Nation Convention on the Law of Non-Navigational uses of International watercourses, Customary International Law and interest of upper riparian states, 33 TEX. INT' L. J. 257 (1998), 270.

[6] Adams Oloo, The Quest for Cooperation in the Nile Water Conflicts: the Case of Eritrea, 11 AFR. SOC. REV. 95, 2007, 96.

[7] Preamble of the Treaty Between Ethiopia and Great Britain on the Delimitation of the Frontier between Ethiopia and Sudan, United Nations, Legislative Texts and Treaty Provisions Concerning the Utilization of International Rivers for Other Purposes than Navigation, United Nations Legislative Series (ST/LEG/SER.B/12), United Nations publication, 115,116 [hereinafter United Nation Publication].

[8] Exchange of Notes Between the United Kingdom and Italy Respecting Concessions for a Barrage at Lake Tsana and a Railway Across Abyssinia From Eritrea To Italian Somaliland, Signed at Rome 14 and 20 December 1925, see United Nation Publication Supra note 7 at 99

[9] OkothOwiro, The Nile Treaty, State Succession and international Treaty Commitments: A case Study of the Nile Water Treaty, http://www.kas.de/wf/doc/kas_6306-544-1-30.pdflast visit 1/4/2012.

reaches Egypt.[10]

Egyptian Scholars argued that the assassination of Sir Oliver Lee Stack, the British governor-general of Sudan, in late 1924 was the reason for concluding such a note.[11] Later, after the 1925 exchange, the British authority in Sudan used this accident to apply pressure on the Egyptian policy in Sudan. It threatened the Egyptian government with increasing areas irrigated with the Nile River water in Sudan, as punishment for murdering Sir Oliver Lee Stack. Hence, the Egyptian government worked to develop a new study of the Nile River water for irrigation purposes.[12] Thus, the notes between Her Majesty's Government in the United Kingdom and the Egyptian Government on the Use of Waters of the Nile for Irrigation were concluded in 1929.

The Note was between the Chairman of the Council of Ministers Mohamed Mahmud Pasha- as a representative of the Egyptian government - and Lord Lloyd from the British government. The first paragraph of the note asserted that "a solution to these problems [irrigation] would not be deferred to a subsequent date when it became possible for the two Governments to come to terms on the status of the Sudan but, regarding the settlement of the present provisions, it expressly reserves every freedom at any negotiations which could precede such an agreement."[13] Egyptian Note sent from the Chairman of the Council of Ministers Mohamed Mahmud Pasha stated that "{t}he present agreement can in no way be considered as affecting the control of the River - this being a problem which will cover free discussions between the two Governments within the framework of negotiations on the Sudan."[14] The second paragraph was reconfirmed later in the 1959 Convention. It asserted its acceptance to the increase of water quantity to Sudan without any "infringement on neither the natural and historical rights of Egypt."[15]

The significance of 1929 Nile water agreement was embedded in three issues. First, Egypt ensured full control of any construction work on the Nile.[16] Based on this fact, the Ethiopian authority was prevented from building a dam on the Lake Tana in 1935.[17] Second, it changed the legal status of the different Nile Basin states. It had fully recognized the principle of equitable utilization.[18] The determination of such utilization is based on finding of a commission.[19] Thirdly, 1929 agreement was a symbol of recognition of "the principle of established rights." Egypt insisted on the recognition of its "natural and historic rights." They have been the most fundamental elements of Egyptian policy approach to the Nile waters.

2.1.5. 1959 Agreement between Egypt and Sudan

The High Dam (1960-1969) was built after months of concluding the agreement between the United Arab Republic and the Republic of Sudan for the full utilization of the Nile Water on November 8th, 1959. This agreement was mainly held for the sake of building the High dam; it determined the Egyptian share in the Nile water regarding the Dam and its lake.[20]

The importance of the 1959 Convention is based on various factors. First, 1959 put a bilateral obligation on both states to negotiate with other riparian states in case of their request to increase their water share. They did not exclude the other riparian states' right to ask for future increases in their own share. Secondly, The Convention was mainly to enhance water utilization for both states. Additionally, it increased the water share of Sudan to compensate for building the high dam; the Egyptian government additionally paid 15 million pounds to the Sudanese government for any damage afflicted on the Sudanese territory form building the dam. Moreover, the Convention helped the two states to form one of the oldest institutional arrangements in the Nile Basin, which is the Permanent Joint Technical Commission for Nile Water (PJTC).

2.2. Institutional Framework of the Nile Basin States

2.2.1. Permanent Joint Technical Commission for Nile Water (PJTC)

The history of institutional arrangement of the Nile Basin started in the early 1950s. In 1959, as a part of the 1959 Convention, Egypt and Sudan formed the Permanent Joint Technical Commission for Nile Water (PJTC). It is considered as one of the oldest arrangements for the Nile Basin. The reason for establishing the Commission was to ensure the technical cooperation for the Nile control projects.[21]

This cooperation tool is a bilateral cooperative one. It did not include any other states from the rest of the Nile Basin except Egypt and Sudan.[22] Both countries stated that for the best interest of the PJTC success, other Nile Basin states shall be involved in another big institutional arrangement. Hence, the result was establishing the HYDROMET project, which paved the road to both UNDUGU and TECCONILE later on.[23]

---

[10]/ see United Nation Publication supra note 7 at 115

[11] /YunanLabibRizk, Adiwan of Contemporary Life, Al Ahram, http://weekly.ahram.org.eg/2000/503/chrncls.htmlast visit, 1/4/2012. See also AZIZA MANSUR FAHMI, supra note4.

[12]/ P. P. HOWELL AND J. A. ALLAN, THE NILE: SHARING A SCARCE RESOURCES; A HISTORICAL AND TECHNICAL REVIEW OF WATER MANAGEMENT AND OF ECONOMICAL AND LEGAL ISSUES, Cambridge University Press, (1st ed.), (1994), 538.

[13]/ United Nation Publication supra note 7 at 101

[14]/ United Nation Publication, Id at 101

[15]/ United Nation Publication Id at 101.

[16]/Supra note 2 at 98.

[17]/ Econ. & Soc. Commission For Western Asia, Assessment of Legal Aspects of the Management of Shared Water Resources in the ESCWA Region, ¶U.N. Doc. E/ESCW A/ENR/2001/3, (Feb. 22, 2001), 14.

[18]/ Id at 16

[19]/ AZIZA MANSUR FAHMI, supra note 4 at 136.

[20]/ Agreement between the United Arab Republic and the Republic of Sudan for Full Utilization of the Nile Waters, see United Nation Publication supra note 12 at 146.

[21]/ Art. 4 of 1959 Convention

[22]/ United Nation Publication Id at 50.

[23]/Id. at 51.

2.2.2. Meteorological and Hydrological Survey on the Equatorial Lakes HYDRO-MET

The HYDROMET project included Egypt, Sudan, Uganda and Tanganyika.[24] Later on, Burundi, Rwanda and Zaire joined the project, while Ethiopia remained as an observer.[25] This project was a survey to the catchments of Lakes Victoria, Kyoga, and Mobutu SeseSeku (Lake Albert).[26] The aim of the project was to help its members in:

a)   Determination of their equitable entitlements to the use of the Nile;
b)   Formulation of national water master plans;
c)   Development of their capacities and basin-wide information system;
d)   Preparation of a basin-wide institutional and legal arrangement;
e)   Enhancement of training procedures;
f)   Environmental impact assessment and water quality management capacity.[27]

Some writers argue that the HYDROMET project is older than the PJTC.[28]They maintain that in 1950, Egypt agreed to work on a meteorological and hydrological survey on the equatorial lakes with the assistance of Great Britain. [28] However, official establishment of the HYDROMET project was in 1967, eight years after the 1959 Convention.[29] The project took about 35 years until it turned into TECCONILE.

2.2.3. Technical Cooperation Committee for the Promotion of the Development and Environmental Protection of the Nile TECCONILE

The Technical Cooperation Committee for the Promotion of the Development and Environmental Protection of the Nile was established in 1992. Rwanda, former Sudan, Tanzania, Zaire and Egypt established the TECCONILE for a fixed period of three years as a transition period until the establishment of a wide institutional arrangement. Some other Nile Basin states participated in the TECCONILE as observers like Ethiopia and Kenya.[30] The main reason of the establishment of the TECCONILE was to address the Egyptian domination in the previous arrangement, especially in the UNDUGU.[31]Brunnee and Toope saw that the Egyptian technical expertise gave it the upper hand in the previous institutional arrangement. This expertise threatened Ethiopia and Kenya. [32] The TECCONILE was supposed to work for only three years as a transition period

before launching the Nile Basin Initiative. However, this period was extended to more than nine years.

2.2.4. UNDUGU: Brotherhood

Another project that many writers did not give due attention was UNDUGU. Egypt was able to convene with Sudan, Uganda and Zaire to form a league called UNDUGU in 1981. UNDUGU means brotherhood in Swahili. The plan was to reorganize this convivial group into a more scientific organization. It was concerned with technical matters that ministers, who were concerned with political affairs, were not very interested in or knowledgeable about.[33]

Mekonnen argued that both the UNDUGU and TECCONNILE paved the road to establish the NBI, which is considered to be a corner stone in the institutional arrangement of the Nile basin. However, many writers challenged this finding. Brunnee and Toope argued that the UNDUGU was an Egyptian initiative as a part of its "hegemonic aspirations." [34] They further stipulated that Egypt "sought to create multi bargaining situations most likely to result in agreement than negotiations purely devoted to water issues."[35]

Additionally, YacobArsano argued that Ethiopia challenged the UNDUGU. He affirmed his argument that Ethiopia declared that UNDUGU had no legal foundation as a legitimate body, and it was ended after the "ministerial meeting in Addis Ababa."[36] However, TakeleSobokaBulto maintained Ethiopia was always against the Egyptian aims of the UNDUGU and TECCONILE, as it was acting in both as an observer. [37] Hence, there is mutual intention from both upper and lower riparian states to take a stand against each other, otherwise, these arrangements would have succeeded.

2.2.5. Nile Basin Initiative

Many writers argued that NBI is the successor of the TECCONILE. [38] They maintained that NBI secretariat is housed in the old TECCONILE buildings. However, there are many differences between the NBI and TECCONILE. First, Ethiopia and Kenya did not join TECCONILE, while both of them are members of NBI. They declared their refutation of the TECCONILE on the bases of it not proviting any "fundamental equitable concerns of water apportionment."[39] Secondly, TECCONILE was to provide states with technical expertise, while NBI is to contribute to poverty alleviation, reverse environmental degradation and

[24]meaningSalman M. Salman, The New State of South Soudan and the Hydir- Politics of the Nile Basin, 36WATER INT'L154, 159. {hereinafter The New State of South Soudan}.37

[25]/ Econ. & Soc. Commission For Western Asia, Assessment of Legal Aspects of the Management of Shared Water Resources in the ESCWA Region, ¶U.N. Doc. E/ESCW A/ENR/2001/3, (Feb. 22, 2001), 14.18.

[26]/The New State of South Soudan Supra note 23 at 37

[27]/ Supra note 25 at 19

[28]/ Id at 18.

[29]/ Id at 19

[30]/ Does the Law Matter, supra note 1 at 133-134

[31]/ Id at 133-134

[32]/ Does the Law Matter, supra note 1 at 133-134

[33]/ Yosef Yacob, From UNDUGU to the Nile Basin Initiative, An Ending Exercise in Futility, Ethiopia TECOLAHACOS, http://www.tecolahagos.com/undugu.htm last visit 21 May 2012.

[34]/ See Does Law Matter supra note 1 at 133

[35]/ Id..at 133

[36]/ YacobArsano, Ethiopia and the Nile: Dilemmas of National and Regional Hydro politics, (2007), (Ph.D. dissertation, University of Zurich) (on file with author)

[37]/ see Ambivalence and Necessity supra note 1 at 318.

[38]/ Does Law Matter supra note 7 at 108.

[39]/ Id at 134.

promote socio-economic growth in the riparian countries.[40]

The Nile Basin Initiative is a cornerstone in the overall Nile Basin relationship among the Nile basin states. They made such joint effort to "achieve sustainable socio-economic development through the equitable utilization of, and benefit from, the common Nile Basin resources."[41] The Nile Basin Initiative was established on February 22, 1999 in Darussalam, by the Ministers responsible for "Water Affairs of each of the nine Member States." [43] These states are Burundi, the Democratic Republic of Congo, Egypt, Ethiopia, Kenya, Rwanda, Sudan, Tanzania, and Uganda.[42] As of yet, there are no available resources about the membership or the position of South-Sudan.

The significance of NBI was manifested in the attempt to reach a legal solution to the pending issues among the Nile Basin states. After one year of its official work, NBI prepared an "Agreement on the Nile River Basin Cooperative Framework."[43] The agreement was based on the scholarly work in the field of the international water law. It will pave the road to form the "Permanent River Nile Basin Organization" or the "Nile Basin Commission." These arrangements will be concerned with the enforcement of any legal arrangement among the Nile Basin states.

2.3. Pending Legal and Institutional Issues

2.3.1. Legal Issues

It should not be forgotten that the NBI had eventually reached a form of legal arrangement, an agreement on the Nile River Basin Cooperative Framework. Ethiopian scholars argued that such a framework would end the Egyptian hegemony on the Hydro-political aspects of the Nile water. Abadir Ibrahim argued that the new agreement would end the Egyptian hegemony unless upper riparian states use a counter hegemonic strategies, which will based on affect on the flow of the Nile to Egypt.[44] However, such a perspective is more imaginary and lacks fundamental reading of the Agreement. Firstly, even though the Agreement did not answer the main question of States' water share or distribution of water among them, it is based, to a great extent, on the international water law principles. Article 4, paragraph 2, about the Equitable and Reasonable Utilization, is a copy of the successive articles regarding

the same issue. There are the works of the Helsinki Rules, the International Law Commission, Draft Articles of the United Nation Convention of Non-Navigational Uses of International Water Course, and the recent development of the International Water Law represented in International Water Law Association conferences.

Furthermore, the new viewpoints have introduced new standards to the principle of equitable and reasonable utilization and participation. Article 4 (2)(h) of the agreement regarding the equitable and reasonable utilization stated that "The contribution of each Basin State to the waters of the Nile River system"[45] as one of the considerable measurement of the equitable utilization principle. This measurement was eliminated during the negotiation of the United Nations Convention on the Law of the Non Navigational Uses of International Watercourses. Additionally, Mekonnen argued that the main reason that kept Egypt and Sudan from joining the Agreement on the Nile River Basin Cooperative Framework was that a new term "water security" had been introduced in the draft.[46] This term led to the suspension of the draft articles, especially that of article 14. [47] Article 14 made the interpretation of both principles of equitable utilization and no harm connected to the water security of the states. Egypt and Sudan did not accept this measure; instead, they proposed to connect the states' water security with "current uses and rights of any other Nile basin state,"[48] which was maintained in article 4 para 2.e "existing and potential uses of the water resources." [49]

Moreover, in April 2010, Egypt maintained – in the *Sharm Al Sheikha* convention among the Nile Basin states - that the new Agreement shall include an article stating the Egyptian "Historical and Natural rights" in the Nile water. The Egyptian Minister of Water and Irrigation Dr. *Hussein El Atafy* made an official statement against the agreement on the Nile River Basin Cooperative Framework. He asserted that this Agreement "violates the agreed upon procedures and does not relieve member states of their commitments to valid previous agreements with Egypt."[50] He further stipulated that " the International Court of Justice considers these rights as enshrined as boarder agreements and those countries cannot change existing and valid agreement under the pretext that they were signed during the era of colonialism."[51]

Thirdly, Article 5 of the Agreement dealt with the principle of 'Obligation not to cause Significant Harm.' It stated that "Nile Basin States shall, in utilizing Nile River

---

[40]/see , Claudia Sadoff and David Grey, Beyond the River: The Benefits of Cooperation on International Rivers, 4 WATER POL' 389, 2002, 401, see Nile Basin Initiative ,http://nilebasin.org/newsite/index.php?option=com_content&view=articl e&id=71%3Aabout-the-nbi&catid=34%3Anbibackground-facts&Itemid=74&lang=enlast visit 10/3/2012 , .

[41]/ Samuel Luzi, Mohamed Abdel, MoghnyHamouda, FranziskaSigrist and EvelyneTauchnits, Water Policy Networks in Egypt and Ethiopia, 17 J. ENV.& DEV. 238, 2008, 239. [43] Nile Basin Initiative ,About the NBI, supra note 51 [44] See Does Law Matter supra note 7 at 108.

[42]/ See Does Law Matter supra note 7 at 108.

[43]/ The Nile Basin Cooperative Framework Agreement, supra note 7.

[44] / Abadier M. Ibrahim, The Nile Basin Cooperative Framework Agreement: The Beginning of the End of Egyptian Hydro-Political Hegemony,18 MO. ENVTL. L. & POL'Y REV 284, 308

---

[45]/ Art.4 par. 2/h

[46]/ See The Nile Basin Cooperative Framework Agreement Negotiations, Supra Note 7 at 428.

[47]/Id at 428.

[48]/ Id at 428.

[49]/Art.4 para 2/e.

[50] Egypt and its Historical Rights in Nile Water, Egypt State Information Service, http://www.sis.gov.eg/En/LastPage.aspx?Category_ID=1144 last visit 30/10/2012.

[51] /Id.

System water resources in their territories, take all appropriate measures to prevent the causing of significant harm to other Basin States." [52] It also recognized the principle of reparation, as it claimed the right of the injured state which has sustained significant harm to ask for compensation for the act. The second paragraph stated that "{w}here significant harm nevertheless is caused to another Nile Basin State, the States ... take all appropriate measures, having due regard to the provisions of Article 4 above, in consultation with the affected State, to eliminate or mitigate such harm and, where appropriate, to discuss the question of compensation." [53]

2.3.2. Institutional Issues (Nile River Basin Commission)

In addition to Nile River Basin, the Commission will succeed the NBI in all its purposes and functions. Article 16 of the Agreement on the Nile River Basin Cooperative Framework dealt with its own new purpose and objective of the Commission. It stated that it has three main objectives:

a) Promote and facilitate the implementation of the principles, rights and obligations of the Agreement

b) Serve as an institutional framework for cooperation among Nile Basin States in the use, development, protection, conservation and management if the Basin and its water

c) Facilitate closer cooperation among states and peoples of the Nile River Basin in Social, economical, and culture fields. [54]

Besides the main objectives and purpose of the Nile River Basin Commission, it was given extra functions in regards to dispute settlements, information exchange, and mutual cooperation. Article 33 of the Agreement on the Nile River Basin Cooperative Framework gives the Nile River Basin Commission a reasonable role in dispute settlement. It urged the states' members to use the Nile River Basin Commission as mediator or conciliator between the quarreled parties.

3. Inadequacy of the Legal Framework of the Nile Basin

3.1. Historical and Acquired Rights

3.1.1. Egyptian Argument

a. Mixing of Historical and Acquired Rights in the Egyptian Legal Literature

Many writers in the field of international water law (Stephen McCafferey, Aziza Fahmy and MufidShehab) have intermixed the historical and the acquired rights. One can say that there is a general confusion in the legal literature of the Nile regarding the Historical and Acquired rights. However, these writers are justified in their

perspective, since most related conventions asserted Egypt's historical rights. On the other hand, it is easy to find other writers confusing the two rights. *Adel Aela* declared that the Egyptian right is "Historical Acquired Rights," as one terminology describes the Egyptian rights. [55] Egypt's position has reached a stage that when talking of Egypt's acquired rights is radically connected to its historical rights.

i. Scope of the Egyptian Argument

The structure of the legal argument related to the specific framework is categorized by opposing claims. Every state based its rights on the refutation of the rights of others. Egypt clings to its historical rights of 7000 years of Nile water utilization, as well as its acquired rights in the successive notes and conventions; conversely, Ethiopia refutes such rights.

The Egyptian government argued that its water rights are based on factual and legal bases. For the factual dimension, *Aziza Fahmi* stated that according to the 1959 agreement Egypt only uses "55.5 milliard cubic meters out of total 200 milliard cubic meters of water resources in the Nile basin." [56] She additionally maintained that Egypt "relies totally on the waters of the Nile for its existence, for its survival because it is an arid desert land." [57] Besides, Fahmi further stipulated that Egypt never used and it will never use the "right of veto". [58] She based her argument on "the principle of abuse of right," [59] the basic principle of State Responsibility that prevents any unreasonable use of the right of veto. [60] Hence, she considered other Nile Basin states' position against the Egyptian Nile share is an "exaggeration." [61]

For the legal dimension, Egypt built its legal argument on the successive legal notes and agreements. All of these maintained that the water should flow to the lower riparian states (Egypt and Sudan). There is no reference to the quantity of water specified to Egypt during such time. In 1929-note exchange, *Mohammed Mahmoud Pasha* asserted the Egyptian historical rights, without any reference to such quantity. The different rules and conventions held a clear position that none of them affect the existing bilateral or other agreements between states by any means. [62] Article 1 of the 1966 Helsinki Accords stated that the "general rules of international law as set forth in these chapters are applicable to the use of the waters of an international drainage basin except as may be provided otherwise by

---

[52] /Art. 5 para 1

[53] /Art. 5 para 2

[54] / Art. 16

[55] / MOHAMED SHAWKI ABDEL AEL, EL ANTFA' EL MONSAF BAMAYAH EL ANHAR EL DAWLAYAH: MA' EL ASHARAH ELA NAHR EL NILE, Motada el Kanwan El Dawli, 2010, 17.

[56] /Id 137.

[57] /Id 137.

[58] /Id 137.

[59] /Id 137.

[60] /Id 137.

[61] /Christina M. Carroll, Supra note 2 at 137.

[62] /Kai Wegerich and Oliver Olsson, Later Developers and the Inequality of "equality utilization" and the Harm of "Do no Harm", 35 WATER INT'. 707, 2010, 709.

convention, agreement or binding custom among the basin States.

Article 3 of the United Nations Convention on the Law of the Non-Navigational Uses of International Watercourses, paragraph 1, states that "{I}n the absence of an agreement to the contrary, nothing in the present Convention shall affect the rights or obligations of a watercourse state arising from agreements in force for it on the date on which it became a party to the present Convention."[63]

The rules, which were mentioned in the previous articles, are concealed with the general rules in the international law. However, any inequitable agreements, from the perspective of any party, will not be affected by the international water law rules. In the Nile case, these provisions will not affect the agreement of 1929 of between Egypt and Great Britain. Accordingly, Egyptian scholars argue that "Nile basin states had no legal ground to ask to modify any of the Nile River agreements or conventions."

ii. Counter Argument of Historical Rights

On the other hand, other riparian states consider the Egyptian historical rights as a naïve excuse to get the lion's share of the Nile Water.[64] They respond to such an argument as it is considered prejudice to their water rights.[65] For Ethiopia, the counter argument was based on its position against the 1902 Convention on the one hand, and other conventions and notes on the other. For the 1902 Convention between Ethiopia and Great Britain, Ethiopia's position can be summarized in three points. First, the Convention of 1902 between Great Britain and Ethiopia was never ratified. Second, all the previous conventions did not mention the Ethiopian share in the Nile water. Hence, these conventions are not mandatory to Ethiopia. Third, the British Declaration of adding the Ethiopian territory to the Italian colony cancelled all the conventions and agreements between Ethiopia and Great Britain.[66]

Besides, scholars advocating the perspective of the lower riparian states have developed a counter argument against the rest of the Note and conventions. For the 1925 and 1929 Exchange notes between Egypt and Great Britain, they argued that none of the Nile basin States were a member. Egypt only signed this Note with the colonist. In addition to the previous argument, they added to the 1959 Convention between Egypt and Sudan another a concrete counter argument. They argued that Egypt and Sudan did not have the right to distribute the Nile water share without referring to other riparian states.

iii. Newly Independent States Unilateral Declaration

The case of the unilateral declaration made by the Newly Independent state was mentioned in Geneva Convention on Succession of States in respect of Treaties. The case that was mentioned in Article 9 only tackles specific case. It deals with affirmative action of newly independent state to accept the provisions of a agreement or convention, but the case of rejecting such an agreement or convention is remain unregulated.

The first paragraph of Article 9 tackles the case of a unilateral declaration made by the successor state, providing the continuity of a treaty or a convention in favor of its territory. It stated that "Obligations or rights under treaties in force in respect of a territory at the date of a succession of States do not become the obligations or rights of the successor State or of other States Parties to those treaties by reason only of the fact that the successor State has made a unilateral declaration providing for the continuance in force of the treaties in respect of its territory." [67]   The second paragraph of Article 9 it consequence of the previous act, it stated "the effects of the succession of States on treaties which, at the date of that succession of States, were in force in respect of the territory in question are governed by the present Convention."[68] This article was reflected in the ICJ judgment in the *Nuclear Test Case*. The ICJ dealt with the unilateral declaration from the French Republic to not participate in any future atmospheric nuclear tests in the South Pacific area.[69] The court stated that "{i}t is well recognized that declarations made by way of unilateral acts, concerning legal or factual situations, may have the effect of creating legal obligations." [71] Then the court further declared that "When it is the intention of the State making the declaration that it should become bound according to its terms, that intention confers on the declaration the character of a legal undertaking, the Statebeing thenceforth legally required to follow a course of conduct consistent with the declaration."

iv. Tanzanian Argument

The Egyptian government claims that the 1929 Agreement is not only binding on Egypt, but also on Sudan, Uganda, Tanzania, and Kenya, on whose behalf the British signed the 1929 Agreement. However, these states are forced to abide by the Nyerere doctrine for state succession. This doctrine is considered as unique theory in the field of state succession.[70] The two years grace period honored all treaty before its termination.[71]

In 1962, the government of Tanzania sent the governments of Great Britain, Egypt, Kenya, Sudan and Uganda a memorandum regarding the utilization of the River Nile water. Mr. Nyerere sent his statement in the form of an exchange note to the Nile Basin States. Many of

/63 /Art. 3 of UN Convention, 1997
/64 /OkothOwiro, The Nile Treaty, State Succession and international Treaty Commitments: A Case Study of the Nile Water Treaty, http://www.kas.de/wf/doc/kas_6306-544-1-30.pdflast visit 1/4/2012.
/65 /Id
/66 /Supra note 2at 543.

/67 /Art. 9 Vienna Convention
/68 /Art. 9 Vienna Convention
/ 69 /Alfred Rubin, The International Legal Effects of Unilateral Declarations, 71 Am. J. Int'l L. 1 1977, 2. 71 Nuclear Test (Astralia v. France), Judgement, ICJ, 20 December, 1974, 267.
/70 /Supra note Ошибка! Закладканеопределена. at 181 184
/71 /Problems of State Succession In Africa: Statement of the Prime Minister of Tanganyika, 11 INT'L & COMP. L.Q. 1169 1962, 1211.

the states remained silent towards the content of the Tanzanian memorandum. While Egypt responded, Kenya, Uganda, and Sudan remained silent.

In 1963, the Egyptian government's response was very simple. It did not argue the legality of the declaration; however, it stated that the provision of Exchange of Notes between Her Majesty's Government in the United Kingdom and the Egyptian Government on the use of waters of the Nile for Irrigation would continue to exist until a new convention is drafted.

3.2. Sovereignty versus Cooperation

3.2.1. Conflict between Sovereignty and Cooperation

a. Absolute Sovereignty

In 1898, the Attorney General of the United States declared in his advisory opinion that "the rules, principles, and precedents of international law impose no duty or obligation upon the United States of denying to its inhabitants the use of the water of that part of the Rio Grande lying entirely within the United States, although such use results in reducing the volume of water in the river below the point where it ceases to be entirely within the United States."[72] These words were, according to most of international water legal scholars, the first pillar for Absolute Territorial Sovereignty.[73] It was named after the American Attorney General Judson Harmon. He denied the riparian states' rights over watercourse to allow the flow of water through its territory to other states. Harmon stated that the state department and the United States held no responsibility "for the substantial reduction in Rio Grande water available to Mexico."[76]

The previous theory about absolute territorial sovereignty was taken as a base to allow the "upstream states complete freedom of action with regard to international watercourses within its territory, irrespective of any consequence that might ensue in other countries."[74] Besides, Ethiopian government adopts absolute territorial sovereignty theory. The Ministry of Foreign Affairs in 1978 issued serious of statements, in which it asserts and reserves "all the rights to exploit her natural resources."[75] Harmon Doctrine had become "a potent weapon in the hands of downstream states accusing an upstream state of acting unreasonable."[76]

The theory of absolute territorial integrity is for the sake of lower riparian states. As Stephan McCafferey stated "{w}hile the doctrine of absolute territorial sovereignty

insists upon the complete freedom of action of the upstream state, that of absolute territorial integrity maintains the opposite: that the upstream state may do nothing that might affect the natural flow of the water in the downstream state."[77]

It has been argued that this theory was never adopted in any diplomatic settlement, convention or court decision.[78] However, in the Nile case, the lower riparian states, especially Egypt, asserted their legal and historical rights to have a veto power over the utilization of the water of the Nile. This is based on the right of the lower riparian states to claim the right of continued, uninterrupted flow of the water to its territory from the upper riparian states. This theory gives a right to the lower riparian states to the water of the river.[79]

This theory was criticized from various reasons. First, it ignores the equal territorial sovereignty of the state.[80] Stephan McCafferey described both theories as "factually myopic and legally anarchic." McCafferey maintained that both theories "ignore other states' need for and reliance on the waters of an international watercourse, and they deny that sovereignty entails duties as well as rights. As freshwater became increasingly precious and nations of the world ever more dependent, both doctrines became increasingly less relevant and defensible."[81]

Second, different courts and tribunals have declined this theory, as they considered it a prejudice against other states' rights.[82] In *Trail Smelter Case,* a claim of water and air pollution was held against Canada from the United States. The court held Canada responsible "for extraterritorial injury existed as a matter of general international law."[83]

Third, in these two theories, harm is inevitable to either the upstream or the downstream states. The international law principles oblige states not to cause any harm to other states.[84] Both theories violate the general legal rule that "one should use his property in such a manner as not to injure that of another,"[85] or *sic uteretuoutalienumnon laedas.* The harm in these theories could mean a change in

---

[72] /U.S. Attorney General Harmon, 21 OP. ATT'Y GEN. 274, 281-282 (1898), 274

[73] /Donald J. Chenevert, Application of the Draft Articles The Non-Navigational Uses of International Watercourses to the Water Disputes Involving The Nile River and the Jordan River, 6 EMORY INT'L REV. 459, 1992, 502 STEPHEN C. MACAFFREY, THE LAW OF INTERNATIONAL WATERCOURSES, NON-NAVIGATIONAL USES, Oxford Univeristy Press, (2nd ed.) (July 2007), 115 [hereinafter Law of International Watercourse]at 115

[74] /Id at 115

[75] /Id at 274.

[76] /Id at 118

---

[77] /Law of International Watercourse supra note 73 at 128.

[78] /BONAYA GODANA, AFRICA'S SHARED WATER RESOURCES LEGAL AND INSTITUTIONAL ASPECTS OF THE NILE, NIGER, AND SENEGAL RIVER SYSTEM, 39, (1985)

[79] /Donald J. Chenevert, Supra note 160 at 502, [83] Donald J. Chenevert, Supra note 160 at 504.

[80] /Donald J. Chenevert, Supra note 160 at 504.

[81] /Law of International Watercourse supra note 73 at 128.

[82] /Margaret J. Vick, International Water Law and Sovereignty: A Discussion of the ILC Draft Articles on the Law of Transboundary Aquifers, 21 PAC. MCGEORGE GLOBAL BUS. & DEV. L.J. 191 2008, 215.

[83] /Trail Smelter Arbitral Decision (United States v. Canada), 33 A.J.I.L. 182 (1939); 3 Int. Arb. Awards 1905, 1963 (1949). Mentioned in A.P. Lester, River Pollution in International Law, 57 AM. J. INT'L L 828, 1963, 836.

[84] /Salman M. Salman, The Helsinki Rules, the UN Watercourses Convention and the Berlin Rules: Perspectives on International Water Law, 23 WATER RES. DEV. 625 (2007), 627

[85] /It is the general rule in national and international law, article 5 of the Egyptian civil law asserted such right.

the natural flow of the basin, which could affect the downstream states, or prevent the development of the international watercourse, which could also affect the upstream states.[86] This has happened in the Nile case. While Ethiopia builds a dam and starts its way of development, it will decrease the amount of water allocated to Egypt. Conversely, when Egypt maintains its share of the Nile water, it will handicap possibilities of development of Ethiopia.

b. *Limited Territorial Sovereignty*

Salman argued that the Limited Territorial Sovereignty principle ensures the equality of all riparian states in the use of the international river.[87]McCafferey reluctantly admitted that it is the dominant theory in the field of international water law in determining rights and obligations. (McCafferey:137) The principle of limited territorial sovereignty is based on the fact that: "all riparian states have the right to fully utilize the water of an international river. Besides, states are obliged to ensure that any use will not cause any significant harm to other riparian states. McCafferey described the theory as "{t}he freedom to swing one's fist ends where the other person's nose begins." (McCafferey:137)

The doctrine of Limited Territorial Sovereignty was strongly supported in many cases. The International Court of Justice case*Gabcikovo- Nagymaros* gave considerable weight to the principal of equitable and reasonable utilization of international watercourse. It stated that "Czechoslovakia, by unilaterally assuming control of a shared resource, and thereby depriving Hungary of its right to an equitable and reasonable share of the natural resources of the Danube ... failed to respect the proportionality which is required by international law."[88] In the *Corfu Channel* case, the International Court of Justice maintained that, "it is illegal for states to use or permit the use of their territories for acts that would constitute harm to persons or to the environment in other countries."[89]

In *Lake Lanoux Arbitration,* France declared that it would consider Spainish interests in the flow of the water to its territory unaffected by its hydroelectric project. Later on, France modified the amount of water used in the project, which Spain refused to accept. The tribunal answered the following question of whether or not the French act was a violation of the governing treaty and its protocol, which is the Treaty of Bayonne of 1866. The court concluded that "in the general accepted principles of international law, a rule which forbids a State, acting to protect its legitimate interests, from placing itself in a situation which enables it

in fact, in violation to its international obligations, to do even serious injury to a neighboring State."

Even with the wide acceptance of the principles of equitable utilization and no harm, major criticism to this doctrine is built on the wide disagreement of the essence of both principles. The detailed relationship between the two principles is complex and challenging.[90] The international failure to reach an agreed text of both principles has deprived the limited territorial sovereignty from its content. As the criticism is directed to the application of the theory in the international water law principles, I shall refer to the next subsection, which deals with these principles.

c. *Community Theory*

Community Theory is based on the assumption that "the entire river basin is an economic unit, and the rights over the waters of the entire river are vested in the collective body of the riparian states, or divided among them either by agreement or on the basis of proportionality."[91] Even though this theory sounds new, its origins go back to Roman law. (McCafferey:149) Many philosophers wrote about the notion that "water is something to be treated as common property," Grotius wrote: "a river ... is the property of the people through whose territory it flows, ... the same river viewed as a running water, has remained common property, so that any one may drink or drain water from it."[92]

Community Theory looks for maximum cooperation among states as a must on one hand; while on the other it overlooks the sovereignty principle.[93] The difference between the Community theory and the Limited Territorial sovereignty theory is that the first theory goes beyond the second, through increasing the rights of the collective body of the river concerned.[94]

The idea of the Community theory was presented in the *Territorial Jurisdiction of the International Commission of the River Oder.* Even though this case was mainly about navigational uses, it is worth being presented for the concept of non-navigational uses. If this theory were applicable navigational uses, it would be also appropriate to present it. In the *Commission of River Oder Case,* the permanent Court of International Justice in its decision in 1929 answered the question regarding the jurisdictions of the Oder Commission under Versailles Treaty, within the Polish territory to include also the Warta and Notze Rivers. The court found that Commission jurisdiction was entitled to both rivers.

/[86]/ *Law of International Watercourse supra note 73 at 136.*

/[87]/ *Law of International Watercourse supra note 73 at 627*

/[88]/ /*(Gabcikovo – Nagymaros) Project (Hungary/ Slovakia), Judgment, ICJ, 25 September 1997, 56*

/[89]/ *Corfu Channel Case ICJ, mentioned in Valentina OkaruBisant, Institutional and legal Frameworks for Preventing and Resolving Disputes Concerning the Development and Mangement of Africa's Shared River Basins, 9 COLO. J. INT'L ENVTL. L. & POL'Y 331 1998, 352.*

/[90] / *Salman M. Salman, The Helsinki Rules, the UN Watercourses Convention and the Berlin Rules: Perspectives on International Water Law, 23 WATER RES. DEV. 625 (2007) at 628*

/[91] /Id at 627*

/[92]/ *Hugo Grotius, On the Law of War and Peace, Chapter II the General Rights of Things, Mentioned in McCafferey:150*

/[93] /Id at 627*

/[94] / *BONAYA GODANA, AFRICA'S SHARED WATER RESOURCES LEGAL AND INSTITUTIONAL ASPECTS OF THE NILE, NIGER, AND SENEGAL RIVER SYSTEM, (1985) 137*

4. Conflict between Sovereignty and Cooperation in IWL Principles

4.1. Principle of Equitable and Reasonable Utilization and Participation

Articles four to eight in the second chapter of the 1966 Helsinki Rules regulated the principle of equitable and reasonable utilization and participation. It holds the basin states responsible for "a reasonable and equitable share in the beneficial uses of the waters of an international drainage basin."[95] Article five; paragraph one defined the principle of Equitable and Reasonable Utilization and Participation, as "it shall be determined in the light of all the relevant factors in each particular case."[96]

Article five, paragraph two stated that factors are considered in determining the reasonable and equitable share. These factors include but are not limited to geography, hydrology, climate affecting the basin, past utilization of the waters of the basin, and the economic, social, and population needs of each basin state. There is also the comparative costs of satisfying various needs, availability of other resources, avoidance of unnecessary waste in the utilization of waters of the basin, and practicability of compensation to one or more of the co-basin states as a means of adjusting conflicts among uses.[97] On the other hand, the third paragraph of article six did not give any superiority to any of the previous factors over the other.[98]

Fairly similar to what Helsinki rules stated in Article V, the principle was mentioned in article 5 of the UN Convention.[99] The International Law Commission tried to solve the problems that resulted from the conflict between the two principles of sovereignty and international cooperation. Article five introduced the concept of the 'equitable participation', the main reason for which was to affirm that a system of equitable and reasonable utilization and participation cannot be achieved solely through one state.[100]

Article 6 stated the factors that affect the equitable and reasonable utilization and participation of the international basin.[101] This article increased the scope of the application of the equitable and reasonable utilization and participation of the international watercourse. These factors include (a) Geographic, hydrographic, hydrological, climatic, ecological and other factors of a natural character; (b) Social and economic needs of the watercourse states concerned; (c) Populations dependent on the watercourse in each state; (d) Effects of the use or uses of the watercourses in one watercourse state on other watercourse states; (e) Existing and potential uses of the watercourse; (f) Conservation, protection, development and economy of use of the water resources of the watercourse and the costs of measures taken to that effect; and (g) The availability of alternatives, of comparable value, to a particular planned or existing use.[102]

4.2. Principle of Obligation Not to Cause Significant Harm (Sic uteretuoutalienumnon laedas)

One can argue that the Helsinki Rules of 1966 did not identify explicitly in their provisions an independent principle of obligation not to cause significant harm. It was only mentioned as part of the principle of equitable utilization and participation. However, the principle of *no harm* is an old and well-recognized principle in international law. In 1948, the International Court of Justice mentioned the *no harm* principle in the *Corfu Channel* case. Even though this case does not deal with the international watercourse or environmental damage, many scholars of international environmental law use it as an example of legal analysis.[103] In this case, the ICJ maintained "every state's obligation is not to allow knowingly its territory to be used for acts contrary to the rights of other States."[104]

The *no harm* principle is the most debatable in international water law. It is connected to articles 5 and 6, which were adopted during the negotiation process by a vote of 38 to 4, with 22 abstentions.[105] The UN Convention on the Law of the Non Navigational Uses of International Watercourses significantly added the principle of obligation not to cause significant harm to its provisions as an independent principle. Article 7, paragraph one stated that: "{w}atercoursestates shall, in utilizing an international watercourse in their territories, take all appropriate measures to prevent cause of significant harm to other watercourse states."[106]

The second paragraph made an important connection between the *no harm* principle and that of equitable utilization principle. These two principles are complementary. This claim is built on two bases: Firstly, McCafferey argued that the significant harm must be, in some cases, tolerated by harmed states. In many cases, the insignificant harm aims to achieve the overall regime of equitable utilization of the international watercourse.[107] Secondly, the determination of compensation shall be in light of two factors. In the case of a significant harm affecting a certain state, the negotiation to remedy such harm shall be based on the balance between the two

---

[95]/ Art.4, Helsinki Rules, 1966,

[96]/ Art.5 Helsinki Rules, 1966,

[97]/ Art. 6/2 of the UN Convention, 1997.

[98]/Art. 6/3 Id

[99]/ Supra note 175 at 631-632.

[100100]/ Stephan McCaffrey, An Overview of the UN Convention on the Law of the Non- Navigational Uses of International Watercourses, 20 J. LAND RESOURCES & ENVTL. L, 57, 2000, 62

[101]/ Art. 6 of UN Convention, 1997

---

[102]/ Art. 6 of UN Convention, 1997

[103]/ Law of International Watercourse supra note 73 185

[104]/ Corfu Channel Case (United Kingdom and Northern Ireland V. Republic of Albania) I.C.J.1949 of (April 9), 22, Judgement.

[105]/ Law of International Watercourse Supra note 73 at, 62

[106]/Art. 7/1 UN Convention

[107]/Law of International Watercourse supra note 73 at 370

principles mentioned in articles 5, 6 and 7.[108]

The *no harm* and equitable utilization principles went side by side in more than five places in the UN Convention. Firstly, they were mentioned in the second paragraph of article 10 (relationship between different kinds of uses). Secondly, article 15 dealt with the reply of notification. Thirdly, article 16 tackled the absence of reply to notification. Fourthly, article 17 dealt with consultations and negotiations concerning planned measures. Fifthly, article 19 regulated the urgent implementation of planned measures), all these articles referred to article 5 (Principle of Equitable and reasonable utilization and participation), and article 7 (Principle of obligation not to cause significant harm) as one unit.

Helsinki Rules addressed the *no harm* obligation through the factors for determining the reasonable and equitable utilization. The UN Convention followed the same approach of Helsinki Rules. It separated the *no harm* principle in one article titled "principle of obligation not to cause significant harm" from the equitable utilization principle. The commentary of Article 12 stipulated that the change in the formulation was to "resolve the most debatable issues in the drafting of the UN Convention: the relationship between the principle of equitable utilization and the obligation not to harm another basin state (Article 16)."[109] The current text reflects the right to an equitable and reasonable share of the water of an international drainage basin, in addition to compliance with the equitable and reasonable utilization with the obligation not to cause significant harm to another basin state.[110]

Article 16 dealt with the "Avoidance of Trans-boundary Harm." Article 16 set the states' obligation to "refrain from, and prevent acts or omissions within their territory that cause significant harm to another basin state having due regard for the right of each basin state to make equitable and reasonable use of the waters." This article is just a reflection of the legal rule to "do not use your property so as to injure the property of another."[111] The Commentary of Article 16 refers to the debates regarding the *no harm* principal, and state liability of harm caused from its actions. The commentary looked at the principles as part of the customary law, withdoubt. It stated that "{d}espite the considerable controversy over the application of the "no harm" rule and its relation to the rule of equitable use found in art. 5 of the UN Convention, there actually is little controversy over whether the principle expressed in art. 7 is (sic) part of customary international law."[112]

4.3. Principle of General Obligation to Cooperate

International scholars consider the cooperation principle as an "umbrella term" rather than a strictly legal duty.[113] Any international river can be a source of good relation and cooperation on one hand; while a source of tension and conflict on the other.[114] Tension could result from the use of sovereign states of the international watercourse. In order to preserve the utility of the international watercourse, states shall participate in a cooperative framework.[115] On the other hand, the general principle of international duty to cooperate among states is just a general obligation, as there are no prescribed or specific obligations.[115] There is struggle between the general principle of international duty to cooperate among states - as an international necessity to preserve the existence of the international society- and the principle of sovereignty. States always need to cooperate, to preserve their existence, while reserving their right of sovereignty. The authority of the state ends at a designated point on land, as well as in the water.[116]

It may be argued that the Helsinki Rules of 1966 and their supplements contained many provisions that encourage states to cooperate in the allocation, management, and preservation of internationally shared waters.[117] Nevertheless, the principle of general obligation to cooperate was first introduced as a separate principle in the UN Convention, as article 8 held a general obligation on all riparian states to cooperate in order to reach the maximum benefit of the Basin. The general obligation of cooperation was based on four factors: sovereign equality, territorial integrity, mutual benefit and good faith.[118] The good faith factor was not introduced in early negotiations of the UN Convention.[119] All the three factors are attached to the sovereign state. One of the major contributions of the special rapporteur Mr. Stephan McCaferrey, was introducing the 'good faith' factor,[120] which is currently embedded in many international cases. The *North Sea Continental Shelf* cases maintained that there is an international obligation on states to resolve their delimitation through justice and good-faith..[121] This obligation mandates reaching a satisfactory result without any prejudice against sovereign states.[122]

Unlike the UN Convention, Article 11 limited the

---

[108] /Stephan C. McCaffrey, *Introduction, Convention on the Law of the Non-Navigational Uses of International Watercourses*, United Nation Audiovisual Library of International Law, untreaty.un.org/cod/avl/ha/clnuiw/clnuiw.htmllast visit 10/3/2012.

[109]/ International Law Association, Berlin Conference, Water Resources Committee, 71 INT'L L. ASS'N REP. CONF. 334 2004, 362

[110]/ Id at 362

[111]/ Id at 362

[112]/ Id at 363

[113]/ Id at 361

[114]/ see *Beyond the River, supra* note 40 at 389 [115] Id at 391

[115]/ *Supra* note 107 at 361

[116]/*Preliminary Report on the Law of the Non-Navigational Uses of International Watercourses, Law of the non-navigational uses of International watercourses,* ¶ U.N.Doc. A/CN.4/393 (July 5, 1985) (prepared by Stephen McCaffrey)

[117]/ *Supra* note at 107 at 361

[118]/ Art. 8 Id

[119]/ *First report on the law of the non-navigational uses of international watercourses, Law of the non-navigational uses of International watercourses,* ¶ U.N.Doc. A/CN.4/367 and Corr.1 (April 19, 1983) (prepared by J. Evensen), 174/108

[120]/ Stephen McCaffrey, *sixth report on the law of the non-navigational uses of international water courses, Special Rapporteur*

[121]/ *North Sea Continental Shelf (Federal Republic of Germany v. Denmark),* 1986, I.C.J. 46/47, Judgement.

[122]/ Id

cooperation framework to only one factor, 'good faith.'[123] It stated that "{b}asin states shall cooperate in good faith in the management of waters of an international drainage basin for the mutual benefit of the participating states."[124]

5. Conclusion

After the analysis of the legal and institutional frameworks of the Nile Basin, it is hard to rely on such frameworks fora working plan. The future convention or even the current agreement should be founded on the basis of needs, identified and expressed by the various states. Egypt has to fully understand that unilateral action will not be efficient, and that Egypt is not the sole decision maker within the basin states, if it wishes to consume the same amount of the Nile share. The problem of the Nile will only be solved through unanimous agreement to negotiate and reach an understanding. Any other suggested solution, other than the previouslystated, will cost Egypt a tremendous amount of money, time and effort. Despite the fact that the conflict looks legal atface value, it is in fact a conflict of interest. Additionally, if Basin countries had really intended to solvethe problem, they would have relentedand sought international courts and tribunals decades ago. Finally the thesis proposed a simple solution to the problem, which was proposed by the various parties of the problem many times in the past.

References

[1] Preamble of the Treaty Between Ethiopia and Great Britain on the Delimitation of the Frontier between Ethiopia and Sudan, United Nations, Legislative Texts and Treaty Provisions Concerning the Utilization of International Rivers for Other Purposes than Navigation, United Nations Legislative Series (ST/LEG/SER.B/12), United Nations publication, 115,116

[2] Exchange of Notes Between the United Kingdom and Italy Respecting Concessions for a Barrage at Lake Tsana and a Railway Across Abyssinia From Eritrea To Italian Somaliland, Signed at Rome 14 and 20 December 1925,

[3] International Law Association, Berlin Conference, Water Resources Committee, 71 INT'L L. ASS'N REP. CONF. 334 2004, 362

[4] Preliminary Report on the Law of the Non-Navigational Uses of International Watercourses, Law of the non-navigational uses of International watercourses, ¶ U.N.Doc. A/CN.4/393 (July 5, 1985) (prepared by Stephen McCaffrey)

[5] First report on the law of the non-navigational uses of international watercourses, Law of the non-navigational uses of International watercourses, ¶ U.N.Doc. A/CN.4/367 and Corr.1 (April 19, 1983) (prepared by J. Evensen), 174/108

[6] North Sea Continental Shelf (Federal Republic of Germany v. Denmark), 1986, I.C.J. 46/47, Judgement

[7] Corfu Channel Case (United Kingdom and Northern Ireland V. Republic of Albania) I.C.J.1949 of (April 9), 22, Judgement

[8] (Gabcikovo – Nagymaros) Project (Hungary/ Slovakia), Judgment, ICJ, 25 September 1997

[9] Trail Smelter Arbitral Decision (United States v. Canada), 33 A.J.I.L. 182 (1939); 3 Int. Arb. Awards 1905, 1963 (1949).

[10] U.S. Attorney General Harmon, 21 OP. ATT'Y GEN. 274, 281-282 (1898), 274

[11] Problems of State Succession In Africa: Statement of the Prime Minister of Tanganyika, 11 INT'L & COMP. L.Q. 1169 1962, 1211.

[12] TesfayeTafesse, The Hydropolitical Assessment of the Nile Question: An Ethiopian Perspective, 26 WATER INT'L 1, 2001,

[13] JuttaBrunnee and Stephen Troope, The Changing Nile Basin Regime: Does Law Matter?, 43 HARV. INT'L L.REV. 105, 106.

[14] Sadat to Ethiopia: Leave Nile alone or it's war, The Gazette Montréal , Saturday June 7, 1980,http://news.google.com/newspapers?nid=1946&dat=1 9800607&id=IYkxAAAAIBAJ&sjid=caQFAAAAIBAJ&pg= 1030,2287901 Last visit 24/12/2011

[15] TakeleSobokaBulto, Between Ambivalence and Necessity: Occlusions on the Path Towards a Basin Wide Treaty in the Nile Basin, 20 COLO. J. INT'L ENVTL. L. & POL'Y 291 2008-2009, 318.[hereinafter Ambivalence and Necessity] See also,

[16] Christina M. Carroll, Past and Future Legal Framework of the Nile River Basin, 12 GEO. INT'L ENVTL. L. REV. 269, 199-2000, 282.

[17] Valerie Knobelsdorf, Note: The Nile Water Agreements: Imposition and Impacts of a Transboundary Legal System, 44 COLUM J. TRANSNATL. L. 634. 635

[18] DerejeZelekeMekonnen, The Nile Basin Cooperative Framework Agreement Negotiations and the Adoption of a Water Security Paradigm: Flight Into Obscurity or a Logical Cul-de-sac? 21EUR. J. INT'L.2, (2010), [hereinafter The Nile Basin Cooperative Framework Agreement Negotiations]

[19] DerejeZelekeMekonnen, Between the Scylla of water security and Charybdis of Benefit Sharing: The Nile Basin Cooperative Framework Agreement- Failed or Just teetering on the Brink?,GO. J. INT' L. 3 (2011),

[20] TakeleSobokaBulto, Between Ambivalence and Necessity: Occlusions on the Path Toward A Basin – Wide Treaty in the Nile Basin, 20 COLO. J. INT'L ENVIRL. L. & POL'Y 291, (2008-2009)

[21] JuttaBrunnee. AZIZA MANSUR FAHMI, WATER MANAGEMENT IN THE NILE BASIN: OPPORTUNITIES AND CONSTRAINTS, http://www.isgi.cnr.it/stat/pubblicazioni/sustainable/133.pdf last visit 11/10/2011.

[22] Aaron Schwachach, The United Nation Convention on the Law of Non- Navigational uses of International watercourses, Customary International Law and interest of upper riparian states, 33 TEX. INT' L. J. 257 (1998), 270.

---

[123] Art. 11 Supra note 195
[124] Art. 11

[23] Adams Oloo, The Quest for Cooperation in the Nile Water Conflicts: the Case of Eritrea, 11 AFR. SOC. REV. 95, 2007, 96.

[24] OkothOwiro, The Nile Treaty, State Succession and international Treaty Commitments: A case Study of the Nile Water Treaty, http://www.kas.de/wf/doc/kas_6306-544-1-30.pdflast visit 1/4/2012.

[25] YunanLabibRizk, Adiwan of Contemporary Life, Al Ahram, http://weekly.ahram.org.eg/2000/503/chrncls.htmlast visit, 1/4/2012.

[26] P. P. HOWELL AND J. A. ALLAN, THE NILE: SHARING A SCARCE RESOURCES; A HISTORICAL AND TECHNICAL REVIEW OF WATER MANAGEMENT AND OF ECONOMICAL AND LEGAL ISSUES, Cambridge University Press, (1st ed.), (1994), 538.

[27] Econ. & Soc. Commission For Western Asia, Assessment of Legal Aspects of the Management of Shared Water Resources in the ESCWA Region, ¶U.N. Doc. E/ESCW A/ENR/2001/3, (Feb. 22, 2001), 14.

[28] Agreement between the United Arab Republic and the Republic of Sudan for Full Utilization of the Nile Waters, see United Nation Publication supra note 12 at 146.

[29] Salman M. Salman,The New State of South Soudan and the Hydir- Politics of the Nile Basin, 36WATER INT'L154, 159. {hereinafter The New State of South Soudan}.37

[30] Econ. & Soc. Commission For Western Asia, Assessment of Legal Aspects of the Management of Shared Water Resources in the ESCWA Region, ¶U.N. Doc. E/ESCW A/ENR/2001/3, (Feb. 22, 2001), 14.18.

[31] Yosef Yacob, From UNDUGU to the Nile Basin Initiative, An Ending Exercise in Futility, Ethiopia TECOLAHACOS, http://www.tecolahagos.com/undugu.htm last visit 21 May 2012.

[32] YacobArsano, Ethiopia and the Nile: Dilemmas of National and Regional Hydro politics, (2007), (Ph.D. dissertation, University of Zurich) (on file with author)

[33] Claudia Sadoff and David Grey, Beyond the River: The Benefits of Cooperation on International Rivers, 4 WATER POL' 389, 2002, 401,

[34] Nile Basin Initiative http://nilebasin.org/newsite/index.php?option=com_content &view=article&id=71%3Aabout-the-nbi&catid=34%3Anbi-background-facts&Itemid=74&lang=enlast visit 10/3/2012 , .

[35] Samuel Luzi, Mohamed Abdel, MoghnyHamouda, FranziskaSigrist and EvelyneTauchnits, Water Policy Networks in Egypt and Ethiopia, 17 J. ENV.& DEV. 238, 2008, 239.

[36] Abadier M. Ibrahim, The Nile Basin Cooperative Framework Agreement: The Beginning of the End of Egyptian Hydro-Political Hegemony,18 MO. ENVTL. L. & POL'Y REV 284, 308

[37] Egypt and its Historical Rights in Nile Water, Egypt State Information Service, http://www.sis.gov.eg/En/LastPage.aspx?Category_ID=1144 last visit 30/10/2012.

[38] /¹/ MOHAMED SHAWKI ABDEL AEL, EL ANTFA' EL MONSAF BAMAYAH EL ANHAR EL DAWLAYAH: MA' EL ASHARAH ELA NAHR EL NILE, Motada el Kanwan El Dawli, 2010, 17.

[39] Kai Wegerich and Oliver Olsson, Later Developers and the Inequality of "equality utilization" and the Harm of "Do no Harm", 35 WATER INT'. 707, 2010, 709.

[40] OkothOwiro, The Nile Treaty, State Succession and international Treaty Commitments: A Case Study of the Nile Water Treaty, http://www.kas.de/wf/doc/kas_6306-544-1-30.pdflast visit 1/4/2012.

[41] Alfred Rubin, The International Legal Effects of Unilateral Declarations, 71 Am. J. Int'l L. 1 1977, 2. ⁷¹ Nuclear Test (Astralia v. France), Judgement, ICJ, 20 December, 1974, 267.

[42] Donald J. Chenevert, Application of the Draft Articles The Non- Navigational Uses of International Watercourses to the Water Disputes Involving The Nile River and the Jordan River, 6 EMORY INT'L REV. 459, 1992, 502

[43] STEPHEN C. MACAFFREY, THE LAW OF INTERNATIONAL WATERCOURSES, NON-NAVIGATIONAL USES, Oxford Univeristy Press, (2nded.) (July 2007), 115

[44] BONAYA GODANA, AFRICA'S SHARED WATER RESOURCES LEGAL AND INSTITUTIONAL ASPECTS OF THE NILE, NIGER, AND

[45] Margaret J. Vick, International Water Law and Sovereignty: A Discussion of the ILC Draft Articles on the Law of Transboundary Aquifers, 21 PAC. MCGEORGE GLOBAL BUS. & DEV. L.J. 191 2008, 215.

[46] A.P. Lester, River Pollution in International Law, 57 AM. J. INT'L L 828, 1963, 836.

[47] Salman M. Salman, The Helsinki Rules, the UN Watercourses Convention and the Berlin Rules: Perspectives on International Water Law, 23 WATER RES. DEV. 625 (2007), 627

[48] Valentina OkaruBisant, Institutional and legal Frameworks for Preventing and Resolving Disputes Concerning the Development and Mangement of Africa's Shared River Basins, 9 COLO. J. INT'L ENVTL. L. & POL'Y 331 1998, 352.

[49] Salman M. Salman, The Helsinki Rules, the UN Watercourses Convention and the Berlin Rules: Perspectives on International Water Law, 23 WATER RES. DEV. 625 (2007) at 628

[50] Hugo Grotius, On the Law of War and Peace, Chapter II the General Rights of Things, Mentioned in BONAYA GODANA, AFRICA'S SHARED WATER RESOURCES LEGAL AND INSTITUTIONAL ASPECTS OF THE NILE, NIGER, AND SENEGAL RIVER SYSTEM, (1985) 137

[51] Stephan McCaffrey, An Overview of the UN Convention on the Law of the Non- Navigational Uses of International Watercourses, 20 J. LAND RESOURCES & ENVTL. L, 57, 2000, 62 .

[52] Stephan C. McCaffrey, Introduction, Convention on the Law of the Non-Navigational Uses of International Watercourses, United Nation Audiovisual Library of International Law, untreaty.un.org/cod/avl/ha/clnuiw/clnuiw.htmllast visit 10/3/2012.

[53] Stephen McCaffrey, sixth report on the law of the non-navigational uses of international water courses, Special Rapporteur

23

Salinity of drinking water and its association with renal failure in Gaza strip, Palestine

Khalid Qahman[1], Eman Abu-afash Mokhamer[2]

[1]Ministry of Environmental Affairs, Palestine
[2]School of Public Health, Alquds University, Palestine

Email address:

kqahman@gmail.com(K. Qahman), emokhamer@yahoo.com(E. Abu-afash Mokhamer)

Abstract: Gaza aquifer is the only natural water source for domestic, agricultural, and industrial purposes in Gaza Strip with a population of about 1.7 million. Current rates of the aquifer abstraction are unsustainable and deterioration of groundwater quality is documented in many parts of the Gaza Strip. The overall aim of this study was to determine salinity of drinking water and its association with renal failure in the southern part in Gaza Strip. Another aim was to explore the relationship between renal failure and socio-economic demographic variables. Descriptive, analytic design was used with survey samples from renal failure patients. A face to face questionnaire for renal failure patients was developed. The sample size for patients was 194subjects, with response rate of 70%. This rate was proportional with respect to its size. Reliability was approved by Cronbach alpha test, and validity was approved by content and face validity method. Analysis of the four quantitative extracted domains that reflected subjectsperception for drinking water salinity level in their localities. All water chemical tests of thesouthern municipal domestic wells have been reviewed since 1987. The tests were fluoride, chloride, nitrate, TDS, and sodium levels in all groundwater wells, which reveal a general trend of increasing from north to south in the southern part. The results show that only 8% of the municipal wells meet the WHO drinking standards in chloride level. Chloride, nitrate, TDS, fluoride and sodium concentration exceed 2-9 times the WHO standards in 92% of the southern wells. The study findings show that there was no association between renal failure prevalence and chloride level, sodium level, TDS level and nitrate level and showed only association with fluoride level, with which there was strong and positive association. There is an urgent need to modify the mixing process according to fluoride level, and initiate public information and awareness programs.

Keywords: Gaza Aquifer, Groundwater, Drinking Water Quality, Salinity, Renal Failure

1. Introduction

This study examined if there is a relationship between renal failure and drinking water salinity (domestic water) in Khan Yunis Governorate. In the following section the researcher intended to explore the relationship between demographical variables, etiological factors, and ecological indicators that affect the level of salinity, as well as to find the relationship between chronic renal failure prevalence and water salinity level, using a comparative study and some domain express some relations, in order to describe the relationship.

This paper presents the results of the statistical analysis of the data characteristic and distribution of the respondents. Itpresents some statistical tests to explore the relationships between the dependent variables and independent variables, and explores the distribution of the subjects' percentage according to the different variables that may have an effect on the renal function. It also describes and discusses the independent variables, demographical variables, water quality, and medical history variables; the historical data of the water resources and the access to it; and the ecological pattern and main salinity indicators that affect the water quality and may be related to increasing or decreasing the renal failure incidence in the southern part. Then statistical methods have been used to express the association and relation for the effect of the dependent variables on the independent variables, by using the independent t-test and one way ANOVA statistical tests, excel as well as other applications.

2. Study Area

The Khanyounis governorate is one of five governorates of the Gaza Strip. Gaza Strip is located in an arid area with scarce water resources. It is a part of the Palestinian coastal plain in the south west of Palestine (Figure 1). Where it forms a long and narrow rectangular area of about 365 km^2, with 45 km length, and between 5 and 12 km width. Nowadays, its five governorates are; Northern, Gaza, Middle, Khanyounis and Rafah. It is located on the south-eastern coast of the Mediterranean Sea, between longitudes 34° 2" and 34° 25" east, and latitudes 31° 16" and 31° 45" north. The Gaza Strip is confined between the Mediterranean Sea in the west, Egypt in the south. Before 1948, it was part of Palestine under the British Mandate. From 1948 to 1967, it was under the Egyptian administration. From 1967 until 1994, the Gaza Strip was under Israel occupation. According to the peace agreement between Israel and the Palestinian, the Gaza Strip has been under the Palestinian Authority control since May, 1994.

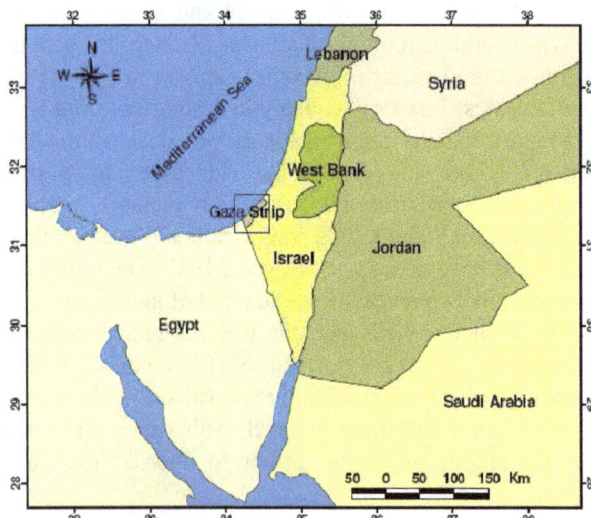

Figure 1: Location of the Gaza Strip (Aish, 2004).

The Gaza coastal aquifer is an important source of water to over 1.7 million residents in Gaza Strip. It is utilized extensively to satisfy agricultural, domestic, and industrial water demands. The extraction of groundwater currently exceeds the aquifer recharge rate. Today, the Gaza Strip is a land under great pressure (Qahman and Larabi, 2006). It is densely populated, with population of more than one million in the year of 1998 and the population increased rapidly up to approximately 1.7 million now in 2013 which means that the environment in Gaza has been under great pressure and as a result most of the people there suffers severely now. In 2006, about 280 thousand inhabitants are living in Khanyounis. The Khanyounis governorate consists of six municipalities: Khanyounis, BaniSuhaila, Abasan El-Kabira, Abasan El-Saghira, Quarrara, Al Fakhari and the Khuza'a.

3. Methods

The study design is cross sectional, descriptive, analytical study. The target group is the total population of renal failure patient in the acute or chronic stage treated or followed by artificial kidney department doctors in Nasser hospital and previously or currently having access to domestic and drinking water from the resources available in Khanyounis Governorate (Municipal, UNRWA, Mekarout, Private wells) and registered in governmental health sector, as renal failure patient in (Nasser hospital). Their total number is 194 subjects conducted to treatment and registered.

Questionnaire data was collected by researcher only with some assistance and co-ordination from the team worker in the artificial kidney department. The patient questionnaire was face interviewed questionnaire. The interview was started by giving the patient explanation about study and its objectives and their importance in giving true answers, a face interview questionnaire was conducted by researcher to the patient, because some of them were illiterate. They were given a complete instruction about the study and how they included in it, there privacy and safety during interview were maintained as the interview was done in the place of work, taking into consideration not to interrupt the work, during the interview. The process of data collection last about 25 full working days.

Patient data for prevalence calculation was obtained and audited by the researcher as there were repletion of cases, the researched apply all the old and new cases on excel sheet and remove the repeated cases.

Drinking water qualitydata was collected by the researcher from many agencies includes (municipalities, coastal water utility in the southern and Gaza region, public health laboratory, and Water Authority).

4. Results and Discussion

4.1. Descriptive Results

4.1.1. Demographical Variables

Regarding gender, the result shows the distribution of gender as male patient represented 58.1%, while female patient represented 41.9% of total sample. The majority of patients weremale; given that male the number is 1.4 times more than female. Characterization of the patient population with chronic kidney disease reveals that the incidence and prevalence rates are universally greater for males than for females. Two thirds of patients in the NAPRTCS CRI registry and in the database of the ItalKid Project are males. This gender distribution reflects the higher incidence of congenital disorders, including obstructive uropathy, renal dysplasia, and prune belly syndrome in males versus females. In fact, in the Italy Kid Project, males continue to predominate even after excluding patients with posterior urethral valves (Bradley A. Warady, 2007). The findings of the study correspond with the studies conducted by the department of nephrology at

Mercy Hospital in Kansas City, MO USA, and reveal that there is a gap between male and female patient percentage, the males being predominate.

4.1.2. Distribution of Patient by Age Group

Regarding patient age, the patients are distributed according to their age and within four age groups.The lowest age group was 0-19 years as it is only 8.8% of the total sample, while the other two age groups of 20-39 and 40-59 years were equal to 30.1%.The fourth age group is from 60+ years and higher, and this group was a little bit higher than the second and the third group with a percent of 30.9%, with (mean 46.01, mode 60, median 47.50). The majority ofpatients are older than 19 years. The study finding corresponds to the study conducted by all three Units of Medicine of Allied and District Headquarters at Allied Hospital Punjab Medical College, (PMC), Faisalabad, a comparative study. The period is January of 1995 to the end ofMay; 1997.The mean age for the study was 58 years.

4.1.3. Distribution of Subjects by Demographical Data

The patients' residences were distributed throughout eight demographic areas within the KhanYounis governorate and the eastern villages in the southern area of Gaza Strip(KhanYounis city, KhanYounis camp, Kuza, Abassan, Banisuhila, Qizan an Najar, Maen, Qarara), according to the demographical localities and according to the study variable when needed.Table (1) shows the distribution of subjects according to locality, the highest percent was being in KhanYunis city as represented with 52.9%. Thethe highest midyear population during 2006 was KhanYunis camp and Baniseihlawith a percent of 11%. KhanYunis camp is 3.7 times BunSuhila by the Mid year population of 2006 Maen, represented with 7.4%, and the lower percentages were found in Kuza, Abassan, Qizan, and Qarara, as each area represented only 4.4% of the sample. The majority of subjects were living in KhanYunis city. The majorities of subjects were married and represent 64.7%. The level of education represented in the table shows that the majority of the subjects' education level is less than Tawjehi (secondary), and more than half of the sample the rest percent distributed between, Tawjehi level with 30%, Diploma 5.9%, Bachelor 4.4%.

Current occupations presented in table (1) show that currently about 89.7% are unemployed due to the political and economical situation in the country, and only 10.3% are currently working.Outof the working percent, agriculture represents 28.5% of the total workers, construction represents 14.2% of the total workers, and all other jobs which include physicans, administrative workers, nurses, managers, teachers, pharmacists, lab technicians, hair dressers, finance professionals, and policemen comprise 57%. Each job represents a percentage of about 7% of the total jobs.Agriculture and construction have the highest and percentages, as well as previous employment presented in the table which show that 63.9% were previously unemployed. Only 36.02 % were previously employed, and

out of the total employed 36.7% were in agriculture, 22.4% were in construction , and 40.8% were in all other jobs that were previously mentioned with percentages not exceeding 5% for each job. The majority of employed subjects either were or still are working in the agriculture sector, and the second largest job category is construction. The study findings correspond with discussed paper by (HenkdeZeeuw, 2000).

Table 1: Distribution of subjects by demographical data

Variable		Frequency	Percent
Residency place	Khan Yunis city	72	52.9%
	KhanYunis Camp	15	11%
	Khuza	6	4.4%
	Abasan	6	4.4%
	Bani Suheila	15	11%
	Qiza an Najjar	6	4.4%
	Maen	10	7.4%
	Al Qarara	6	4.4%
Marital Status	Single	34	25%
	Married	88	64.7%
	Widow	12	8.8%
	Divorce	2	1.5%
Academic certificate	Less than Tawjehi	80	58.8%
	Tawjehi	42	30.9%
	Diploma	8	5.9%
	Bachelor	6	4.4%
Current employment	unemployed	122	89.7%
	employed	14	10.3%
Current job	Agriculture	4	28.5%
	Construction	2	14.2%
	Others	8	57.1%
Previous employment	Unemployed	87	63.9%
	Employed	49	36.02%
Previous job	Agriculture	18	36.7%
	Construction	11	22.4%
	Others	20	40.8%
Family Type	Nuclear	98	72.1%
	Extended	38	27.9%

Regarding the family type of the subjects, 72.1% of the subjects represented nuclear family and 27.9% represented extended. The first type is most common.

4.2. Distribution of Subject by Medical Data (Variables)

4.2.1. Distribution of Subjects by Incidence Age Group

Regarding patient age, the incidence age group for the renal failure patients vary in percent. The patients are distributed within four age groups, and the highest age groups for incidence were the third age group of 40-59 with percent 37.5% of the total sample; the second age group of 20-39 years with a percent of 28.7%; and the third, ages 0-19, with a percent of 20.6%.The lowest is 60+ with percent 13.2%. The majority of renal failure incidence occurs at the age 40-59. Figure (2) shows the distribution of renal failure incidence age.

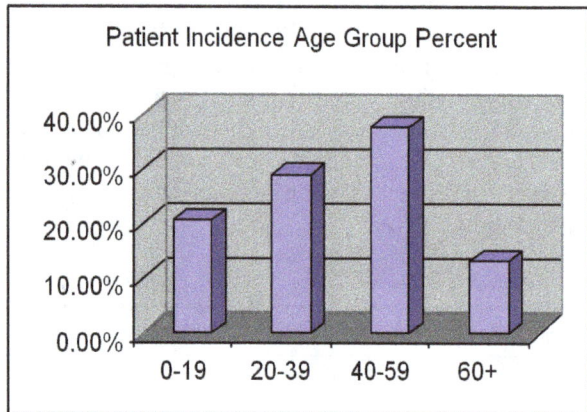

Figure 2: *Distribution of Subjects by incidence age group*

4.2.2. Distribution of Subject by Etiological Variables (Diseases)

The medical history for the subjects is represented in table 2 below. The table shows that only 20.6% of the total subjects do not suffer from chronic disease, but the majority of subject are suffering from chronic diseases as represented by 79.4% of the subjects.The major diseases are diabetes (40.7%), hypertension (33%), congenital disorder (13% - five of the subjects who are suffering from congenital disorder having only one kidney, two of them having right kidney and three having left kidney),heart disease (7.4%), gland disorder (3.7%), and the minority suffering from both diabetes and hypertension and they (3.8%). Two studies (K. Amin and Tufai Muhammad, 2000)and (Arrigo et al, 2000) showed the prevalence of diabetes to be 25%35%respectively.

Table 2: *The medical history for the subjects.*

Variable		Frequency	Percent
Chronic disease	Suffering from chronic disease	108	79.4%
	Don't suffer from chronic disease	28	20.6%
Type of disease	Diabetus	44	40.7%
	Hypertension	35	32.4%
	Hypertension and Diaabetus	3	2.8%
	Heart disease	8	7.4%
	Congenital disorder (five subjects having only one kidney)	14	13%
	Gland disorder	4	3.7%

Mellitus is a quite significant risk factor in the people of our area. The high prevalence of diabetes in patients of chronic renal failure was due to poor glycemic control and lack of knowledge about the hazardous effects of diabetes.

Hypertension is also one of the important risk factors in chronic renal failure. The study corresponds with another study conducted by Allied Hospital Punjab Medical College, (PMC), Faisalabad. 25 patients out of a total 300 patients, i.e. 8.33%, were having hypertensive nephropathy.

Table 3: *The cause of renal failure*

Variables		Frequency	Percent
Cause of renal failure	Glomerulonephritis	58	42.6%
	Renal atrophy	56	41.2%
	Renal stone	22	16.%
Type of stone	Calcium	12	52.2%
	Exalate	11	47.8%

4.2.3. Distribution of Patient by Cause of Renalfailure

The majority of subjects who responded to the study suffer from glomerulonenephritis represented by 42.6% out of the total subjects, and 41.2% of the subjects have renal atrophy. The rest (16.2%) developed renal stone, while 52.2% developed calcium stones.The others (47.8%) developed exalate type of stone.Table 3 represented the cause of renal failure.

4.2.4. Distribution of Subjects by Severity of Disease

As presented in table 4, all the subjects in the study were suffering from chronic renal failure and 59.6% are still severe but don't need to conduct hem dialysis.The rest of the subjects (40.4%) are at the highest level of severity and conduct hemi dialysis.The majority of these subjects (52.7%) need 2 sessions per week and around 3-4 hours per session.These subjects stay around seven hours per week in the department of artificial kidney in the hospital for treatment.The second highest group represented (40%) needs three sessions per week with average treatment 10.5 hours per week.The third group represented (5.5%) need one session per week with average 3.5 hours.The fourth group represented (1.8%) need four sessions per week, with an average of 14 hours per week. Seven subjects (5.2% of the total subjects) previously conducted renal transplantation; five out of seven with (3.7% of the total subjects) conducted transplantation once, and two subjects (1.5%) conducted transplantation twice.

Table 4: *Distribution of subjects by severity of disease*

Variables		Frequency	Percent
Level of servility	Chronic and Conducted hem dialysis	55	40.4%
	Chronic Not conducting hem dialysis	81	59.6%
Frequencies of hemi session per week dialysis	1 session	3	5.5%
	2 sessions	29	52.7%
	3 sessions	22	40%
	Four sessions	1	1.8%
Renal transplantation	No	129	94.8%
	Yes	7	5.2%
Repeat ion of transplantation	once	5	3.7%
	twice	2	1.5%

The cost of medical treatment for most of the acquired kidney disease has been expensive. Renal transplantation is limited because of the shortage of donors not only in Palestine but worldwide.

4.2.5. Distribution of Subject Relatives by Renal Failure Prevalence (History)

The majority of the subjects(76.5%) do not have relatives suffering from RF, and 23.5% do have relatives suffering from RF.For the previous percent, the researcher notices that some of the subjects answer with no even if they have a relative visiting the hospital and suffering from RF.This may refer to the nuclear type of family, or because of weak relationships within the extended family, so the percent may be a little higher than the mentioned percent.Regarding their relative, 43.8% of the subjects with a «yes» answer knew that one of their relatives were affected by Renal Failure; 37.5% knew that two of their relatives were; 12.5% knew that three relatives were; 3.1% knew of five relatives; and 3.1% knew six relativesthat were suffering from renal failure.

The majority of affected relatives were first degree with 59.4%; 28.1% of their relatives were second degree, and the minority had both first and second degree relatives affected by RF. Regarding their relatives' residency, 62.5% are living in the same demographical area (city, camp).The majority of relatives had the same source of water, but the rest of relatives (28%)lived in the same governorate but maybe did not share the source of water. Twenty nine relatives out of thirty two are living in KhanYunis governorate or the eastern villages with a percent of 90.6% of the whole relatives, and only 9.3% are living outside the governorate. Table 5 shows the distribution of subject relatives.

Table 5: *Distribution of subject relatives by renal failure prevalence (history)*

Variables		Frequency	Percent
Relative history RF	Yes	32	23.5%
	No	104	76.5%
Number of relatives	1	14	43.8%
	2	12	37.5%
	3	4	12.5%
	5	1	3.1%
	6	1	3.1%
Level of relation	First degree	19	59.4%
	Second degree	9	28.1%
	Both first and second degree	4	12.5%
Relative Residency	Living in the same area (district)	20	62.5%
	Live in the same governorate but not in the district	9	28%
	Outside the governorate	3	9.3%

4.2.6. Distribution of Subject by Domestic Water Source

The subjects have access to domestic water through one of the four suppliers in the governorate.The majority have municipal access (77.9% of the total subjects) 13.2% have access to private well , 7.4% have Makarout access and 1.5% only have UNRWA access as shown in table 6.

There is a wide range of differences between previous and current domestic water treatment before ten years.Only 2.9% of the total subjects treated domestic water before using for drinking purpose, but recently only 7.4% of the subjects didn't treat domestic water before using it for drinking purposes.Many do not treat because they have access to Mekarout water, which has WHO approval for drinking water.Sometimes this water has a lesser chloride level than the WHO standard, meaning that 92.6% of the total subjects don't use domestic water without treatment.

Table 6: Distribution of subject by domestic water source

Variables		Frequency	Percent
Domestic water source	Municipality	106	77.9%
	UNRWA	2	1.5%
	Macarout	10	7.4%
	Private wells	18	13.2%
Previous domestic water treatment (since the last 10 years)	No	132	97.1%
	Yes	4	2.9%
Treatment tool	Home filter	4	100%
Current domestic water treatment	Yes	126	92.6%
	No	10	7.4%
Current treatment tool	Desalinate (Sold water)	66	52.4%
	Home filter	38	30%
	Mekarout	22	17.5%
Treatment time	Before incidence	32	25%
	After incidence	94	75%

This could be reflectedby the Gaza strip society as all have the same access for the saline ground water with different level of salinity. The majority (52.4%) of subjects who treat water buy water from desalination stations a using home filter, while the rest take water from Mekarout recently as they have access. 75% of the subjects treated water after renal failure incidence while the other 25% only treated water before renal failure incidence.

Table 7: Distribution of subjects by water consumption pattern

Variables		Frequency	Percent
Total amount consumed for drinking per day	Less than three liter	102	75%
	Three liter	0	0
	More than three liter	24	17.4
	uncertain	10	7.4%
Using treated water for cooking	Yes	104	76.5%
	No	32	23.5%

As shown in table 7, regarding the total consumed drinking water by subjects, 75% of the total subjects consumed less than three liters per day, 17.4% drink more than three liters, 7.4% don't know the total daily amount, and none drink exactly three liters per day. Most of the subjects (76.5%) used treated water for cooking, but 23.5% still use untreated domestic water for cooking.

4.2.7. Subjects Sub-Scale Domains

It was difficult for the researcher to study each item of the Likert scale presented in the questionnaires alone. The researcher classified the items into four domains through the questionnaire to make it easy and applicable for analysis.The first domain is domestic water quality, the second concerns practices and attitudesaccess to safe drinking water, the third knowledge and awareness, and the fourth satisfaction.

Table 8: Distribution of subject's domain by mean and percentage

Domain	Mean	Percent
Domestic water quality (Physical characteristics, level of salinity)	3.62	72.4%
Practice " keeping access to safe drinking water	2.88	58%
Knowledge about salinity problem	2.91	58.5%
satisfaction " domestic water"	2.91	58.5%
Over all domains	3.08	62%

The overall domain is the summation of all factors. The highest mean was for the water quality and the lowest mean was subjects' practice.Table 8 shows the distribution of all subjects by domains of mean and percentage

4.2.7.1. Domestic Water Quality Domain

The subjects perceive this domain as the highest positive (72.4%), more so than other domains. This could be explained by the fact that the majority of subjects that have access to saline water, not pure, have changeable characteristics like color, odor, and poor even for cooking. This meets the study finding that more than 90% percent have another source of drinking and cooking water through either a home filter or by buying desalinate water.

4.2.7.2. Practice and Attitude

This reflects the ability of subjects to keep continuous access to safe water twenty for hours per day for drinking and cooking.These subjects represented 58% from the whole, and 42% percent are not able to keep a continuity of safe drinking water. It is expected that the last have a higher prevalence of water disease than the others.Access to safe water is very vital and critical for public health, and there should be serious thought and efforts to supply the citizenswith infrastructure access for safe water.This will reduce diarrhea, typhoid, skin sepsis, ulcers.

4.2.7.3. Knowledge and Awareness

This domain reflects the subject's knowledge and awareness about both the salinity problem and the renal failure problem; if there is direct link or effect between the two; if there is any relation between salinity and renal

failure; and if they have attended any workshops or awareness sessions. The percentage of this domain reported 58.5%, which is considered moderate in between other domains. On the other hand, the majority of subjects agree about the positive role of knowledge and awarenessin decreasing the prevalence of renal failure.Eventhough the majority did not join any session (T test statistically significt with this domain for female),there is a need to focus on both health and environmental awareness for the public.This may assist health improvement.

4.2.7.4. Subjects Satisfaction

These items included subject's level of satisfaction due to their usage of saline domestic water.This domain represented 58.5%. The majority of subjects absolutely agree that domestic water is very poor and cant even used for cooking. This correspond with the subjects agreement to change the source of domestic water, the use water treatment tools at home,and access to clean water.Nearly all subjects in all areas accept this domain positively except (Kuza an Absan) as all subjects in this area are satisfied and don't agree to change domestic water.

4.2.7.5. Overalldomains

Refer to summation of the four domains. The percentage of overall perception of domain was 62%, and its mean 3.08.

4.3. Ecological Variables (Historical Review)

The researcher goes deeply into the history of water quality and review all the tests was carried out by the Public Health lab for all the wells in Khanyounis Governorate and the eastern villages since 1987and distribute subjects living areas according to water quality in each area.

The mathematical method is used to calculate the average for each well within the following periods (1987-1992 , 1993-1998, 1999-2004,and 2005-2009).Each of the previous periods have been studied separately (within the same period a major change occured with nitrate levels, which increased 20% -25% in the spring test in eastern villages and Al Qarara , but in other areas nitrate level increased from 10-15%, and the level came down from 20-15% during autumn test). During the same period the well number doesn't changed.The changing amount of produced water from each well doesn't exceed 10%.The average for each well water quality during the period has been calculated by summation of the all tests carried divided by the number of tests, then the average multiply by the production percent for each well (the level for each area is the summation of the well's production percent for supply water for the area).This calculation has been done for each well within the four separate periods. As example figure 3 shows the variation of TDS level by time and locality.

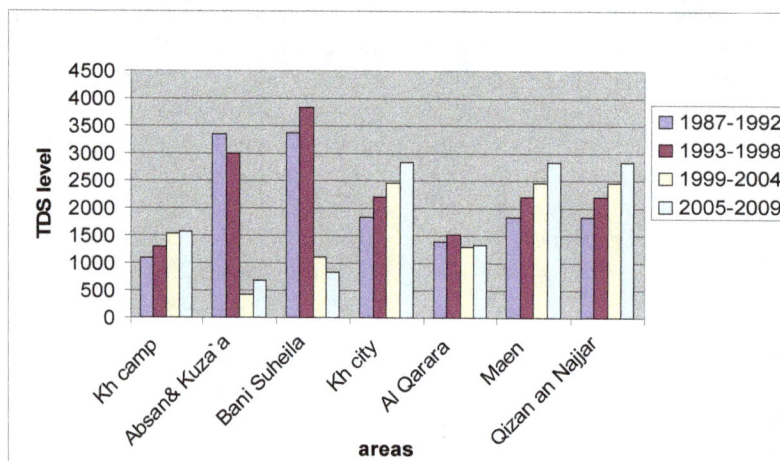

Figure 3: Variation of TDS level by time and locality

The following were observed:

1. There are different levels of water quality as well as salinity; the main sources of water is Municipal wells since 1987, Private Wells that are owned by the Municipality, and the UNRWA wells. There are 22 wells that save all the domestic water supply for KhanYnis city, and KhanYuniscamp.Six of the wells have worked since 1987, and others were drilled since 1991-2000.

2. KhanYunis Camp water supply conducted by nine wells with total amount produced at 290190 per month.Different percentages of water production to the camp were (Al Sada well 11.3% , Al Ahrash 14.2%, Al Amal 11.9%, Al Amal Al Jaded 24.5%, Rashwan B 10.4%, Al Tahady Al Jadied 8.6% , Istad El Ryady 13.8% , Al Wakalla Al Shamaly 2.4%, Al Wakalla Al Janoby 2.5%).

3. KhanYunis city water supply conducted by thirteen different wells for the city with total amount 329560 m3 per month.Even the mid year population for KanYunis city is three times than the camp. (Al Janoby 10.7%, New Janoby 9.5%, Aya 20. % , Al Shargy7% , Al Madina Al Ryadia 10% . Al satar al shamaly 2.9%, Al Najar 16%, Maen 6.5%, Al Markaz Al Thagafy 10%, Al Bahar Al Jadied 1.7%, Al satar al Jadid 5.7% ,).

4. During1987 there was only one well suppling water for Qarara and owned by its municipality, but another well was producing during the period from1991-1998 and has been closed due to elevated nitrate level.The third well was drilled in1998, and now the two wells are supplying water for Qarara (with equal percent).

5. Since 1987 to 2001, the eastern villages (Khuza, Abasan, BaniSuhila) were getting water from the different wells (Khuza and Absan) from one source, but BuniSuhila from another source.The wells were closed and not being used for Khuza and Abasan, but one of the closed wells is still supplying water for Bunsuhial in summer if there is shortage of

water.This happened because BuniSuhila need larger amount of water than both Khuza and Absan. But after 2001, both Absan and Khuza are having directly from Mekarout and Merage three (which met the WHO standard for drinking), but Banisuhila has from the same source and another covers about 30% of Bunisuhiala daily supply. The other well is owned by Eastern villages municipality, so Banisuhila is having 14 hours/day only from the same source and the other ten hours/day having mixed water from Merage and another well (Al Najar for eastern villages).

During the period from 87-92 and by comparing the TDS, Nitrate, Chloride, Fluoride and sodium BaniSuhila is the highest level (for all the fifth elements,) then (Kuza and Absan), Maen and KhanYunis city , Qarara (is highest with fluoride level), and KhanYunis camp. Within 1993-1998 the highest level found for the fifth parameters is BuniSuhialla , Khuza and Absan, KanYunis city an Maen, Qarara(only higher with fluoride level), Khan Yunis Camp. During 1999-2004 the highest level for all except fluoride wasKhanYunis city and Maen, Qarara , BuniSuhila ,and both Khuza and Absan met the WHO standard for drinking water. Between 1999-2004 the highest level found for the fifth parameters wasMaen, KhanYunis city, KhanYunis Camp, Qarara, BuniSuhila, Kuza andAbsan met the WHO standard for TDS, Chloride, Fluoride, Sodium, but higher with Nitrate level. According to the data, all wells within theKhanYounis governorate exceed the WHO standard (TDS, Nitrate, Chloride, Fluoride, and Sodium) for drinking water within the period from 87-2008, but only the eastern villages Absan and Kuza met or fell below WHO standard (TDS, Cl, Nitrate and sodium), but fluoride is a little bit higher during the period 99-2004 .

The researcher noticed that tests are carried out during the spring and autumn seasons, but not for all wells, and some wells are tested more than twice a year, and fluoride level is not measured in all tests. A noticeable elevation of the nitrate level fromautumn to spring in the same year may be explained byfertilizers or manure storage, which are common causes for nitrate pollution for underground water in the Khanyounis area and eastern villages. Chloride is an indicator ion that if found in elevated concentration, points to potential contamination from septic systems, fertilizer, landfills, or road salt. And another sort for chloride is the seawater intrusion.

5. Inferential Statistic Parts

This part discuss the relationship between the dependent and independent variables for subjects by using some statistical tests, and the researcher provides an explanation and opinion regarding the findings of this study. The dependent variable is the subject's domains to explore the relation between domain and subject acceptance.Theindependentvariable is demographical data such as gender, residency place, marital status, main job, and level of education.

5.1. Subjects Relationships Part

5.1.1. Demographic Characters for Subjects

Age, marital status, and level of education all showed no statistically significant differences in the overall domains.

5.1.2. Differences in Domains by Gender

Gender comparison with domains was done by using an independent t test. Table (9) shows that males and females had no statistical significant differences variation in the mean scores in overall domains (P = .802). Through knowledge and awareness only femaleshad more positive perception than male for the domain (statistical significancewasobserved between the two groups).

Table 9: Differences in domains by gender

Dependent variable " Domain"	Ind. var. "Gender"	N	Mean	SD	t	Sig.
Satisfaction	Female	57	3.02	.517	2.257	.448
	male	79	2.83	.462	2.216	
Practice	F	57	2.85	.787	.343	.471
	M	79	2.90	.853	.347	
Knowledge	F	57	3.02	.355	2.530	.007*
	M	79	2.82	.530	2.692	
Water quality	F	57	3.64	.601	.331	
	M	79	3.60	.507	.332	.374
Over all	F	57	3.13	.316	1.661	.802
	M	79	3.04	.334	1.667	

(*) Statistically significant

5.1.3. Differences in Domains by Employment

Employment comparison with domains (Table 10) show that unemployment and employment had nosignificantstatistical differences in the mean scores in overall domains (P= .123) similarly all sub-scale domains hadno statistical significant differences except in water quality domain. The test shows that unemployed subjects have more positive response to water quality domain than the employed.

Table 10: Differences in domains by employment

Dependent variable	Independent"employment"	N	Mean	Std. Deviat ion	t	Significance
Satisfaction	Unemployed	122	2.90	0.491	0.745	
	Employed	14	3.00	0.491	0.708	0.469
Water quality	Unemployed	122	3.66	0.502	2.442	
	Employed	14	3.29	0.790	1.714	0.002*
Practice	Unemployed	122	2.87	0.840	0.653	0.063
	Employed	14	3.00	0.667	0.677	
Knowledge &awareness	Unemployed	122	2.91	0.478	0.168	0.976
	Employed	14	2.89	0.455	0.174	
Overall	Unemployed	122	3.08	0.339	0.423	0.123
	Employed	14	3.04	0.222	0.589	

5.1.4. Differences in Domains by Level of Severity

By comparing level of severity with domains by using an independent t test Table (11) shows that subjects who conduct hem dialysis and subjects who don't conduct hem dialysis have a small variation in the mean scores in overall domains. The results show no statistical significance between the two groups within the four domains and the overall domain (p=.903)

Table 11: differences in domains by level of severity

Dependent variable	Conduct dialysis or not	N	Mean	SD	T	Sig
		55	2.96	.501	.988	
	Yes	81	2.87	.488	.983	
satisfaction	No	55	3.61	.599	.150	.955
Water quality	Yes	81	3.62	.513	.145	.392
Practiceandattitude	No					.787
Knowledgeawarness	Yes	55	2.91	.823	.311	.817
	No	81	2.86	.828	.312	
	Yes	55	2.85	.460	.190	
		81	2.95	.482	.201	
Over all	No					
	Yes	55	3.08	.329	.075	.903
	No	81	3.08	.331	.075	

5.1.5. Differences of Domain Scores Regarding to Water Treatment

By comparing subjects who treat water before drinking and who don't treat with response to domains shows that subjects who treat water before drinking and subjects who don't treat water before drinking had no statistical significant differences in the mean scores in all domains (P= .213) similarly practice, knowledge, except satisfaction, and water quality. The study shows that subjects who treat water before drinking had more positive perception to satisfaction and water quality domains than subjects who don't treat water before drinking.

5.1.6. Differences of Domainscores by Watersource

One way anova used the water source of participant shows that there were differences in theoverall perceptions of domains with significant statistical difference (P=.007). The respondents show different statistical significance in satisfaction, water quality, and practice. No significant statistical differences shown in knowledge and awareness. Scheffe test shows that the subject who has Municipal access had more positive perception, than subjects who have UN, and Private well, and Mekarout. (Ascending quality arrangement)

5.1.7. Differences of Domainscores by Residency

One way anova used to the residency of participant shows that there were differences in the overall perceptions of domains with statistical significant difference (P=.000). The respondents shows different statistical significance in satisfaction, characteristics, & and knowledge. No statistical significant in practice and attitude. Scheffe test shows that KhanYunis city subjects had more positive perception.

Age group, marital status, and academic certificate all show no statistical significance in overall domains with some variation in the subscale domain.

5.2. Distribution of Renal Failure Prevalence and Water Salinity Level (from 2005 -2009)

The total population of the renal failure patients registered in Nasser Hospital have been classified according to their water source area (salinity level).The classification was done by the artificial kidney staff.Upon the researcher's request the old and new cases were included, and the prevalence of each area has been calculated (Prevalence per thousand = old and new cases/Mid year population *1000). The Mid year population for all areas have been conducted from " Palestinian Central Bureau of Statistics" except Man the Mid year population for 2006, which was obtained from KhanYunisMunicipality, Absan include (Absan Al Jadida , Absan Al Kabira, Kuza`a). A comparison between renal failure prevalence and level of salinity including , TDS level , Nitrate level, Chloride level , Fluoride level and sodium level during the period from 2005-2009, shows that the highest prevalence for renal failure is found in Qezan An Najjar representing 1.28 renal failure patients per 1000 in Qizan an Najjar , then Maenwith 92 renal failure patients per, then KhanYunis City with75 renal failure, then Bunisuhila , Absan and Kuza ,and KhanYunis camp.

5.2.1. Comparison between the Distribution of Renal Failure prevalence (Dependent Variable) and TDS Level (Independent Variable) in the Different Areas

All areas haveTDS levels higher than the level recommended by the WHO standard (1000mg/l) for drinking water except Kuza,Absan&Bunisuhila. Figure 4 links the relationship between TDS and the prevalence of renal failure.The study findings show that when the TDS increased, the renal failure prevalence increased in one and in another decreased.The highest TDS level were in Maen but the highest prevalence was in Qizan andNajjar. The lowest TDS levels were in Kuza&Absan while the lowest prevalence was in KanYunis Camp. Thepeak of TDS level was not in the area with the peak prevalence or off peak so there is no association between TDS level and renalfailure prevalence.The study findings show that more than 90% of the subjects don't accept domestic water before treatment

due to high level of TDS, and 95% prefer to change domestic water source due to the negative effect of TDS on public health as well as taste and odor. The life of home hot water heaters decreases by approximately one year for each additional 200 mg/l of TDS in water above the typical household level of 220 mg/l.The secondary maximum contaminant level (SMCL) of 500 mg/l for TDS is reasonable because it represents an optimum value commensurate with the aesthetic level to be set as a desired water quality goal(New Jersey secondary drinking water regulation, 1992).

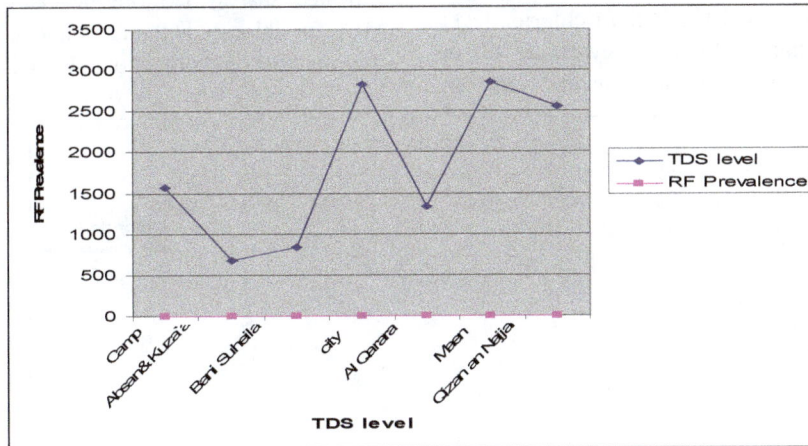

Figure 4: Distribution of renal failure prevalence and TDS level

5.2.2. Comparison between the Distribution of Renal Failure Prevalence (Dependent Variable) and Nitrate Level (Independent Variable) in the Different Areas

All areas haveNitrate levels higher than the level recommended by the WHO standard for drinking water (50mg/l). Figure 5 links the relationship between Nitrate, and the prevalence of renal failure. The study findings show that when the Nitrate level increased, the renal failure prevalence once increased and another decreased.The

highest Nitrate level was found in KhanYunis camp and the lowest Nitrate level in Kuza&Absan.The peak nitrate level is not in the area with the prevalence peak or off peak, so there is no association between nitrate level and renal failure prevalence. The study findings show that the peak nitrate level is during spring, and the highest increase occurred in Kuza, Absan, BuniSuhila, and Al Qrara. This could be due to soil washing in these agricultural areas during winter and excessive unplanned used offertilizers and manure.

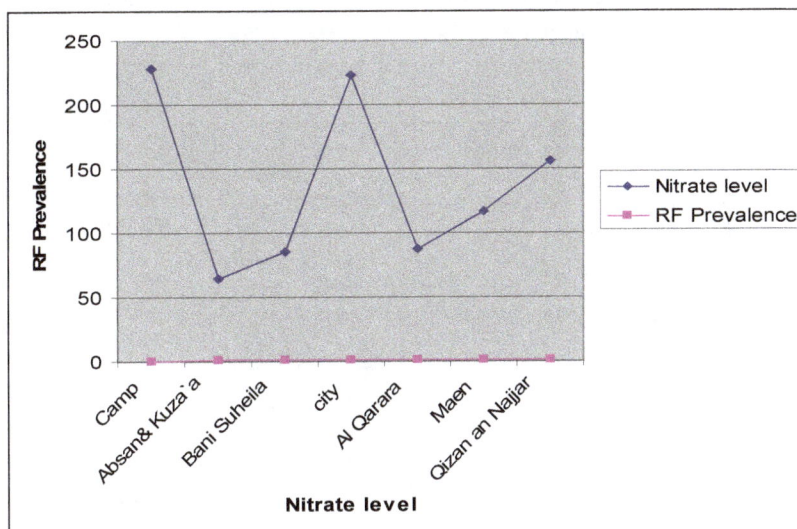

Figure 5: Distribution of renal failure prevalence and nitrate level

5.2.3. Comparison between the Distribution of Renal Failure Prevalence (Dependent Variable) and Chloride Level (Independent Variable) in the Different Areas

All areas havechloride levels higher than the level recommended by the WHO standard (250mg/l) for drinking water except Kuzaand Absan which have chloride level less than the WHO standard for drinking water. Figure 6 links the relationship between chloride level, and the prevalence of renal failure. The study findings show that when the chloride increased the renal failure prevalence once increased and another decreased.The highest chloride level was in KhanYunis city but the highest prevalence was in QizanaNajjar. The lowest chloride level was in Kuza&Absan(moderate) while the lowest prevalence was in KanYunis Camp. Thepeak of chloride level was not in the area with the prevalence peak or off peak so there is no association between chloride level and renal failure prevalence. The study findings show that more than 90% of the subjects don't accept domestic water taste without treatment even for cooking but Kuza and Absan accept water taste without treatment and 90% of the subjects in these areas don't treat water, and the same percent refused to change domestic water source. The study findings correspond to the Federal safe drinking water which confirmed that if chloride levels exceed 250mg/l the SMCL for chloride is the level above which the taste of the water may become objectionable to the consumer.

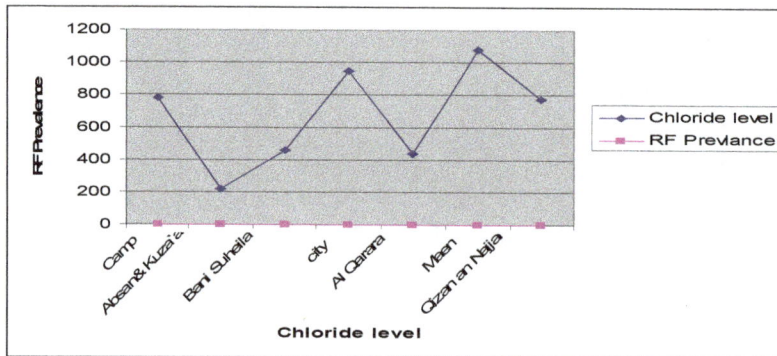

Figure 6: *Distribution of renal failure prevalence and chloride level*

5.2.4. Comparison between the Distribution of Renal Failure Prevalence (Dependent Variable) and Fluoride Level (Independent Variable) in the Different Areas

All areas hadfluoride levels higher than the level recommended by the WHO standard for drinking water (1.5 mg/l). Figure 7 links the relationship between Fluoride, and the prevalence of renal failure. The study findings show that when fluoride level increased the prevalence of renal failure increased.The highest fluoride level was inQezanaNajjar and the highest prevalence was in the same area. The lowest fluoride level wasKhanYunis camp and the lowest prevalence was in the same area.In all areas when the fluoride level increased the prevalence increased so there is a positive association between the fluoride level and renal failure.The diagram links a semi linear relationship.The study findings correspond with a new study, to be published in the journal Environmental Research, and adds further support to recent conclusions on fluoride toxicity by the National Academy of Sciences (NAS).

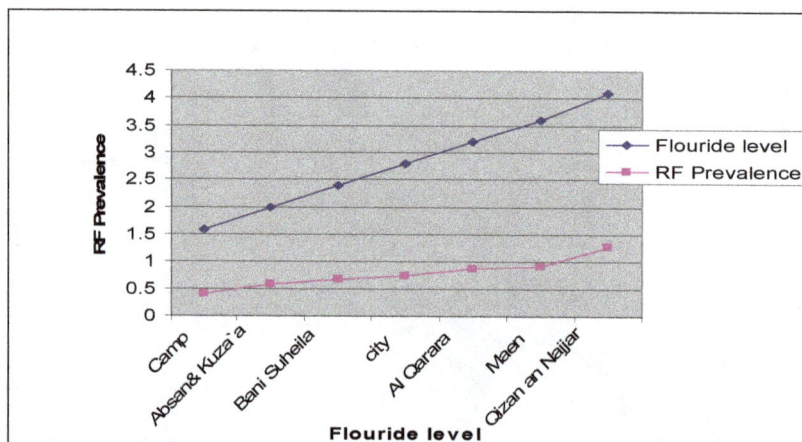

Figure 7: *Distribution of renal failure prevalence and fluoride level*

5.2.5. Comparison between the Distribution of Renal Failure Prevalence (Dependent Variable) and Sodium Level (Independent Variable) in the Different Areas

All areas hadSodium levels higher than the level recommended by the WHO standard (200mg/l) for drinking water except Kuzaand Absan which have sodium level less than the WHO standard for drinking water. Figure 8 links the relationship between sodium level, and the prevalence of renal failure. The study findings show that when the sodium level increased the renal failure prevalence once increased and another decreased.The highest sodium level was in Maen but the highest prevalence was in QizanaNajjar, and the lowest sodium level was in Kuza&Absan(moderate) while the lowest prevalence was in KanYunis Camp. Thepeak sodium level was not in the area with the prevalence peak or off peak so there is no association between sodium level and renal failure prevalence.

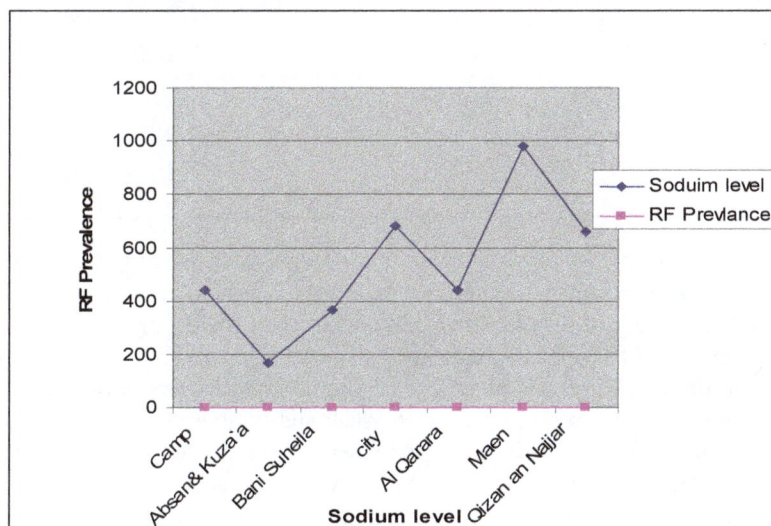

Figure 8: *Distribution of renal failure prevalence and sodium level*

6. Conclusions

1. Before ten years the majority of subjects used domestic water access for drinking directly without treatment and these represented 97.1% of the total subjects, that's mean that nearly all subject having one access which fit their needs & requirement for domestic and drinking.

2. At present time majority of subjects treating domestic water before drinking these presented 92.6% of the total subject, and only 7.4% don't treat domestic water before drinking these subjects has direct access to mirage storage tank.

3. The results showed that only 8% of the municipal wells meet the WHO standards for drinking in chloride level. Chloride, nitrate, TDS, fluoride and sodium concentration exceed 2-9 times the WHO standards in 92% of the southern wells.

4. The study findings showed that there was no association between renal failure prevalence and chloride level, sodium level, TDS level and nitrate level and showed only association with fluoride level, there was strong and positive association.

References

[1] Aish A, 2004. Hydrogeological Study and Artificial Recharge Modeling of the Gaza Coastal Aquifer Using GIS and MODFLOW. PhD Dissertation, Holland, VrijeUniversiteit Brussels VUB, Belgium.

[2] ArrigoSchieppati, Norberto Perico and Giuseppe Remuzzi "preventing end stage of renal disease NDTNephro Dialysis Treatment" http://ndt.oxfordjournals.org/cgi/content/full/18/5/858

[3] Bradley A. Warady and VimalChadha, R. (2006): "chronic kidney disease in children, the global perspective". Journal ofNursing Management, 14(8). P. 610–616.

[4] Henk de Zeeuw, (2000)Co-coordinator of the Resource Centre on Urban Agriculture and Forestry (RUAF), ETC, the Netherlands; and Karen Lock, Visiting research fellow, London School of Hygiene and Tropical Medicine, UK. Yahoo, http://www.fao.org/urbanag/Paper2-e.htm

[5] Khalid Amin and Tufail Muhammad. (2000) Etiological prevalence of chronic renal failure paper No. 04 http://search.msn.com/results.aspx?q=renal+failure+prevela nce+by+gender&FORM=MSNH11

[6] Qahman K, Larabi A. 2006. Evaluation and numerical modeling of seawater intrusion in the Gaza aquifer (Palestine). Hydrogeology Journal 14: 713–728.

24

Assessment of institutional and asset-related functions in the urban water sector in Libya

Khaled A. Rashed

Associate Professor, Department of Civil Engineering, Faculty of Engineering, University of Tripoli, Tripoli, Libya

Email address:

k65rashed@yahoo.co.uk

Abstract: The total urban water supply in Libya is about 600 million cubic meter per year, of which more than 90% came from groundwater. Surface water resources are minimal and Libya relies on wells, desalination and transported water for urban water supply. Transported water supply is essentially targeted towards the agricultural sector; however, the share of the urban sector has been increased to cope with increasing demand. This paper focuses on the organization of urban water sector with regards to institutional level and asset level, trying to high light the problems facing urban water sector and proposes solutions. In addition to the newly re-established Water and Wastewater Company, there are six main players in the urban water sector today that duplicate institutional and asset-based functions. Three of them deal with supply side (Desalination Company, Water Authority, and Man-made River Authority) and the other three deal with demand side (Ministry of Utilities, Project Execution Authority, and Environment Authority). After assessing the current situation of the urban water sector mainly around organizational consideration, one can conclude that in terms of institutional setup there is no clarity around supply/demand decisions. On the asset-related side, water and wastewater operations have been confused by frequent re-organizations. Apart from fragmentation, overstaffing of Water and Wastewater Company is significant, especially in light of the level of service provided to consumers. In order to build a professional urban water sector, in terms of institutional and asset-related levels, clear key functions for both levels have been proposed.

Keywords: Urban Water Supply, Institutional Functions, Asset-Related Functions

1. Introduction

Libya, which accommodates about 6 million people, relies on groundwater, transported water and desalination for urban water supply. Groundwater supply figure, which is based on data gathered by the Water Authority (WA) from the municipalities through field visits and interviews (no flow metering was used), is estimated to be around 282 million cubic meter per year. Transported water supply, which comes from deep aquifers in the south of Libya via the Man-Made River (MMR), is estimated to be around 286 million cubic meter per year (Rashed, 2004). Desalination supply figure, which is based on actual water deliveries made by the operators to the urban water network, is estimated to be around 35 million cubic meter per year. These three sources together provide a total of about 600 million cubic meter per year, of which more than 95% came from groundwater. These quantities of water (600 million cubic meter per year) are delivered to consumers through network of pipes, but the Libya's water network

coverage rates are not exactly known. In order to estimate the water network coverage rates, interviews with key stakeholders and published reports lead us to an estimate of 65%, based on a 70% average coverage for users living in cities and a 50% average coverage for users living in rural areas. Our estimate of 65% average coverage for Libya is in fair agreement with the World Bank estimation (World Bank, 2007), which estimates a national average of 70% for Libya and average water coverage in the Middle East and North Africa (MENA) region stands at 75% (figure-1). According to the population density, the country can be divided into fifteen water regions. Population by water region are presented in figure-2 (GIA, 2010). Using the quantity of water provided for urban supply and the percentage of people connected to the water networks lead to an initial national daily supply per capita estimate of 420 litre. This figure of water consumption includes domestic and non-domestic consumptions and unaccounted for water. In fact there is a wide regional variation in water supply as presented in figure-3. This paper focuses on the organization of urban water sector with regards to institutional level and asset level, trying to highlight the problems facing urban water sector and proposes solutions.

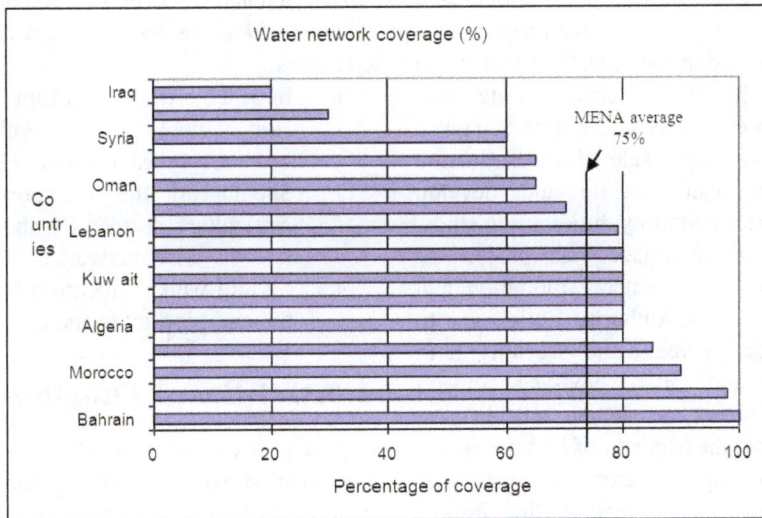

Figure-1: *Water network coverage in the MENA countries*

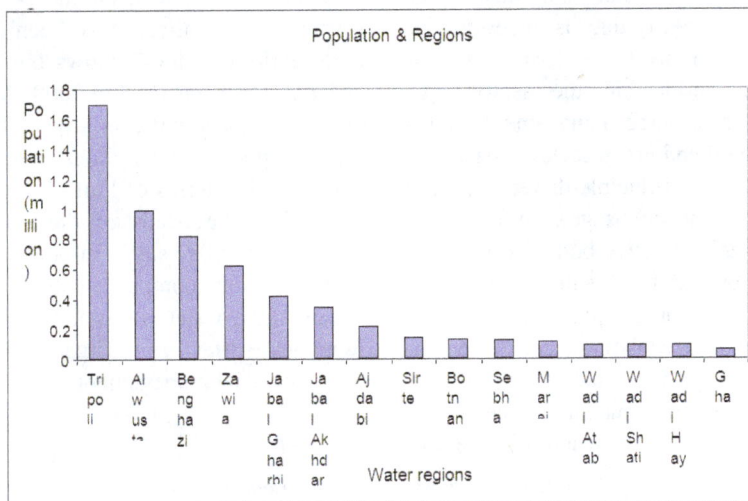

Figure-2: *Population by water regions*

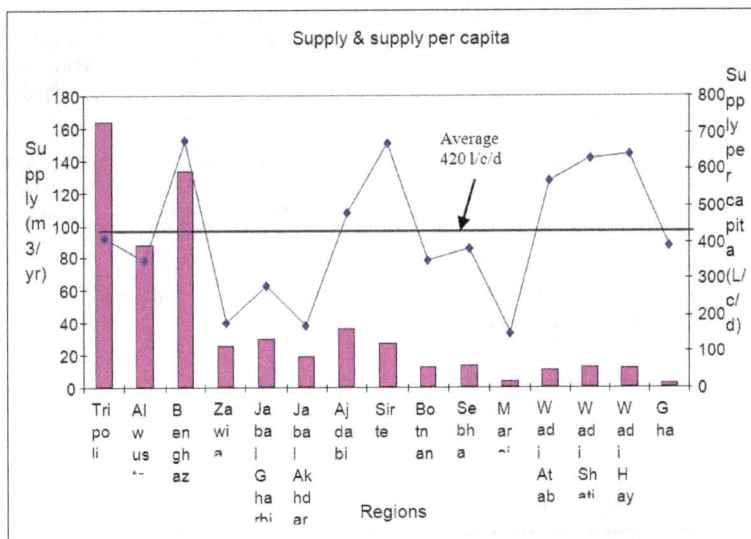

Figure-3: *Regional variation in water supply Institutional functions*

In addition to the newly re-established Water and Wastewater Company, there are six main players in the urban water sector today that duplicate institutional and asset-based functions. Three of them deal with supply side and the other three deal with demand side. The Water and Wastewater Company duties are to operate water and wastewater assets; test water quality and bill and manage customers. With respect to supply side; the Desalination Company duties are to plan, operate and develop desalination plants; the Water Authority duties are to study, plan and implement ground and surface water projects as well as to recommend on groundwater exploitation and allocation; the Manmade River Authority duties are to determine the supply rates, develop and set long term strategy, identify current and forward looking demand requirements, strategic priorities, investment and benefits. With respect to demand side; the Ministry of Utilities duties are to perform urban planning of water and wastewater infrastructure and monitor water quality; the Project Execution Authority duties are to execute projects related to expansion and refurbishment of water and wastewater networks; the Environment Authority duty is to set quality standard (Rashed, 2010). Other institutional roles are not clearly allocated between entities in the sector. For examples; in terms of water resource management, water allocation between agricultural and urban sector is based on negotiations instead of a principle-driven policy (economics, environment, domestic usage); in terms of water quality control, no single effective body for ensuring legal compliance with water and wastewater legislations. Some institutional roles are not in place at all. For examples; in terms of sector regulating, tariffs set by central government, without regard to value of water to end user, production cost or demand management needs; in terms of customer service, no national authority setting standards for customer service. A key element that is missing in the institutional setup is clarity around supply/demand decisions. According to the current situation there are no clear answers to the following key questions:

1. In which order should supply sources be utilized?
2. How quickly should withdrawals from ground water resources be reduced?
3. Should all the transported water go to the Agriculture sector? Or should an increasing share feed the urban network?
4. Should water allocation be based on the value of the end product to the economy?

2. Asset-Based Functions

Assets can be classified into five classes; (1) Desalination plants; (2) Well pumps; (3) Storage tanks; (4) Transmission pipes; and (5) Meters (WWC, 2009). Description and performance of these assets are summarized in table-1. On the asset-related side, water and wastewater operations have been confused by frequent re-organizations. Table-2 shows the recent changes in urban water sector, from 1998 to 2009. Apart from fragmentation, overstaffing is significant, especially in light of the level of service provided to consumers. Staff functions are unbalanced in terms of Full Time Equivalent (FTE) and do not reflect the needs of the operation. Figure-4 shows obvious surplus staff in the Water and Wastewater Company when compared with other countries (UN, 2000). In summary, water assets are in poor condition and asset functions suffer from organizational fragmentation and weak financial management.

Table-1: Assets description and performance

Assets class	Description	Performance
Desalination plants	- 12 desalination plants in 8 of the 15 regions	- Low utilization rates - Network bottleneck - Geared towards electricity - Lack of spare parts
Well pumps	- Spread throughout the country, depending on regional needs and aquifer locations - 9 out of 15 regions rely almost entirely on local wells for supply	- Ageing pumps - Poor maintenance - Frequent interruptions and shut-downs - Lack of spare parts
Water tanks	- Located at the outskirt of large urban areas, feeding into network - Aggregate water supplies - Regulate water flow and act as buffer supply	- Old tanks - Lack of proper maintenance
Transmission/ Distribution pipes	- Connect water tanks to consumers - Diameters (100 to 1200 mm) - Cast iron and galvanized steel; increasing use of PPE	- Old and requires maintenance - Leakage is poorly understood
Meters	- Visual reading - Low penetration	- Low meter penetration - Require maintenance and are either out of operation or not properly working

Table-2: *Changes in urban water sector in Libya*

Years	1998	2003	2006	2009
Scope	Water, wastewater	Water, wastewater	Water, wastewater, Garbage collection, Gardening	Water, wastewater
Level	National company	32 local companies	32 local companies	National company
Reporting to	MoS *	Municipality	HUC **	MoU ***

* Ministry of Services; ** Housing & Utilities Corporation; *** Ministry of Utilities

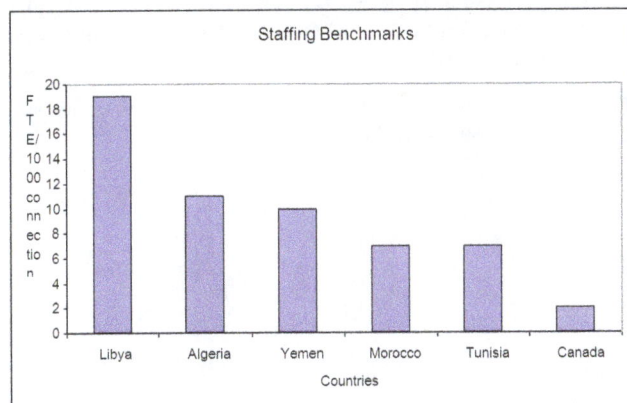

Figure-4: *Water and Wastewater staffing Benchmarks*

3. Recommendations

Typically, we would expect to see two broad functions (institutional and asset-related) in the water and wastewater sector, each with a specific set of activities;

Firstly, six key functions at institutional level can be recommended;

1- Perform Water Resource Management:
1.1- Develop long term supply/demand requirements.
1.2- Develop policies and regulations for water demand, supply allocation and sustainable development.
2- Define sector strategy:
2.1- Establish targets (service levels, and coverage).
2.2- Recommend new tariff mechanisms.
2.3- Recommend degree and form of private sector involvement.
2.4- Recommend pricing changes.
3- Regulate sector:
3.1- Ensure sector participants comply with water laws while financially viable.
3.2- Ensure that monopoly status is not abused (price caps and control over efficiency).
3.3- Define consumer standards.
4- Control water quality:
4.1- Inspect and monitor water resources.
4.2- Enforce quality standards.
5- Issue licences and manage water right:
5.1- Define terms and conditions of licenses and water rights in line with water resource management policies.

5.2- Enforce terms of licensing.
6- Address consumer complaints:
6.1- Receive user's complaints.
6.2- Pursue and investigate complaints.

Secondly, five key functions at asset-related level can be recommended;

1- Plan assets and investments:
1.1- Long-term strategic plans for new asset requirements.
1.2- Budget and authority to develop an asset strategy.
2- Finance asset development:
2.1- Water companies should not depend entirely on government funding.
2.2- Revenues should cover cost of operation.
2.3- Understanding of cost of production and operating cost.
3- Operate and maintain assets:
3.1- Network maintenance and renewal should be the responsibility of Water Company.
3.2- Engineering, operational and financial capabilities require strengthening.
4- Billing and collection:
4.1- Setting tariffs is the Water Company right.
4.2- Revenues are essential to funding company operations.
4.3- Ability to enforce bill payment.
5- Provide customer service:
5.1- Complaints are dealt with by customer service centres.
5.2- Meter for every customer so that customers can get information to reduce consumption and billing.

4. Conclusion

Libya (6 million people) relies on groundwater, transported water and desalination for urban water supply. These three sources provide 282 million cubic meter per year, 286 million cubic meter per year, and 35 million cubic meter per year respectively. In this paper, a comprehensive study has been conducted especially with regards to institutional and asset-related functions and activities. After assessing the current situation of urban water sector mainly around organizational consideration, one can conclude that in terms of institutional setup there is no clarity around supply/demand decisions. On the asset-related side, water and wastewater operations have been confused by frequent re-organizations. Apart from fragmentation, overstaffing of Water and Wastewater Company is significant, especially in light of the level of service provided to consumers. Six key functions at institutional level have been proposed and five key functions at asset-related level have been proposed.

Acknowledgement

The author would like to tank Mr. A. Elmetsho (Water and Wastewater Company) for allowing access to the company's data.

References

[1]   GIA. 2010. National Statistical Book for the year 2010.

[2]   Rashed, K. A. 2004. A 2045 vision for Municipal water supply in Tripoli. Proc.

[3]   Symposium on Challenges Facing Water Resources Management in Arid and Semi-Arid Regions, Lebanon, October, 7-9.

[4]   Rashed, K. A. 2010. Challenges facing urban water sector in Libya. Proc. BHS international hydrology symposium, UK, July, 19-23.

[5]   United Nations. 2000. We the Peoples: The Role of the United Nations in the 21st

[6]   Century. UN Report 10017, New York.

[7]   World Bank. 2007. Making the Most of Scarcity: Accountability for Better Water Management in the Middle East and North Africa. MENA Development Report 41113, Washington, DC.

[8]   WWC. 2009. Water Problems in Tripoli Municipality and Proposed Solutions: Internal Report.

25

The potential of rainwater harvesting: A case of the city of Windhoek, Namibia

Festus Panduleni Nashima[1, *], Martin Hipondoka[1], Inekela Iiyambo[2], Johannes Hambia[3]

[1]University of Namibia, Private Bag 13301, Windhoek, Namibia
[2]Rossing Uranium of Namibia, Private Bag 5005, Swakopmund, Namibia
[3]University of Namibia, Ogongo Campus, Private Bag 5520, Oshakati, Namibia

Email address:

fnashima@unam.na (F. P. Nashima), mhipondoka@unam.na (M. Hipondoka), inekelaiiyambo@yahoo.com (I. Iiyambo),
jhambia@unam.na (J. Hambia)

Abstract: Windhoek, Namibia's capital has experience for more than 25 years of novel approaches in integrated water management largely driven by the scarcity of water in the area. Notably absent in their approaches however, is the rooftop rainwater harvesting which is regarded as one of the viable alternative sources of water for domestic use. This paper assesses the potential economic benefits for rooftop rainwater harvesting for the City of Windhoek. The rooftop areas from four representative formal suburbs of Okuryangava (low income) in the north, Academia (middle income) in the south-central, Pioneers Park (middle income) in the south-west and Ludwigsdorf (high income) in the east, were estimated from high resolution satellite images captured from Google Earth. These estimates were used to extrapolate for the potential amount of rainwater that can be harvested in an average rainy season (i.e. December to April) in the study area. The estimated harvestable amount for each residential area was developed using a simple model that incorporates total rooftop area and estimated rainwater. The derived figure was then expressed in terms of cost per unit prices charged by the City of Windhoek. Pioneers Park attests to have the highest (134 m^3) potential harvestable rainwater per household, while Okuryangava is estimated to harvest the least amount of 36 m^3 per raining period. Given the high density of erven, however, Okuryangava has a potential to harvest approximately 920484 m^3 of water per hectare, ranking this suburb second after Academia, which stands at 1120716 m^3. This is a significant amount of water effectively taken from rainwater rooftop that could also provide justifiable saving to residents if used instead of tap water. It is therefore recommended that the City of Windhoek actively promotes rooftop rainwater harvesting for the benefit of residents and also as a measure to reduce storm-water runoff due to urban development.

Keywords: City of Windhoek, Rooftop Area, Rainwater Harvesting, Water Saving Cost

1. Introduction

The earth is covered by approximately 71% of water but only a total of 3% is fresh water (Bhandari, 2003). Fresh water is largely stored in polar ice sheets, rivers, lakes, streams, below ground as soil- and groundwater and the atmosphere. Only 0.003% of freshwater is accessible to humans, however (Bhandari, 2003). This signifies the scarcity of freshwater resources on earth. According to UNFAO (2007) around 1.2 billion people, or almost one-fifth of the world's population, live in areas of physical water scarcity, and 500 million people are approaching this situation. Additionally, 1.6 billion people, or almost one quarter of the world's population, face economic water shortage.

Global freshwater resources are facing dramatic changes as a result of global climate change, high water demands, population growth, industrialisation and urbanisation. As climate change leads to more extreme variations, water rationalization schemes through rain harvesting has been identified as an effective intervention measure towards meeting Africa's Millennium Development Goals (MwengeKahinda et al., 2007). This paper aims to assess the potential amount of rainwater that can be harvested from rooftops in the City of Windhoek.

1.1. Background

Rainwater harvesting is a viable alternative freshwater source, particularly for arid countries. It is also valuable in this age were freshwater resources are under increasing pressure from pollution and over-utilization. This practice is currently taking place in many countries across the globe, especially developing countries in Africa and Asia (HRDC, 2007; Baker *et al.*, 2007). Rainwater harvesting is relatively cheap and pollution-free. This is mainly due to the fact that as the rain falls, rainwater is collected before it is allowed to interact with the ground, where the higher potential for pollution exists. Therefore, the likeliest source of pollution and contamination of the harvested water would be the pollution inherent in the atmosphere or the unhygienic conditions of the collecting devices (HRDC, 2007).

Windhoek, Namibia's capital, has a population of approximately 300 000 residents, with a growth rate of 4.44% per annum (City of Windhoek, 2010). Like much of the country, the area receives summer rainfall, mostly during the months of December to April; the average annual rainfall is approximately 360 mm (NORIT, 2002). Water supply for Windhoek comes from a combination of three sources.

Since the early 1970s, the main water source is surface water collected in the dams around Windhoek (mainly Von Bach). The second source is the water acquired from recycling of waste water at Goreangab water reclamation plant, whilst the third source is borehole water (Louw, 2013). All these sources are delicate and vulnerable to environmental or economic factors. Water recycling, for example, has many economic implications due to the expensive nature of the recycling of waste water to potable quality, whilst the aquifers in Windhoek are said to be very vulnerable to pollution because they are shallow, and unprotected (Murray & Tredoux, 2004; Mapani, 2008).

With increasing growing population, demand for water will rise proportionally. Rainwater harvesting could therefore be another source of water for the residents of Windhoek. The economic benefits could arise from the fact that residents will save on potable water supplied by the city, through the usage of rain water on gardens, laundry as well as on most activities taking place outside the household. The city will also benefit by saving on the money needed to recycle the water required to meet the demand of the

Windhoek residents. Despite its tremendous potential, rainwater harvesting has not received adequate recognition in Namibia (New Era, 2012). Though rainwater can be harvested from many surfaces, rooftop harvesting systems tend to be the most commonly used (Baker *et al.*, 2007; Verlag, 2002). The technique has proven to work well in most countries and this can be an excellent strategy for adoption to Namibia.

2. Methodology

To estimate the potential amount of rainwater that can be harvested from rooftops within the four selected suburbs of Okuryangava, Academia, Ludwigsdorf and Pioneers Park, data were collected as follows: three sub-tiles of satellite images covering part of the City of Windhoek were onscreen captured from Google Earth 2009. These sub-tiles cover a 3.1 ha area of Okuryangava, 13.98 ha section of Ludwigsdorf and a 14.57 ha area of Pioneers Park and Academia combined. These suburbs were selected based on income classification per suburb, where low income suburb was represented by Okuryangava, middle income suburb constitute of Pioneers Park and Academia, while Ludwigsdorf correspond to a high income suburb.

The onscreen-captured image tiles were imported into ILWIS 3.3 (ITC, the Netherlands), an object oriented, raster-based Remote Sensing/GIS package. The images were then geo-referenced to a UTM coordinate system, yielding a re-sampled spatial resolution of less than one meter. The process of on-screen digitizing focused on roofs of the largest, contiguous building in each erven. This was necessitated under the assumption that installations for water harvesting would be more beneficial from one large surface area. Due to rainwater intercept and stem flow, only the unobstructed roof area was digitized where no tree(s) covered the roof. In essence, the average roof area for each erven would thus be marginally under-estimated under this approach. The digitized roof area for each suburb was then converted to polygons and sizes were calculated using the same GIS program. For the estimation of the potential amount of rainwater that can be harvested per house or hector in each suburb, the following formula was used:

Water volume (L) = Average annual rainfall (mm) x coefficient of runoff x roof area (m²)   (Equation 1)

A coefficient of rooftop runoff of 0.9 applicable to most buildings structure and used in literature (Romsey, 2010) was used to obtain a rough estimate. (Note: $1m^3 = 1000L$).

The cost of potable domestic tap water per m³ as provided by the Municipality Table 1, was used to estimate the

saving amount (N$) from the municipal water tariffs per average rainy season if this rainwater is collected and used instead of tap water. For the estimation of the saving amount (N$) to residents per average house the following formula was used:

Saving amount (N$) = Water volume collected from rooftop area (m³)* Cost of one cube water (N$)   (Equation 2)

3. Results

Pertinent measurements and derivatives from the four

selected suburbs attest that Okuryangava, Ludwigsdorf, Academia and Pioneers Park had the following average roof areas sizes for house as follows: 111.51 m², 258.40 m², 394.88 m², 413.31 m², respectively. While per hector, the roof area sizes were 2841 m^2, 1737 m^2, 3457 m^2, 2633 m^2 respectively. The number of roof structure per hector estimates are 25.5, 6.7, 8.8 and 6.4, respectively.

3.1. Potential Harvestable Rainwater

The estimates of the average potential amount of rainwater that can be harvested from rooftop area per household (m^2) are shown in Figure 1. These calculations are based on the annual average total rainfall of Windhoek (360 mm).

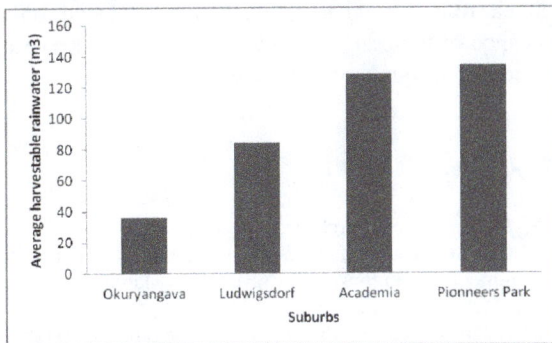

Figure 1. Average harvestable rainwater per household in each suburb per annum

The average amount of rainwater that can be harvested per household in each suburb per year is highest in Pioneers Park followed by Academia with estimated water volume of 134 m^3 and 128 m^3, respectively. Ludwigsdorf is estimated to yield an amount of 84 m^3 while the minimal

volume of harvestable rainwater for Okuryangava is 36 m^3.

The comparison of suburbs with regards to the estimated average amount of rainwater that can be harvested per hectare is showed in Figure 2. Academia suburb tends to have the greatest rooftop area per hectare in contrast to all other suburbs, followed by Okuryangava, Pioneers Park and the least Ludwigsdorf with an estimated rainwater amount of 1120716 m^3, 920484 m^3, 853092 m^3, and 562788 m^3, respectively.

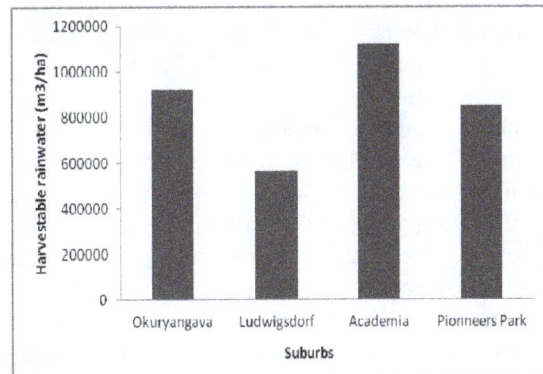

Figure 2. Potentially harvestable rainwater per hectare in each suburb

3.2. Saving Amount from Municipal Water Tariffs (N$)

For the City of Windhoek the potable water consumption tariffs for domestic use is shown in Table 1.

Table 1. Water consumption tariffs for the Municipality of Windhoek (Source: City of Windhoek, 2010).

| Tariff Code | Consumer Description | WATER CONSUMPTION TARIFFS- POTABLE | | |
		Tariff per m^3	VAT	Total
WC 10	Domestic			
	0 – 0.200 m^3 per day (0-6 m^3 p.m.)	8.08	0%	8.08
	0.201 – 1.50 m^3 per day (6-45 m^3 p.m.)	13.44	0%	13.44
	More than 1.50 m^3 per day (>45 m^3 p.m.)	24.76	0%	24.76

The potential amount of water that can be harvested from rooftop areas of building structure is calculated in terms of its cost as tap water to estimate the probable amount of money to be saved instead of using tap water. The calculation on cost saving was based on the quantity of water usage consumption as set up by the City of Windhoek, Municipality (Table 1).

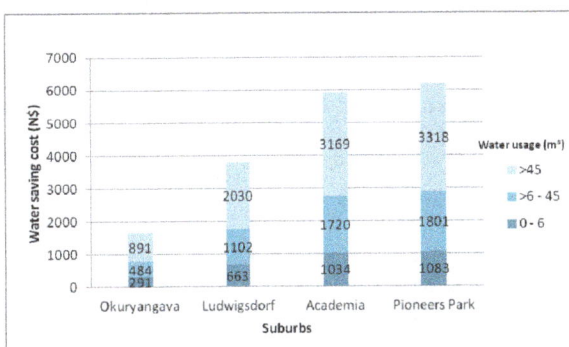

Figure 3. Average water saves per house in each suburb per rainy season

The saved amount per rainy season was determined based on the water volume collected from the rooftop area. It is evident from Figure 3, that the Municipal water consumption tariffs are based on water usages (m^3). The comparison of saved amount in each suburb based on municipal water tariffs (Table 1) shows that more saves is to be obtained at water usage of > 45 m^3 in Pioneers Park and Academia, with N$3318 and N$3169, respectively. Whereas, residents for Ludwigsdorf and Okuryangava can save about N$3169 and N$891, respectively, when assessed at the maximum tariff price for water usage (>45m^3).

4. Discussion

The GIS analysis employed in this study was essential for a systematic evaluation of roof rainwater harvesting in the City of Windhoek. Through the application of GIS it was possible to estimate the total amount of water harvestable at the household level as well as per hector in each suburb. As a result, low income areas with limited roof sizes per household, such as Okuryangava, emerged to have relatively less harvestable amount of 36 m^3, whereas middle income suburbs per household was dominated by larger single story buildings, such as Pioneers Park and Academia, and have a potential to harvest as much as 130 m^3 of water per rainy season. High income suburbs had mostly double story buildings covering relatively a smaller surface area, such as Ludwigsdorf can be expected to harvest an amount of approximately 80 m^3 per rainy period. Okuryangava area constitute of high density of erven per hector in comparison to all other suburbs. Thus, the water collection potential surpassed Ludwigsdorf and Pioneers Park with an amount of 920484 m^3. Though, Academia had the highest water collection potential which amount to 1120716 m^3 per hector. This is a significant amount of potential rain water that could be channeled to good use.

Considering the calculated water save per household if rainwater is collected and used, a justifiable saving can be deduced per rainy season by individual resident. These values can be escalated if measurement is calculated per hector, though not considered in this study as these are mere estimation. This quantity of water would make considerable savings towards watering of gardens, car washing and laundry, which are listed as some of the most water consuming activities in Windhoek (Uhlendahl *et al.*, 2010) and may not require purified water.

Currently, only a few residents located in formal areas of the city are reported to harvest rainwater from roofs (New Era, 2012). This limited usage of rainwater harvesting was in part attributed to economic factors and lack of awareness. The initial cost, which includes the acquisition of storage tank and installation, may far exceed the annual expenditure (averaging approximately N$3000 per household) for water consumption at present. Valued in short term perspective, individual household may therefore not justify investing in roof water harvesting. A general misconception in Namibia that residents in urban areas are not allowed to harvest rainwater has been recently dispelled by the City of Windhoek (New Era, 2012). The only condition attached to such practice is that harvested water shall not be connected to any of the municipal supply systems and can only be used for private consumption. Residents are thus allowed to liberally make use of water harvested from roofs at a household level.

5. Conclusion

The City of Windhoek has tremendous technical potential for rooftop rainwater harvesting. Based on the estimated results from this study it is envisaged that Windhoek residents can potentially collect substantial amount of rainwater and further more save on water cost, particularly towards usage that does not require purified water. Whilst it is unlikely that all of this potential will be developed shortly, it is evident that rainwater harvesting can provide justifiable amount of water to Windhoek residents.

Recommendation

In order to facilitate the implementation and swift operationalization of rainwater harvesting practices, the City of Windhoek need to provide support and assistance to its resident which can include mandatory installation of water tanks on all new houses or inclusion of rainwater harvesting designs to all future settlements. Furthermore, voluntary (with incentives) installation of water tanks on existing houses is also recommended. Significantly, water saving campaigns should be necessitated by the City of Windhoek to ensure that residents are educated on the significance of adopting strategies for rainwater harvesting as an additional source of water for their household use.

References

[1]   S. Baker, E. Grygorcewicz, G. Opperman and V. Ward. "Rainwater Harvesting in the Informal Settlements of Windhoek, Namibia". Retrieved 31 March 2011, from website: http://www.wpi.edu/Pubs/E-project/Available/E-project-051207-15 2911/unrestricted/report.pdf, 2007.

[2]   B.B. Bhandari. "What is happening to our Freshwater Resources? Institute for Global Strategies. Environmental Education project". Tokyo, 2003.

[3]   City of Windhoek. "City Development and Planning". Retrieved 7 April 2010, from website: http://www.windhoekcc.org.na/default.aspx?page=42, 2010.

[4]   VV. Berlag. "Rainwater harvesting facilities part 1: design, construction, operation, and maintenance". Berlin, Deutsches Institut für Normunge (DIN), 2002.

[5]   Habitat Research & Development Centre (HRDC). "Rainwater Harvesting in Namibia". Newsletter of the Namibia National habitat committee Vol. 3, No. 3, 2007.

[6]   D. Louw. "The Windhoek Aquifer: An important source in the water supply to the City of Windhoek". Namibia Scientific Society. Windhoek. 2013. (Presentation).

[7]   B.S. Mapani and U. Schreiber. "Management of city aquifers from anthropogenic activities: Example of the Windhoek aquifer, Namibia". Journal of Physics and Chemistry   of the Earth, volume 33, 674-686, *2008.*

[8]   E.C. Murray and G. Tredoux. "Planning Water Resource Management: The Case for Managing Aquifer Recharge". Proceedings of the 2004 Water Institute of Southern Africa (WISA) Biennial Conference, 2 – 5 May 2004. Cape Town. 430-437, 2004.

[9]   J. MwengeKahinda, A.E. Taigbenu and R.J. Boroto. "Domestic rainwater harvesting to improve water supply in rural South Africa". Physics and Chemistry of the Earth, 32: 1050-1057, 2007.

[10] New Era. "Harvesting rainwater in Windhoek a cheaper option". Retrieved 10 November 2012, from   website: http://www.infrastructurene.ws/2012/07/03/harvesting-rainw ater- in-windhoek-a-cheaper-option/, 2012.

[11] NORIT. "New Goreagab Water Reclamation Plant". Retrieved 7 April 2010, from website: http://www.norit.com/import/assetmanager/5/5565/02-CASE_HISTORY_GoreangabB.pdf\, 2002.

[12] Romsey. "Rainfall harvesting calculator". Retrieved 12 February 2011, from website: http://home.iprimus.com.au/foo7/tank2.html, 2010.

[13] T. Uhlendahl, D. Ziegelmayer, A. Wienecke, L. Mawisa and P. du Pisani. "Water consumption Windhoek 2010". Institute of Cultural Geography, Berlin: Albert-Ludwigs University of Freiburg, 2010.

[14] United Nations Food and Agriculture Organization (UNFAO). "Coping With Global Water Scarcity". Rome, FAO, 2007.

26

Experimental and numerical investigation on wave interaction with submerged breakwater

Md. Ataur Rahman, Silwati Al Womera

Dept. of Water Resources Engineering, Bangladesh University of Engineering and Technology, Dhaka, Bangladesh

Email address:

mataur@wre.buet.ac.bd(Md. A. Rahman)

Abstract: Experimental studies are carried out in a two-dimensional wave flume (21.3 m long, 0.76 m wide and 0.74 m deep) to investigate the performance of rectangular type submerged breakwater. A set of experiments are carried out at 50 cm still water depth with fixed submerged breakwaters of three different heights (30 cm, 35 cm and 40 cm) for five different wave periods (1.5 sec, 1.6 sec, 1.7 sec, 1.8 sec and 2.0 sec) in the same wave flume. For fifteen run conditions, water surface elevations are collected at six different locations both in front of and behind the breakwater. Also the type of wave breaking and position of wave breaking are simultaneously recorded with a digital video camera. Effects of breakwater height and length along the wave direction on wave height reduction are analyzed. It is found that both the relative structure height (h_s/h) and relative breakwater width (B/L) have strong influence in reducing transmitted wave height. Experimental analysis prevails that the reduction of transmitted wave height are 50%, 58% and 68% for relative structure height (h_s/h) of 0.6, 0.7 and 0.8 respectively, for a particular value of relative breakwater width (B/L =0.35). Also, the reduction of transmitted wave height is 32% and 50% for relative breakwater width (B/L) of 0.25 and 0.4 respectively, for a particular value of relative structure height (h_s/h =0.6). A two-dimensional numerical model based on the SOLA-VOF method has been developed in this study to investigate the wave interaction with fixed submerged breakwater. The developed model can simulate time series water surface profiles, water particle velocity field, VOF function F, pressure around a breakwater. The water surface profiles and wave breaking positions in various wave conditions simulated by the developed numerical model show good agreement with the experimentally measured values. The numerical model developed in this study is expected to serve as tool to analyze wave deformation due to submerged breakwater and will be important for designing submerged breakwater as a coastal protection measure.

Keywords: Experimental Investigation, Submerged Breakwater, Wave-Structure Interaction, SOLA-VOF

1. Introduction

Submerged breakwaters, a special type of breakwaters distinguished from other emerged offshore ones, are built with their crests submerged in the water. With this advantage, they avoid the generation of significant reflected wave that affect the nearby shoreline. Although it might take some disadvantage for navigation, they may be used efficiently as a mean of erosion control as they provide an inexpensive measure of protecting beaches exposed to small or moderate waves and offer fast installation for temporary offshore works. Fixed submerged breakwaters are more effective in reducing wave heights and are also less susceptible to structural failure during catastrophic storms. This nature-conscious coastal protection work has become increasingly popular due to their multiple functions, which are to protect shoreline or harbor and to prevent beach erosion by providing a safe and agreeable environment in coastal areas (Mizutani et al., 1994). This structure scarcely harms the coastal scenery nor obstructs the utilization of sea area; thus, by creating a calm sea, it facilitates the utilization of the sea for recreational and residential developments.

For safety design, many researchers have investigated the interaction between waves and the fixed submerged breakwater experimentally and numerically. Several reports on submerged breakwaters have appeared in the recent past (Rufin et al., 1996; Mizutani et al., 1998; Golshani et al., 2003; Hur et al., 2003), and the state-of-the-art literature review for complete review of the performance characteristics of various types of submerged breakwaters is done (Cheng et al., 2003). Dong-Soo Hur (2004) has investigated the wave deformation of multi-directional random waves passing over an impermeable submerged breakwater installed on the slope. The experiments are carried out in the three-dimensional wave basin which is equipped with a multi-directional random wave generator with a segmented wave-maker installing the wooden submerged breakwater on a sloping bed. A study on application analysis of submerged breakwaters to compare the reflection and transmission characteristics with other kinds of breakwaters has been done by Cheng (2003). He has studied the performance of different types of submerged breakwaters both experimentally and numerically.

Al-Banna and Liu (2007) have conducted a numerical study on the hydraulic performance of submerged porous breakwater under solitary wave attack based on solving the Reynolds-Averaged Navier-Stokes (RANS) equations. Lee et al. (2007) have studied the transformation of irregular waves propagating over a submerged breakwater. By providing the incident irregular waves with repeatable amplitude and phase for each wave component, effects of the height and width of the breakwater on the wave transformation have been studied systematically. Hur and

Mizutani (2003) have developed a numerical model, combining the VOF model and porous body model, to estimate the wave forces acting on a three-dimensional body on a submerged breakwater. They have examined wave induced deformation on the permeable submerged structure making use of the porous body model (Sakakiyama and Kajima, 1992) to express the governing equations of fluid motion. Rahman et al. (2006) has developed a two-dimensional numerical model combining the SOLA-VOF model and porous body model, to estimate the wave forces acting on a pontoon type submerged floating breakwater.

In this study, two-dimensional experimental studies are carried out in the laboratory to investigate the performance of submerged breakwater in wave breaking as well as dissipating the incoming wave energy. Also a two-dimensional numerical model based on SOLA-VOF method is developed in this study for simulating wave interaction with submerged breakwater.

2. Experimental Investigation

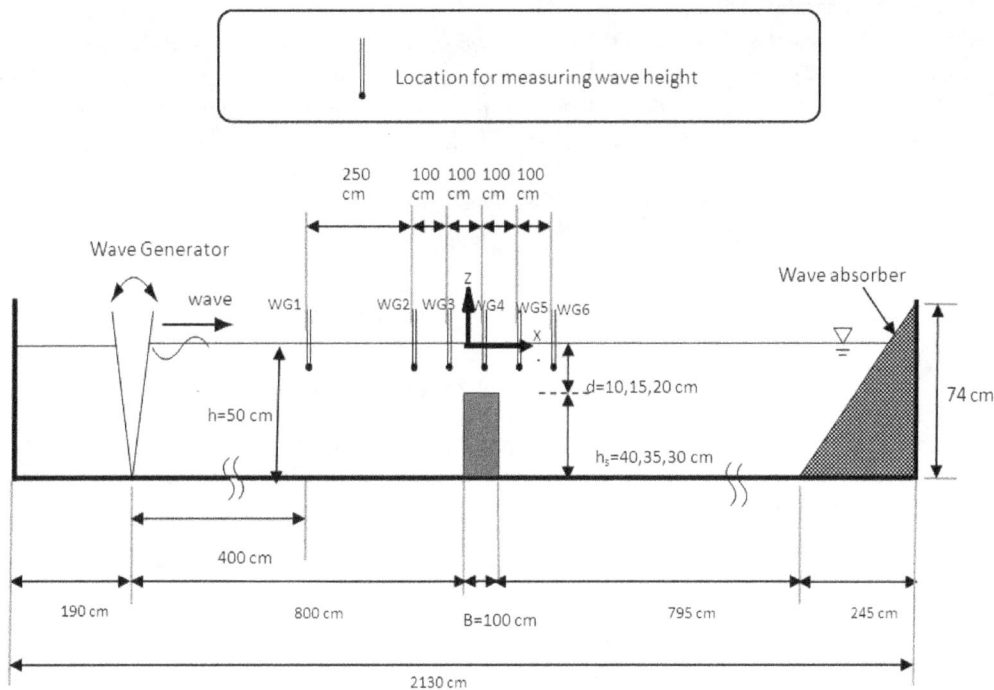

Figure 1. Detail of the experimental setup

To investigate the performance of proposed rectangular type submerged breakwater, experimental studies are carried out in a two-dimensional wave flume (21.3 meters long, 0.76 meter wide and 0.74 meter deep) at the Hydraulics and River Engineering Laboratory of Bangladesh University of Engineering and Technology. Breakwater (length 76 cm, width 100 cm and three different heights of 30 cm, 35 cm and 40 cm) made of cement-sand mixture was installed in at almost middle position of the wave flume containing 50 cm still water depth. Regular waves with five different wave periods (T=1.5 s, 1.6 s, 1.7 s, 1.8 s and 2.0 s) were generated

from a flap type wave generator installed at the upstream end of the wave flume. To damp the transmitted wave after passing the breakwater a wave absorber was installed at the end of the wave flume. For each of fifteen different run conditions, time series water level data were measured at six different positions - three were in front of the breakwater, one was over the breakwater and two positions were behind the breakwater. Moreover, still photographs and video recordings were taken during each run. The detail of experimental setup is shown in Figure 1.

3. Numerical Model

3.1. Governing Equations

The free surfaces are modeled with the Volume of Fluid (VOF) technique, which was first reported in Nichols and Hirt (1975), and more completely in Hirt and Nichols (1981). The basic equations used for VOF method are the continuity equation, the Navier-Stokes equation for incompressible fluid and the advection equation that represents the behavior of the free surface. Because the wave generation source is placed within the computational domain, these equations involve the wave generation source. The continuity equation is,

$$\frac{\partial u}{\partial x} + \frac{\partial w}{\partial z} = q(x, z, t) \tag{1}$$

$$q(x, z, t) = \begin{cases} q^*(z, t) \dots\dots\dots\dots x = x_S \\ 0 \dots\dots\dots\dots\dots\dots x \neq x_S \end{cases} \tag{2}$$

where u and w are the flow velocity of x and z direction respectively, q is the wave generation source with q^* as the source strength which is only located at $x = x_S$ and t is the time. The wave generation source q^* is defined as follows so that the vertically integrated quantity of q^* is equal to that in the non-reflection case (Ohyama and Nadaoka, 1991). q^* is also gradually intensified for the three wave periods from the start of wave generation in order to guarantee a stable regular wave train, as mentioned by Brorsen and Larsen (1987), shown in Eq. (3).

$$q^* = \begin{cases} \left\{ 1 - \exp\left(-\frac{2t}{T_i} \right) \right\} .2U_0 \frac{\eta_0 + h}{\eta_0 + h} / \Delta x_s & : t/T_i \leq 3 \\ 2U_0 \frac{\eta_0 + h}{\eta_0 + h} / \Delta x_s & : t/T_i > 3 \end{cases} \tag{3}$$

where t is the time from the start of wave generation, T_i is the incident wave period, h is the still water depth, and η_s is the water surface elevation at the source line ($x = x_s = 0$). Δx_s is the mesh size in the x-direction at $x = x_s$, and is required in order to apply the non-reflective wave generator to the finite difference method. U_0 and η_0 are the time variation of horizontal velocity and water surface based on third-order Stokes wave theory, respectively. The Navier-Stokes equations are,

$$\frac{\partial u}{\partial t} + u\frac{\partial u}{\partial x} + w\frac{\partial u}{\partial z} = -\frac{1}{\rho}\frac{\partial p}{\partial x} + \upsilon\left(\frac{\partial^2 u}{\partial x^2} + \frac{\partial^2 u}{\partial z^2} \right) + uq \tag{4}$$

$$\frac{\partial w}{\partial t} + u\frac{\partial w}{\partial x} + w\frac{\partial w}{\partial z} = -\frac{1}{\rho}\frac{\partial p}{\partial z} + \upsilon\left(\frac{\partial^2 w}{\partial x^2} + \frac{\partial^2 w}{\partial z^2} \right) + wq + \frac{1}{3}\upsilon\frac{\partial q}{\partial z} - g - \beta w \tag{5}$$

where p is the pressure, υ is the kinematic viscosity, ρ is the fluid density, g is the gravitational acceleration and β is the wave dissipation factor which equals 0 except for the added dissipation zone. The advection equation of VOF function F is derived by considering conservation of mass of the fluid in each cell. The advection equation of VOF function F is,

$$\frac{\partial F}{\partial t} + \frac{\partial uF}{\partial x} + \frac{\partial wF}{\partial z} = Fq \tag{6}$$

3.2. Computational Procedure

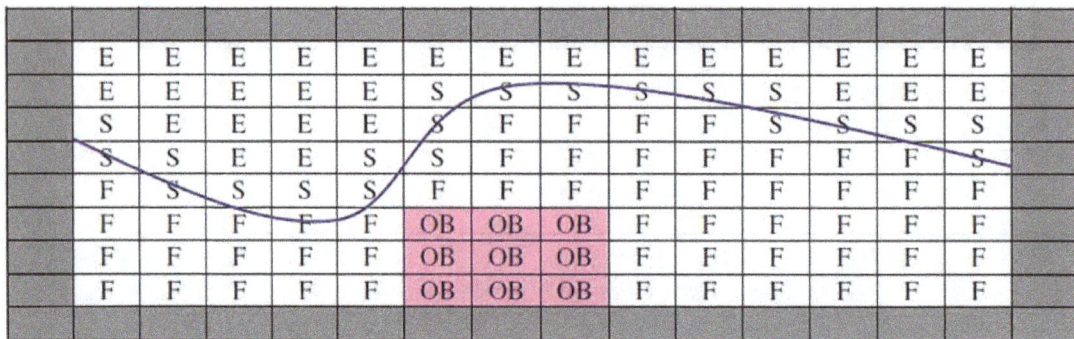

E = Empty cell, S = Surface cell, F = Fluid cell, OB= Obstacle cell

Figure 2. Free surface geometric model of VOF method

The equations (1) to (6) are calculated by a finite difference method using a staggered mesh. The free surface geometric model of VOF method is shown in Figure 2. The grid size of $\Delta x = 2$ cm and $\Delta z = 1$ cm has been used in the model. On the staggered mesh, the flow velocities u and w are put on the cell boundary, and the pressure p, wave generation source q and VOF function F are set on the center of each cell. The cell is classified into four types; a full cell filled with fluid, an empty cell occupied by air, a surface cell containing both fluid and air and an obstacle cell that represents the structure. The SOLA scheme is employed to calculate the pressure and flow velocity in each time step. And a type of donor-acceptor flux approximation is used to calculate the advection of the VOF function F computing the free surface. The advections are calculated by velocities of the adjoining cell using a donor cell which transports a fluid and an acceptor cell which receives an advect fluid. The physical characteristics of the cell are defined by the values of VOF function F. The cell in air, in the surface and in the water are denoted with $F=0$, $0<F<1$, and $F=1$ respectively.

3.3. Boundary Conditions

There are two boundary conditions for water particle velocity, that is, a boundary condition for the velocity parallel to the free-surface and a boundary condition for the velocity normal to the free-surface. In the first boundary condition, the velocity on the surface cell is set equal to the velocity on the interface in contact with the adjacent full cell, which can be calculated by the governing equations. In the second boundary condition, the velocity is determined so that the continuity equation is satisfied in surface cells. An added dissipation zone method (Hinatsu, 1992) is used to treat the open boundaries. The waves are damped by numerical dissipation effects due to the coarse grids and the fictitious damping forces based on the Stokes damping law. Sommerfeld radiation condition is applied for the open boundaries. And, non-slip condition is applied on the sea bed.

3.4. Calibration of Model

At first the developed numerical model is run for incident wave period, T= 0.8 sec, incident wave height, H_i= 4 cm and still water depth, h= 40 cm without any breakwater in the computational domain. The model simulated water surface profiles are compared with the waves generated from Stokes 3rd order wave theory, which shows very good agreement. Then the model is run for five different wave periods ranging from 1.5 sec to 2.0 sec and three different heights of breakwater as 30 cm, 35 cm and 40 cm to simulate water surface profile, velocity profile, VOF function F and pressure along the computational domain.

4. Results and Discussions

4.1. Experimental Results

4.1.1. Effect of Relative Structure Height on Wave Height Reduction

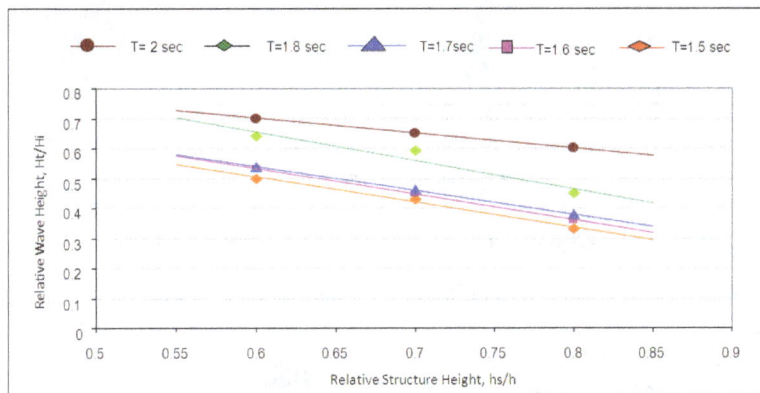

Figure 3. Effect of relative structure height on wave height reduction

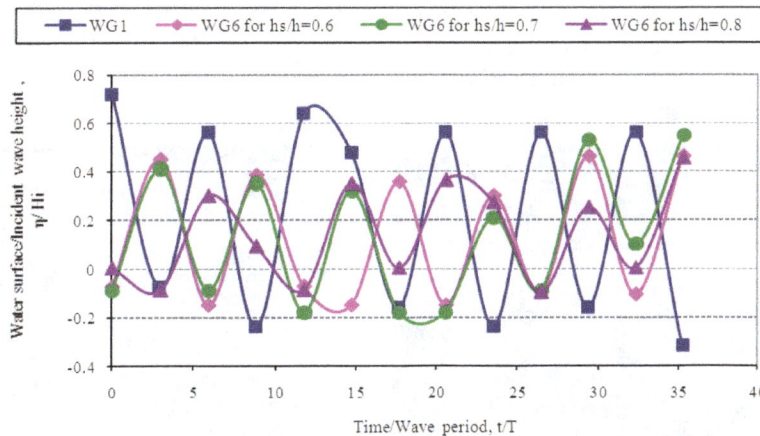

Figure 4. Variation of η/H_i with t/T for $T= 1.7$ sec, $H_i= 13cm$

In Figure 3 the effect of relative structure height, h_s/h (h_s is the height of breakwater and h is the still water depth) on wave height reduction, H_t/H_i (H_t is the transmitted wave height measured at wave gauge 6 and H_i is the incident wave height measured at wave gauge 1) is shown for five different wave periods. From this figure it prevails that as the relative structure height, h_s/h increases, the transmitted wave height, H_t decreases more with respect to the incident wave height, H_i for any particular wave period. For wave period T=2 sec, transmitted wave height reduces about 30% when $h_s/h=0.6$.

For h_s/h=0.7 and 0.8, the reduction of transmitted wave height occurs up to 35% and 40% respectively. Moreover, for a particular value of h_s/h, the reduction of transmitted wave height occurs more for lower wave periods than for the higher wave periods. When h_s/h=0.8, the wave height reduces 40% and 65% for T= 2 sec and T=1.5 sec respectively. For h_s/h=0.6 and 0.7, the variation of wave height reduction follow the similar trend. The variation of η/H_i with t/T at wave gauge 1 and wave gauge 6 for different relative structure height is shown in Figure 4. Wave gauge 1 and wave gauge 6 were placed to measure the incident and transmitted wave respectively. In Figure 4 it is seen that the incident wave damps when it passes over the breakwater. Moreover, it prevails from this figure that the wave damping

increases with increasing relative structure height (h_s/h).

4.1.2. Effect of Relative Structure Width on Wave Height Reduction

Figure 5 shows the effect of relative structure width, B/L (B is the width of breakwater along the wave direction and L is the wavelength) on wave height reduction, Ht/Hi (Ht is the transmitted wave height measured at wave gauge 6 and Hi is the incident wave height measured at wave gauge 1) for three different relative structure heights of hs/h=0.6, 0.7 and 0.8. For a particular relative structure height, it is seen that as B/L increases, the reduction of transmitted wave height also increases.

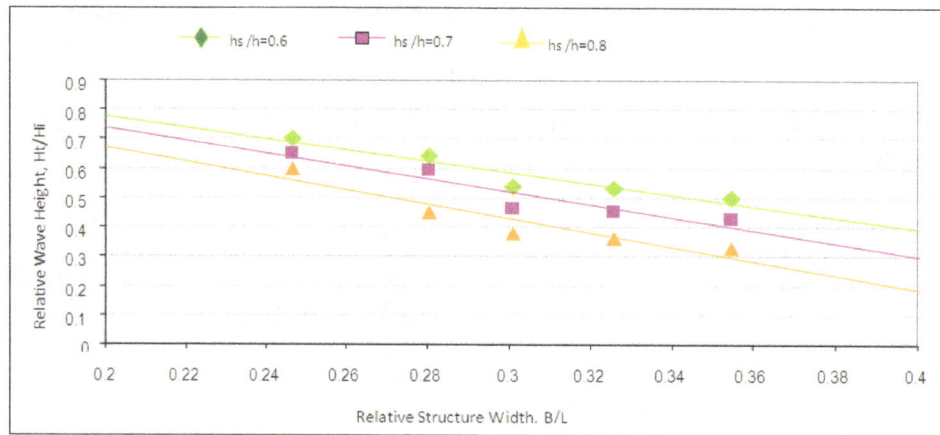

Figure 5. *Effect of relative structure width on wave height reduction*

For hs/h =0.6 and B/L= 0.25, reduction of transmitted wave height is 32%, whereas for the same breakwater height, wave height reduces 50% as the relative structure width, B/L becomes 0.35. The scenario is similar for hs/h= 0.7 and 0.8. Again for a particular value of B/L, the transmitted wave height, Ht decreases more with respect to the incident wave height Hi for higher value of hs/h. When the ratio of relative structure width, B/L is 0.2, wave height is reduced to 32%, 27% and 22% for hs/h= 0.6, 0.7 and 0.8 respectively. For any value of B/L the trend is almost similar. But as discussed

earlier, when breakwater width increases, the reduction of wave height occurs more for any value of hs/h.

4.1.3. Wave Breaking

From the still photographs and video recording data during each experimental run, the positions of wave breaking are analyzed. Wave breaking over and behind the breakwater during two experimental runs are shown in Figure 6.

Figure 6. *Wave breaking over the breakwater*

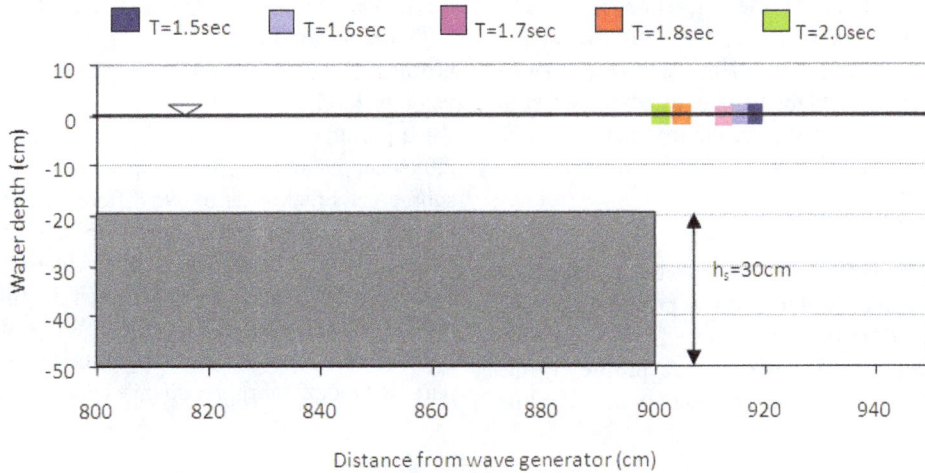

Figure 7. *Position of wave breaking for 30 cm height of breakwater*

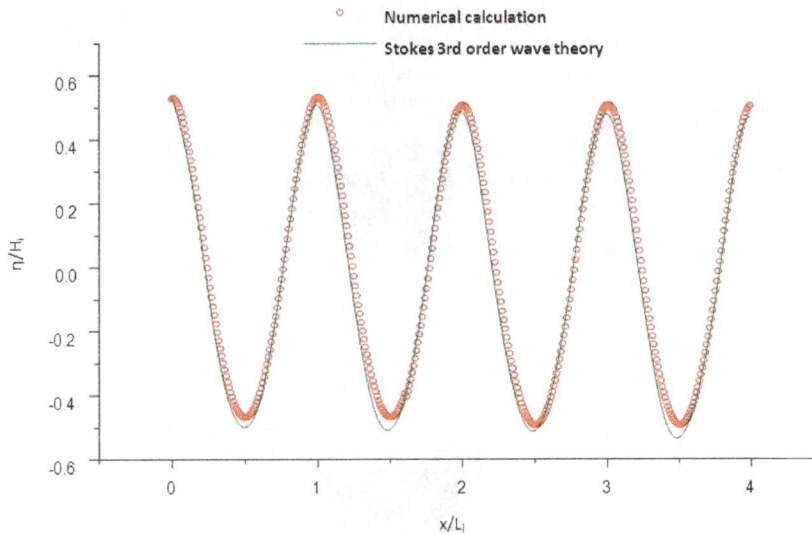

Figure 8. *Comparison of numerical and 3rd-order Stokes wave theory results of dimensionless water surface profiles (H_i=4cm, T=0.8 sec, h=40 cm)*

Figure 7 shows the position of wave breaking behind the breakwater for breakwater height hs = 30 cm. The breaking positions are seen just behind the breakwater. For 35 cm and 40 cm height of breakwater the breaking positions are seen at the onshore and at the middle position over the breakwater respectively. For a particular height of breakwater, as the wave height increases, the breaking position comes toward the offshore side. For hs/h=0.6, the variation of wave breaking positions are within 20 cm behind the breakwater for all wave periods. For the same wave periods, breakwater having hs/h= 0.7 and 0.8 wave breaks with 25 cm and 15 cm variation in length over the breakwater respectively. Moreover, when the breakwater height increases keeping the same still water depth, the waves tend to break more quickly than that for lower height of breakwater.

4.1.4. Performance of the Submerged Breakwater

Analysis of laboratory experimental data reveals that both the height and width of the submerged breakwater has strong influence in reducing the wave height propagating over the breakwater and transmitted to the shore. As energy of wave is direct propotional to the square of the wave height (E ∝ H^2), so damping of incident ocean wave height of 50%, when it propagates over the breakwater, results 75% dissipation of wave energy trasmitted to the shore. This performance of the submerged breakwater makes it very effective in beach restoration project. It is revealed from Figure 5 that the breakwater having height of 70% of the water depth (h_s/h=0.7) and width of 30% of incident wave length (B/L=0.3) damps 50% of wave height.

It should be mentioned that the emerged breakwaters are designed to attenuate the whole wave action and are submitted to the direct impact of wave breaking, resulting in larger structures that often eliminate water circulation at the lee side (in the protected area). Consequently degradation of water quality and of natural habitats in the lee-side is a frequent phenomenon. The environmental result is obviously not highly appreciated, due to its big visual impact, together with the strong erosion phenomena noticed at the gaps between barriers. On the other hand, in submerged breakwaters, as these structures are constructed below a specified design water level, some overtopping is permitted, allowing he circulation along the shoreline zone. The submerged breakwaters have a height of 50-70% of the water depth may permit sufficient water exchange preserving the healthiness and the bathing use of the water in the protected area and resulting an environment-friendly, natural and low price stable beach.

4.2. Numerical Model Results

The developed numerical model is verified first by comparing the simulated water surface profile with that should be as Stokes 3rd order wave theory. In Figure 8, it is seen that the dimensionless water surface profile by the developed model show very good agreement with that of the Stokes 3rd order wave theory.

4.2.1. Simulation of Water Surface and Velocity Profiles

Figure 9 shows the numerical simulation of water surface profiles along the channel length and the water particle velocity field around the breakwater. The wave height, the wave period and the water depth are considered as 15 cm, 2.0 seconds and 50 cm respectively. Rectangular fixed submerged breakwater of two different heights as 20 cm, and 40 cm are considered here for the numerical model simulation. For 20 cm high breakwater no wave breaking occurs and no disturbance is seen in the water surface profiles as well as in velocity field. When 40 cm breakwater is installed, the wave breaks over the breakwater and water particle velocity abruptly changes due to breaking.

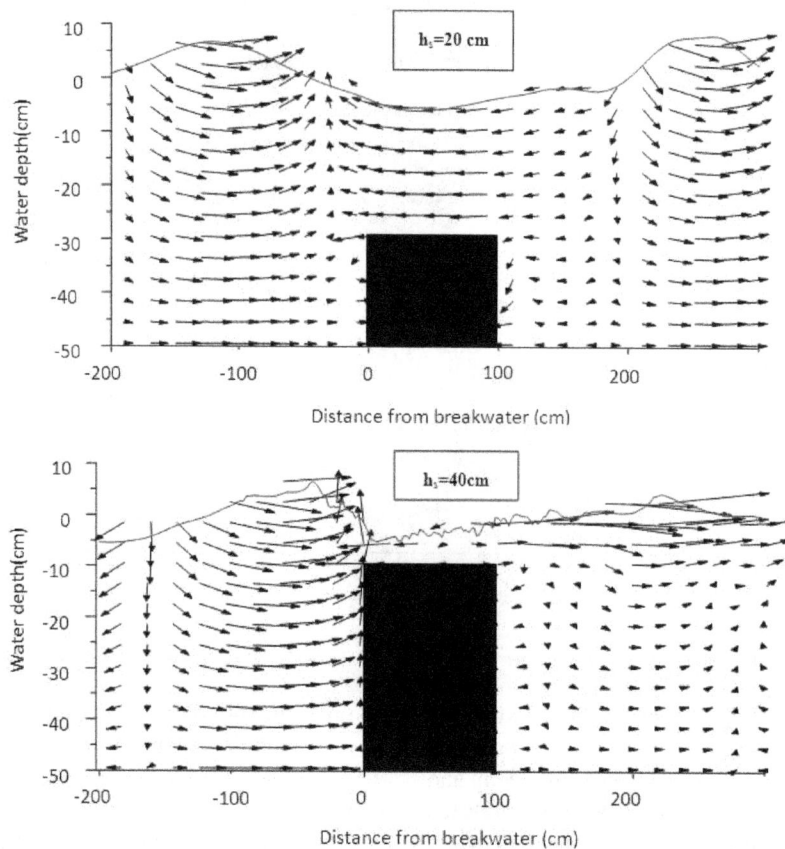

Figure 9. Numerical model simulation of water surface profile and velocity profile for both breaking and non-breaking condition ($H_i=15$ cm, $T=1.5$ sec, h=50 cm)

4.2.2. VOF function F around the Breakwater

Figure 10 shows the numerical simulation of the contour map of the VOF function F, which ranges from 0 to 1 at 15 sec after starting the simulation. The wave height, wave period and the water depth are considered as 10 cm, 1.5 seconds and 50 cm respectively.

The solid portion in the middle of this figure represents the breakwater containing the obstacle cell having F=0. The deep gray color around the breakwater represents the water containing the fluid cells having F=1. The F value of the top surface of the water surface profile is seen less than 1 (F<1), that represents the surface cells. On the right side of the top

surface of the breakwater, it is seen that the wave front becomes light gray color having F<1. It prevails that the breaking of wave occurs here and the air-bubble entrained in the corresponding numerical mesh cells due to wave breaking reduces the water volume less than the full volume of a fluid cell. For this reason the numerical model calculates F value of these cells less than 1. Also, the cells having F<1 are seen in both offshore and onshore side of the breakwater. This may happen due to the reason that the higher water particle velocity in vertically downward direction may cause partial void at some cells near the bottom forming vortex in this zone.

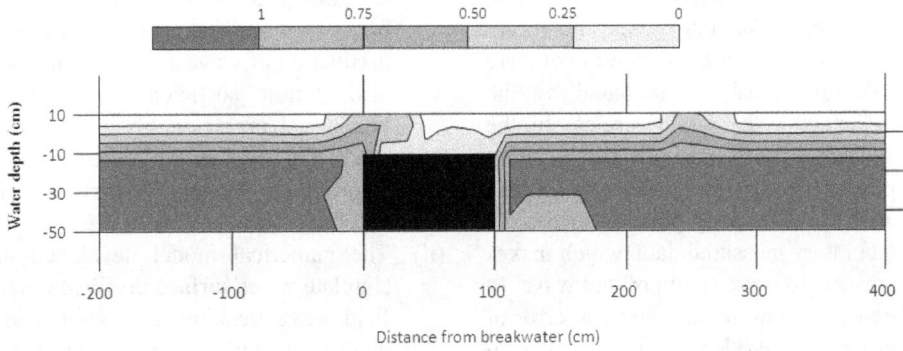

Figure 10. *Numerical model results of contour map VOF function value F around the submerged body with height 40 cm (H=10cm, T=1.5sec,h=50cm)*

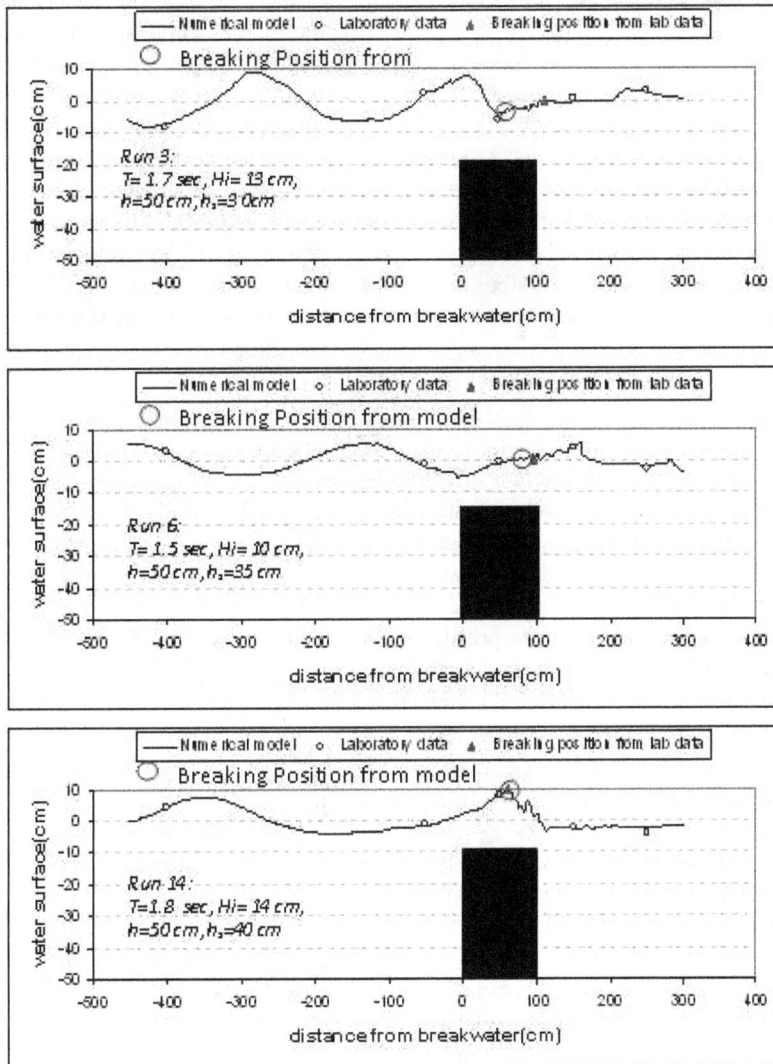

Figure 11. *Comparison between numerical and experimental results of water surface profiles and breaking position*

4.3. Comparison between Numerical and Experimental Results

The performance of the developed two-dimensional numerical model has been verified by comparing the model simulated results with experimentally measured data. The model simulated water surface profiles for three laboratory run conditions (Run No. 3, 6 and 14) are compared with the experimentally measured data for the respective run conditions and are shown in Figure 11. From the figure it is seen that the developed numerical model can track the free surface of the passing wave very well with little error less than 5%. It can track the free surface of the passing wave not only before interaction with the breakwater, but also during breaking due the interaction with breakwater and also during transmitted to the shore. Wave breaking positions measured during Run No. 3, 6 and 14 of the laboratory experiments are also shown in this figure with triangular symbol to compare with the locations of wave breaking simulated by the numerical model marked with the larger circles. In the model simulated water surface profiles the breaking position is considered at the point where the wave collapses. In Run No. 3, model simulated wave breaking position differs 55 cm from that of the laboratory measured data, which makes an error of 12% with respect to wave length of that wave. In Run No. 6, this difference is 8 cm, which makes an error of less than 3% with respect to wave length of that wave. In Run No. 14, it is seen that the numerical model simulates the exact position of the wave breaking position as of laboratory data with no error.

5. Conclusion

Submerged breakwater is a nature-conscious coastal protection work that prevents beach erosion and provides a safe and agreeable environment in the coastal areas. In this research work, the interaction between wave and rectangular fixed submerged breakwater has been investigated both experimentally and numerically to find out the effective size of this protection structure for the reduction of wave height. In a two-dimensional wave flume, fifteen experimental runs have been conducted with solid submerged body of three different structure heights (h_s= 30 cm, 35 cm and 40 cm) in fixed still water depth h=50 cm for five different wave periods as T= 1.5 sec, 1.6 sec, 1.7 sec, 1.8 sec and 2.0 sec. Moreover, a two-dimensional numerical model using SOLA-VOF is proposed in this study to simulate the wave interaction with fixed rectangular shaped submerged breakwater. The model can simulate water surface profile, velocity profile, water pressure all through the flume length including wave breaking over and around the breakwater. The experimentally measured water surface profiles and wave breaking positions are compared with the model simulated results. The key findings from this study are as follows.

(i)   From the experimental investigations, it is seen that for any particular wave period the relative structure height (h_s/h) and the relative structure width (B/L) are the important parameters for damping the wave height passing over the breakwater. Increase of relative structure height (h_s/h) for a particular value of relative breakwater width (B/L) or increase of relative breakwater width (B/L) for a particular value of relative structure height (h_s/h), both can damp the incoming wave height to a significant level, which plays an important role in controlling the beach erosion and restoring the beaches. Considering the disadvantages of emerged breakwater resulting degradation of water quality and of natural habitats in the lee-side, this study suggests submerged breakwater having height of 70% of the water depth (h_s/h=0.7) and

width of 30% of incident wave length (B/L=0.3) which can damp 50% of wave height passing over the breakwater and transmitted to the shore. It results 75% dissipation of wave energy trasmitted to the shore and also permit sufficient water exchange over the breakwater preserving the healthiness and the bathing use of the water in the protected area and resulting an environment-friendly, natural and low price stable beach.

(ii)   The numerical model developed in this study can simulate water surface profiles, water particle velocity field, wave breaking positions, dynamic water pressure due to interaction between the wave and submerged breakwater. Comparison of model simulated water surface profiles with that of laboratory measured data shows very good agreement with little error less than 5%. Moreover, the developed numerical model can simulate the wave breaking positions resulting from wave-structure interaction with error limit up to 12%. Hence the developed numerical model can be used for interaction of submerged breakwater of any height and width exposed to various wave conditions. The developed model can help the coastal engineers to fix the position and to select the dimension during design of submerged breakwater under various wave conditions.

References

[1]   Al-Banna, K. and Liu, P. (2007). "Numerical study on the hydraulic performance of submerged porous breakwater under solitary wave attack". *Journal of Coastal Research*, Vol. 50, pp. 201-205.

[2]   Brorsen, M and Larsen, J. (1987) "Source generation of nonlinear gravity waves with the boundary integral equation method," *Coastal Engineering*, Elsevier, Vol. 11, Amsterdam, pp. 93–113.

[3]   Cheng, S., Liu, S. and Zheng, Y. (2003) "Application Study on Submerged Breakwaters used for Coastal Protection," *Proceedings of International Conference on Estuaries and Coasts, China.*

[4]   Golshani. A., Mizutani. N and Hur. D. S. (2003) "Three-dimensional Analysis of Non-linear Interaction between water waves and Vertical Permeable Breakwater", *Journal of Coastal Engineering*, World Sientific, Vol. 45, pp. 329-3451-28.

[5]   Hinatsu, M. (1992). "Numerical simulation of unsteady viscous non-linear waves using moving grid system fitted on a free surface." *J. Kansai Soc. Naval Arch. Japan* 217: 1-11.

[6]   Hirt, C.W. and Nichols, B.D. (1981). "Volume of fluid (VOF) method for the dynamics of free boundaries," *J. Comp. Physics*, Vol. 39, pp. 201–225.

[7]   Hur, D.S. (2004). "Deformation of Multi-directional Random waves passing over an impermeable Submerged Breakwater installed on Sloping bed", *Journal of Ocean Engineering*, Elsevier, Vol. 31, pp. 1295-1311.

[8] Hur, D.S. and Mizutani, N. (2003). "Numerical estimation of wave forces acting on a three-dimensional body on submerged breakwater", *Journal of Coastal Engineering*, Elsevier, Vol. 47, pp. 329-345.

[9] Hur, D.S., Kawashima. N. and Iwata. K. (2003). "Experimental study of the Breaking Limit of Multi-directional Random waves passing over an Impermeable Submerged Breakwater", *Journal of Ocean Engineering*, Sciencedirect, Vol. 30, pp. 1923-1940.

[10] Lee, C., Shen, M. and Huang, C. (2007). "Transformation of irregular waves propagating over a submerged breakwater", *Proceedings of the 12th ISOPE Conference.*

[11] Mizutani, N., Rufin, T.F. and Iwata, K. (1994), "Stability of Armor Stones of a Submerged Wide-Crown Breakwater", *Proceedings of the 24th International Conference Coastal Engineering*, Kobe-Japan, 1994, pp 1439-1453.

[12] Mizutani. M, Mostafa, A. M. and Iwata, K. (1998). "Non-linear Regular wave, Submerged Breakwater and Seabed Dynamic Interaction", *Journal of Coastal Engineering*, Elsevier, Vol. 33, pp. 177-202.

[13] Nichols, B.D. and Hirt, C.W. (1975), "Methods for Calculating Multidimensional, Transient Free Surface Flows Past Bodies", *Proc. of the First International Conf. On Num. Ship Hydrodynamics*, Gaithersburg, ML.

[14] Ohyama, T. and Nadaoka, K. (1991). "Development of a numerical wave tank for analysis of nonlinear and irregular wave field." *Fluid Dynamics Research* 8: 231-251

[15] Rahman, M.A., Mizutani, N. and Kawasaki, K. (2006). "Numerical modeling of dynamic responses and mooring forces of submerged floating breakwater", *Journal of Coastal Engineering*, Elsevier, Vol. 53, pp. 799-815.

[16] Rufin, T.M., Mizutani. N. and Iwata. K. (1996) "Estimation method of Stable Weight of Spherical Armor unit of a Submerged Wide-crown Breakwater", *Journal of Coastal Engineering*, Elsevier, Vol. 28, pp. 183-228.

[17] Sakakiyama, T. R. and Kajima, R. (1992). "Numerical simulation of nonlinear wave interacting with permeable breakwaters," *Proc. 22nd Int. Conference on Coastal Engineering*, ASCE, Venice, pp. 1517–1530.

27

Feasibility study of AL-Masab AL-Aam water drainage in ThiQar and treatment for irrigation

Kadhim Naief Kadhim, Abbas Yasir Hussein

College of Engineering, University of Babylon-Iraq

Email address:

altaee_kadhim@yahoo.com(K. N. Kadhim), aljawei@yahoo.com(A. Y. Hussein)

Abstract: The Irrigation sector in different part of world including Iraq is a major water consumer to produce adequate food for increasing high population growth and meeting the MGD food goal. The challenges for the Iraq agriculture sector are to increase food production through effective management of the available and potential water sources including drainage and treated waste water and at the same time conserve and protect its environmental.(Ayers, R. S., and D. W. Westcot. (1985). In Iraq, the water users in different districts and policy makers are showing increasing interest in increasing the reuse drainage water as means of augmenting dwindling useable water supplies. Waterhowever its quality must meet crop tolerance to achieve optimal production and reduce environmental impact. To evaluate feasible option this study is concerned with assessing the suitability of drainage water of Al-Masab Al-Aamlocated for irrigation with or without treatment. Al-Masb Al-Aam drainage passes through the territory of governorate of ThiQar for part of the course covering 180 km out of the565 km and the discharge of the total 220 m3 / . The approach is to evaluate the chemical and physical properties of drainage water from the nearby irrigated field and and the water of Al Gharraf river in the Al-Fajr city northern of Nassiriyah, and the water of the Euphrates River south of Nassiriyah. The drainage water was to be blend fresh water of the nearest river through revaluating different ratios of blending starting with R1 which represents 90% drainage water blended with 10% river water, R2 (with 80% drainage water and 20% river water andand followed by blending ration up (R3 - R9)to nine trials. Monthly water samples were taken from four locations: two from drainagewater and two from rivers over the period from June 2011, to July 2012.Physically and chemically analyzed for EC,TDS, PH, Ca++, Mg++, Na+ , K+ , Cl- , SO4 , NO3 , Turb. , PO4 and T.Hwere carried out on 48 samples. The analyses indicated that the Sodium Adsorption Ratio (SAR) for drainage water was less than 12, which is acceptable for irrigation use. In terms of salinity, the drainage water of Al-Masab Al-Aam its acceptable for irrigation because the halophytes can be irrigated with EC) less than 8000 Micro Siemens/cm. The blending between the drainage water and the fresh water of Al Gharraf river showed showed god blending ratio (R7) having the EC) less than 3000 Micro Siemens/cm.

Keywords: Drainage, Water, Blending, Almasab Al-Aam

1. Introduction

The world's irrigated area is currently estimated to be 260–270 million hectares. The past averaged annual growth was estimated 2% and has fallen to less than 1%.While only about 17 percent of the world's cultivated land is irrigated, it produces one-third of the world's fresh food harvest and about half of its wheat and rice production. It is predicted that at least half of the required increase in food production in the near-future decades must come from the world's irrigated land. In view of the role of irrigated agriculture as the "world's food machine", competition for water cannot be allowed to result in even lower food

production growth rates, or an absolute reduction, of the world's irrigated area. The challenge is therefore clearly to produce more food by enhance management measures and none conventional sources such as drainage and adequately treated wastewater source.(FAO. (1992)).

The most feasible options to meet the challenge to enhance the fresh water management are to reduce the amount of irrigation water applied and to reuse the non-consumed fraction of the irrigation water already diverted. It is well documented (Hill 1994; Frederiksen 1992) that, at the field level, a large part (typically half) of the applied irrigation water is not actually consumed by a given crop and therefore ends up as drainage water. Since much of the drainage water commonly becomes the source of the water for downstream irrigation schemes and for other uses, the water use efficiency computed at the basin level is usually much higher than it is at the field or irrigation scheme level. In many irrigated areas, however, there is ample scope for planned reuse of drainage water supported by increasing interest of decision makers and water users and both water users as means of augmenting dwindling useable water supplies.(Gupta, I.C.1979.)

2. Study Area

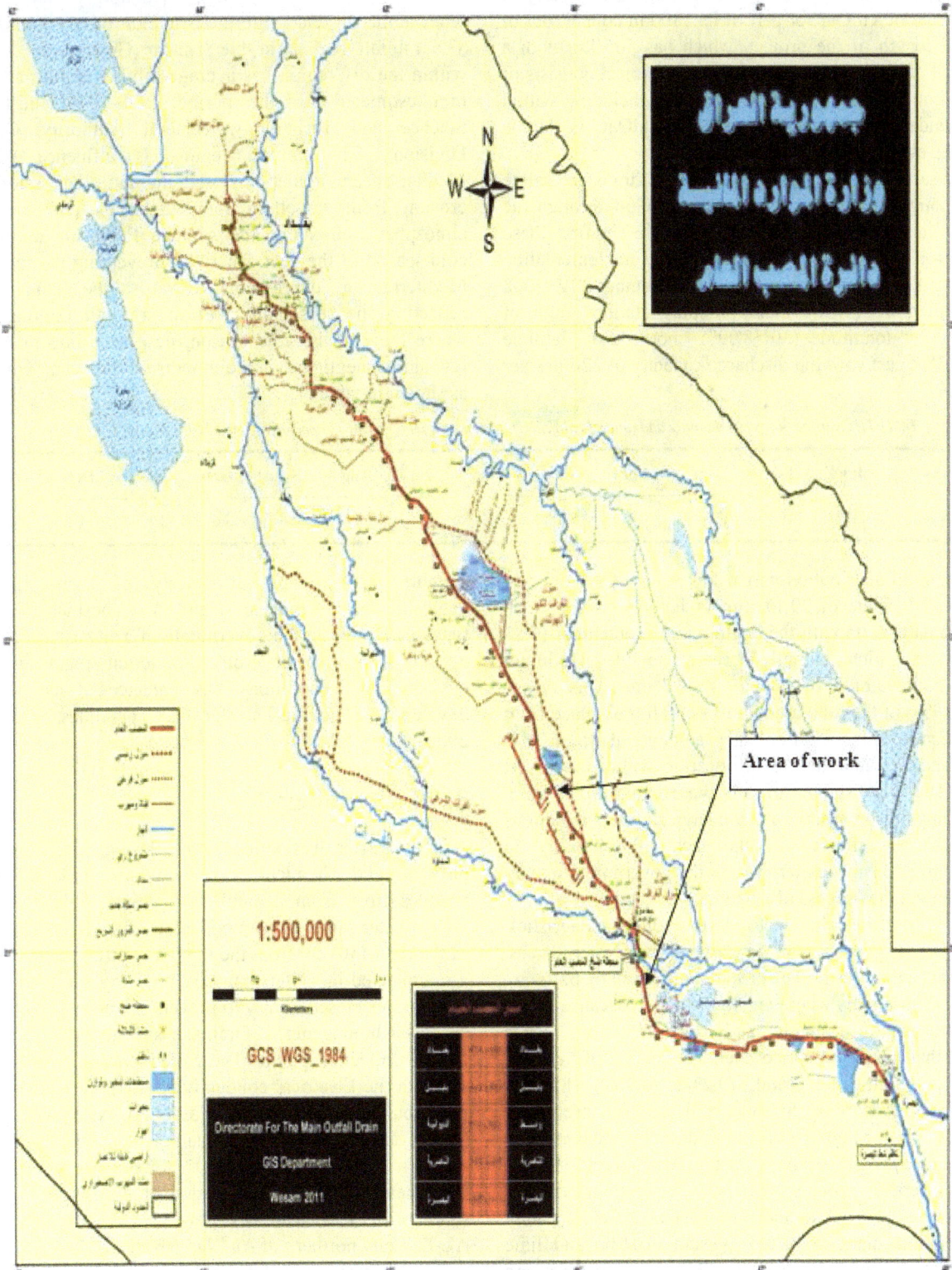

Fig (1): AL-Masb AL-Aam map (Fahad K. 2006)

The project Al-masab Al-aamis one of the major development projects in Iraq aimed at removing salt water collected from the reclamation lands in the central and southern Iraq byinterconnected drainage network . The idea of setting up AL-masabALaam came in the Fifties, when a U.S. company worked to prepare a map of networks drainages between the rivers Tigris and Euphrates.Al-Masb Al-Aam shown in Figure (1) passes through the territory of governorate of ThiQar for part of its 180 km course out of the total length of the project, which has 565 km with a discharge of the total 220 m3 / s and the project consists of several important stages in the operation, including within the boundaries of the province of ThiQar is Home drainages estuaries.

Al-masab AL-aamdrainage networks transfer saline water though the project to the Arabian Gulf through the Home of its parts (North and East andSouth. The first phase of the project started in 1973 followed by implementation of different phases and completed on December , 12 ,2008 by installation of the main pumping station south of Nasiriyah at longitude (30°58'09") north and latitude (46°20' 38") east with the discharge capacity of 220 m3 per second.

The Project area is influenced by the prevailing climatic regime and soil characteristics in relation to crops productions. The climatic condition crop impact influence the yield, the surrounding the process of agricultural production such as soil and water resources, and includes the activity of workers in agriculture and vitality.

The most dominate climatic factor is the rainfall region as influence soil moisture availability and water renewability in relation to the irrigation water requirement. The rainfall regime at the ThiQar Governorate located within the dry region has average rainfall depth of 127.48 mm estimated for the period 1970-2010(Ministry of Science and Technology, General Authority, Climate Division). The rainfall regime is influence by the Mediterranean circulation system due toair depressions and erosion from travelling long distances and the high atmospheric air above the Arabian Peninsula during the cold season of the year. There by preventing the incursion Mediterranean depressions towards the study area, reflecting the seasonality of the rainfall regime, Only where they hit the cold season, despite falling rates very few almost negligible in the months of June and September as shown in table (1)

Table(1): Average 40 years monthly and annual rainfall depth (mm) at DhiQar, for the period (1970-2010) (Najim A.R. 2007)

Month	Jan	Feb	Mar	Apr	May	Jun	Jul	Aug	Sep	Oct	Nov	Dec	Tot
Depth(mm)	29.2	19.2	20.9	10.5	5.0	0.2	-	-	0.8	5.6	15	21.2	127.6

The rainfall distribution varied low value of 0.19 mm in June to high value of 29.17 mm in January as shown in table (1) data starts with the beginning of September (0.83 mm), after which rainfall increases to the maximum amountin the month of January (29.17 mm).This means that (90.9%) of the total amount of rainfall took place from November-April.In addition, the amount of rainfall is erratic and irregular due to its location within the desert region. The small amount of rain, the irregular distribution of the rainy season months and the large annual fluctuations, are problems reflected in the lack of reliability in agriculture whichmakes farming in the governorate mainly dependent on the process of irrigation.

The other influencing factor is the soil characteristics. The soil is of gypsum type in parts of the western areas, and desert sand and gravel in the south western parts are generally loose soils where erosion is active because of the scarcity of natural vegetation and drought. Analysis results showed that the physical and chemical content of the soils were clay, silt and sand (9.2%) (5.7%) (85.1%), respectively. According to triangle tissues soil are sandy soils this mix of rough tissue. Either density virtual reached (1.67 g / cm3).(Al-Zubaidy A. 1990 & Al-Dooryetc 1989)

The average soil properties were estimated for; theporosity (36%), while ,moisture content at field capacity (11%),organic matter (0.2%) gypsum (4.18%) and lime (11.9%) while for the (PH) at (7.7). he soil types are suitable for growing of variety of crops specially vegetables in Um eenayj) and Al-Gabeshiah. The soil types variability in the governorate of ThiQar has provided option of diverse of crops production and favorable environmental conditions has enhanced its potential development for various farming practices. (Fahad K. 2006).

3. Methodology

The analysis was implemented through determiningthe physical and chemical content of the collected water samples from from a number of points along the study areas. Water samples were collected from the surface layer of the river (30) cm from the top, and bottles made ofWater was collected in 5 liters capacity polyethylene bottles with the Nozzles closed tightly to prevent the entry of air after it has been homogenized. Drainage samples were collected from 4 collection points. The collection points samplingto estimate the Electrical conductivity (EC) were at; near the main pump station of AL-Masab AL-Aam southern of Nassiriyah designated by the simple D1,the AL-Fajr city northern of AL-Nassiriyah as D2, for Euphrates river in the point near the main pump station of AL-Masab AL-Aam southern of Nassiriyah as C1 and AL-Gharraf river in the AL-Fajr city northern of AL-Nassiriyah as C2.

The monthly sampling was one sample per station

covering the period July 2011 to June 2012.The physical analyses focused on estimating Total Dissolved Solids,Temperature,Turbidity and Electrical Conductivity while the chemical analysis covered   pH, Nitrate, Phosphate , Total Hardness, Potassium, Sulfate, Calcium, Magnesium, Chloride, and Sodium. The water sample of drainage water blending with water of nearest river with different ratios is shown in table (2)

Table (2): Blending ratio percentage % of the drainage and river water

Blending ratio	Drainage water %	River water %
R1	90	10
R2	80	20
R3	70	30
R4	60	40
R5	50	50
R6	40	60
R7	30	70
R8	20	80
R9	10	90

4. Discussion

The analyses of the Electrical conductivity at the 4 observation points (EC) ranged from 6500 to 8000 Micro Siemens /cm .Higher and almost identical values were observed at two reaches D1 and D2 at near the main pump station of AL-Masab AL-Aam southern of Nassiriyah and the AL-Fajr city northern of AL-Nassiriyah The lowest values of 1000 was at point C2 for AL-Gharraf river in the AL-Fajr city northern of AL-Nassiriyah as shown in figure (2). Low monthly values were observed during the month November for D1.D2 and C1. The high EC values indicate the salt content at these reaches. In case of salinity, the (EC) results showed for ALMasab AL-Aam water drainage a range between 6000 and 8000 Micro Siemens /cm. Thus it can be used for irrigating Eucalyptus, date-palm, barely, wheat (semi-dwarf) and cotton.(AbawiS.and Hassan M.1990 &Al-MosawiA.1977)

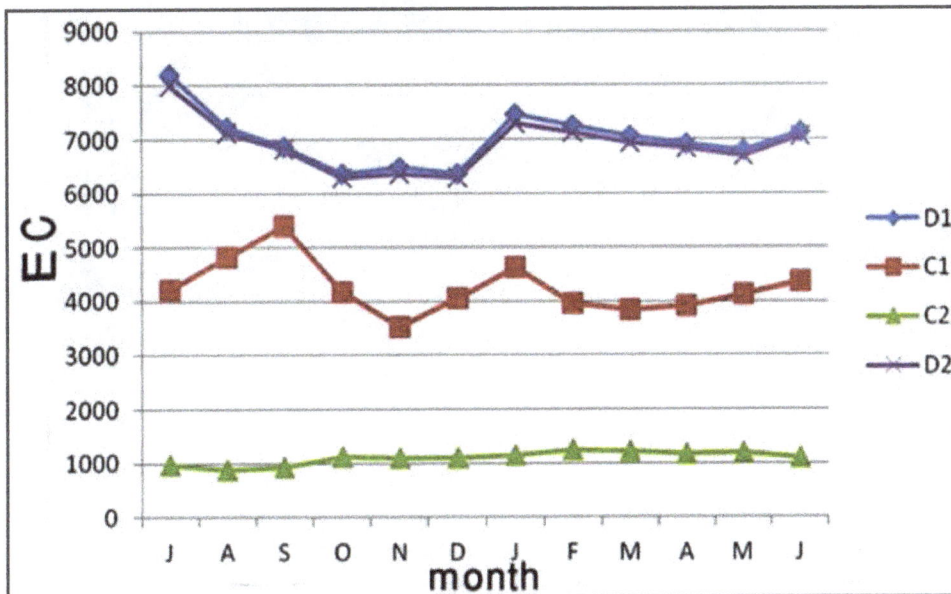

Fig.(2):Variation of the Electrical conductivity at the 4 observation points (C1,C2, D1 and D2)

Also analyses of the Sodium Adsorption Ratio (SAR) at C1 , C2 , D1 and D2 as was done of the Electric Conductively as shown in figure(3). The SAR values ranged from 2.5 to 13 with the higher values at point D1 near the main pump station of AL-Masab AL-Aam southern of nassiriyah, D2 at the AL-Fajr city northern of AL-Nassiriyah . The result of testing showed that the (SAR)

value for the Euphrates is between 6.89 and 11.2, for AL-Garraf river it is between 2.42 and 3.87 and for AL-Masab AL-Aam water the value of (SAR) is between 9.16 and 11.92. From the test results and comparison with the U.S. division (1985) we can deduce that the water of these stationsits acceptable for irrigation according to the (SAR) indicator where the (SAR) its less than 12.

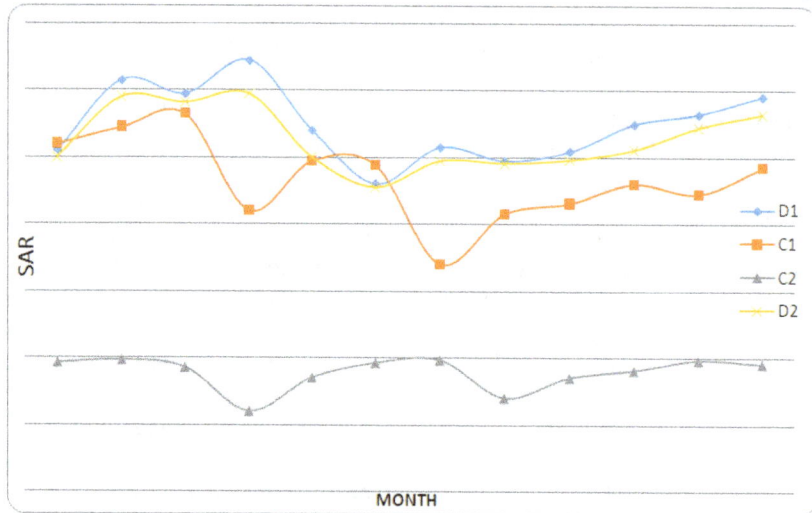

Fig.(3): *Variation of the Sodium Adsorption Ratio (SAR) at the 4 observation points (C1,C2, D1 and D2)*

Figure (4) shows the monthly results for Electrical conductivity (EC) for each blending ratio between the drainage water of AL-Masab AL-Aam and the water of Euphrates river , in the point near the pump station of AL-Masab AL-Aamsouthern of niassiriah .

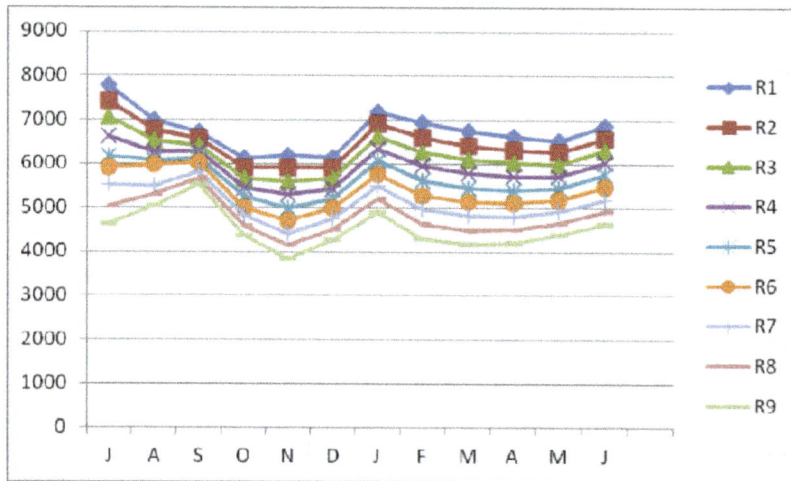

Fig.(4): *Variation of Electrical conductivity (EC) for different blending ratio of the drainage water of AL-Masab AL-Aam and the water of Euphrates River.*

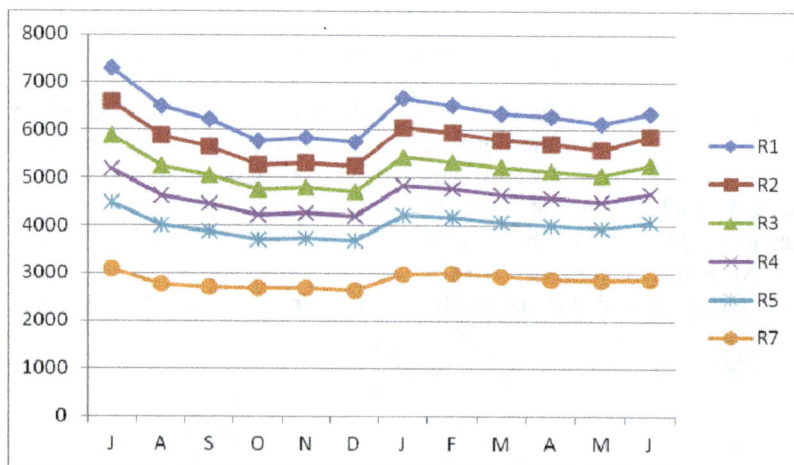

Fig (5): *the monthly results for Electrical Conductivity (EC) for each blending ratio between the drainage water of AL-Masab AL-Aam and the water of AL- Gharraf river in AL-Fajr city.*

The blending between the Euphrates river water and AL-Masab AL-Aam water showed that for blending ratio (R1) and (R2) it is greater than 6000 and less than 7000 (μs/cm) so it can be used to irrigate Sorghum. The blending ratio (R4) would work for wheat irrigation and the blending ratio (R7) for Sunflower

The blending ratio between AL-Masab AL-Aam and AL-Garraf river showed that for the blending ratio (R1) the result can used for irrigate Eucalyptus, date-palm ,barely, wheat(semi-dwarf) , cotton and sorghum , the blending ratio (R4) for sunflower irrigation, the result of blending ratio (R7) is good water for irrigation with electrical conductivity less than 3000 (μs/cm)

5. Conclusions

The conclusions drawn from this study are summarized as follow:

-- The analyses indicated that the Sodium Adsorption Ratio (SAR) for drainage water was less than 12, which is acceptable for irrigation use. In terms of salinity, the drainage water of Al-Masab Al-Aam its acceptable for irrigation because the halophytes can be irrigated with EC) less than 8000 Micro Siemens/cm. The blending between the drainage water and the fresh water of Al Gharraf river showed showed god blending ratio (R7) having the EC) less than 3000 Micro Siemens/cm.

-- Water from location D1 represents the water of AL-Masab Al-Aam water drainage . It may be suitable for

growing Date-palm , Barely , and Wheat(semi-dwarf) .

References

[1] Ayers, R. S., and D. W. Westcot. (1985). Water quality for agriculture. Irrigationanddrainage. No. 29. Roma, Italy. FAO.

[2] Gupta, I.C.1979.Anew classification and evaluation of quality of irrigation water for arid and semi- arid zones of India.Trans. Isdt and Ueds.

[3] FAO. (1992). the use of saline water for crop production. Irrigation and Drainage Papers. No. 48. Rome, Italy.

[4] الدوري ، وليد محمد ،ابراهيم شعبان السعداوي ، مؤيد العاني ، سعد المشهداني. (1989) . مقارنة الملوحة لاربعة تراكيب وراثية من الشعير . ألمجلة العراقية لبحوث علوم الحياة.

[5]) ملوحة التربة ، وزارة التعليم العالي والبحث 1990) الزبيدي ، احمد حيدر العلمي ، جامعة بغداد - بيتالحكمة.

[6] .)تأثير ملوحة التربة على انبات (1977) الموسوي ،عبد الله حمد .)رسالة.(Helianthus   annuus L.)ونمووانتاجية نبات عباد الشمس ماجستير جامعة بغداد.

[7] ، الهندسة العملية للبيئة 1990) عباوي ، سعاد عبد الله وحسن، محمد سليمان فحوصات الماء، وزارة التعليم العاليوالبحث العلمي - جامعة الموصل .

[8]) .تقييم بيئي لنهر الغراف احد الأفرع الرئيسة لــنهر 2006)() فهد ,كامل كاظم دجــلة ضمن قاطع مــدينة الناصرية, أطروحة دكتوراه , كلية الزراعة ,جامعة البصرة .

[9] () نجم عبد الله رحيم ، الخصائص الفيزيائية والكيميائية لتربة محافظة ذي قار وتأثيراتها في الإنتاج الزراعي، اطروحة دكتوراه ، كلية الآداب ، جامعة البصرة.2007 غير منشورة ،

28

Effective safety culture modelling in the maritime industry; an assessment of the Tanker ship subsector

Chinedum Onyemechi[1, *], Lazarus Okoroji[2], Declan Dike[2]

[1]Dept. of Maritime Management Technology, Federal University of Technology, Owerri, Nigeria
[2]Dept of Transport Management Technology, Federal University of Technology, Owerri, Nigeria

Email address:

c_onyemechi@yahoo.com (C. Onyemechi), okoroji_lazarus@yahoo.com (L. Okoroji), declanuba@yahoo.com (D. Dike)

Abstract: The work reviewed the several approaches to safety culture procedures applied in the maritime industry with a view to creating the best model for the sector. In this study, the importance of the international safety management code otherwise known as the ISM code were reviewed vis a vis the development of a proactive safety management culture by the entire tanker sub sector of the maritime industry. Comparison were made between different sources of maritime incidents such as design errors, human errors and organizational commitment to safety in a bid to model an effective safety culture for the entire maritime industry. Finally, the need to evolve safety measurement metrics best suited to analyze the safety demands of the maritime sector was emphasized. The developed model emphasizes an approach of safety orientation as part of the organization's safety philosophy. The adaptability of the model in the entire maritime industry as well as measurement procedures was also proposed.

Keywords: Safety Culture Modeling Safety Orientation, ISM Code, Maritime Incident Reporting, Human Errors

1. Introduction

Modeling of safety culture in the maritime industry to date has been based on the safety criteria recommended by the ISM code. The safety criteria set up by the ISM code provides three indicators presently used for safety analysis and evaluation. These criteria includes

i.   An active and established working process of continuous improvement.

ii.   Commitment from the top management of the company towards safety improvement.

iii. Motivated and encouraged personnel on board to actively initiate safety improvements.[1][2]

Some works stated above have found out barriers existing in different cultures which prevent the execution of the safety process. Another suggestion in the sector includes the demand for a move from a reactive approach to a proactive approach in the tanker sub sector of the maritime industry[3] This work had argued that most decisions of the International Maritime Organization (IMO) was based on lessons learned from previous accidents beginning from the Titanic which created SOLAS even up to the case of Herald Enterprise, which created the ISM code. See also [4]. Finally,

(Havold, 2007)[5] argued in his doctoral thesis for the need for the maritime industry to move from the reactive approach to the approach of safety orientation . All of the above are pointers to the need for the right modeling approach to safety culture in the maritime industry.

1.1. Objective

The objective of this paper is to review the safety modeling approach in the entire maritime industry and thus create an effective safety model for the sector.

2. Literature Review

Assessment of safety culture in the maritime industry has generally lacked instruments of measurement over the years. [5] The safety orientation approach is actually a practical safety culture assessment instrument that indicates the degree of orientation by a group or an organization towards safety.

Three regimes/cultures of development have been associated with safety regulation in recent times.[2] They include firstly, the regime of punishment where responsible parties are assigned by law to pay compensation for their safety liabilities. A good example is American's Oil Pollution Act of 1990 (OPA '90). The second regime has been described as the regime of compliance. Under this regime prescriptive rules concerning ship construction and other safety matters are required to be complied with. The concept of continuous improvement was established by the "quality gurus" in the fifties and sixties of which the American businessman Philip Crosby was one of the best known.

He stated that in order to reach the objectives of a quality management system it is of importance to look into the task as a continuous process where nonconformities continuously are reported and corrected. This process was described in his work Quality is free where he wrote about this process as the 14-step program for continuous improvement. This program was implemented in the American company ITT which by then had revenues of USD 15 billion and 350.000 employees, which made them to be one of the largest companies in the world .The theory of continuous improvement as a never ending process as a system for Quality/Safety Management is well established within all kind of industries and one of the best known is the Kaizen model developed by Toyota in Japan during the fifties. Kaizen is the Japanese word for continuous improvement and is a policy of constantly introducing small changes in order to improve quality/safety. This was assumed, because of their presence, that it is the people within the business who are the ones to best identify where there is room for improvements.

The system can be operated at the individual level or by Kaizen Groups or Quality Circles which are groups for identification of improvements. One issue of importance in order to make Kaizen effective, is the culture of trust between staff and manager including good communication both ways, and an open-minded and democratic view of the employee .

This can be related to the cornerstones of the ISM Code, "Commitment from the top" and continuous improvement by reporting, analyzing and implementing corrective actions as described in section 9 of the ISM Code.

In the work of Sagen [4] he describes a theory developed by Heinrich who stated that the relation between serious accidents and minor accidents is 1 to 30 and minor accidents and near accidents (near misses) is 30 to 300, i.e. 300 near accidents results in up to 30 minor accidents and 30 minor accidents results in 1 serious accident. This theory is commonly accepted within the science of safety research (Sagen, 1999)[4].

Good examples of safety regulations made during this regime include: International Convention for Safety of Life at Sea (SOLAS) and the International Convention on Standards of Training, Certification and Watch keeping for Seafarers (STCW) [6].The third regime or culture is known as the regime or culture of self regulation, this regime is based on standards established by the industry itself. The

ISO 9001 Quality standard and the ISM code are good example of this safety culture regime. [6]

The ISM code itself did not specify statistical measures of performance like its counterpart ISO 9001. However; it literally demands the implementation of the process of continuous improvement regularly. [2] In handling this problem, (Mejia, 2001)[7] suggested the use of qualitative assessment instruments that measures system effectiveness. Mejia defined effectiveness to mean`` the issues of whether desired results are actually achieved". Coming from the backdrop of Policy Management, an evaluation of the ISM code was made categorizing them into outputs and outcomes. The outcomes are the desired goals of the policy. For the ISM code these include:

(i)   The requirement to provide safe practices in ship operation and a safe working environment and the requirement to establish and safeguard against all identified risks.

(ii)   The requirement to continually improve safe management skills of the personnel ashore and aboard including preparation for emergencies and environmental protection.

(iii) The requirement to develop a safety culture. [7].

Next, Mejia set to define outputs as those set of policies which attempt to ensure that the safety management systems of the shipping companies and vessels are compliant with the ISM code. According to him, output measures of the ISM code will include:

- Port sate control detention due to non-conformities and deficiencies in regard to the requirements of ISM code.
- ISM related port inspections carried out by the Flag state Re-Inspections due to major non-conformities observed in connection with external audits performed by the administration.
- ISM deficiencies and non-compliance reported by the ships personnel.[7]

3. Methodology

The method of principal component analysis was applied to measure the output side of the ISM code as presented in the tanker output report and barge output report for the years 2007 and 2008.The total score factor coefficients was calculated and the output result presented. Factor coefficients of the inspections carried out between 2007 and 2008 were computed as a basis for analyzing safety improvement in the tanker ship sub sector.

4. Result Presentation

Analysis of safety in the industry has focused mainly as exposed by the diagrams outlined below. Designs of safety in the system are usually controlled by policies that will eliminate the occurrence or likely occurrence of these errors. Safety result as analyzed by the diagram below reported casualties in percentage. Other measures are revealed subsequently by subsequent diagrams.

An assessment of the output aspects will be shown in this work applying the principal component analytical method to the next table.

4.1. ISM Code Output Analysis Using Principal Component Analysis

The table below contains output reports of inspections carried out on tankers and barges in two subsequent periods, the years 2007 and 2008.The variations in the number of reports requested and those submitted are reflected in the table as differences. The reports from barges were also reflected in the table.

Table 1. *SIRE Statistics (Source[8],)*

	Jan-Dec 2007	Jan-Dec 2008	Difference
Tanker reports submitted	15,730	16,452	+722
Tanker reports requested	52,527	59,736	+7,209
Total tanker vessels in the system	6,222	6,553	+331
Reports per tanker vessel per annum	2.6	2.6	
Barge reports submitted	4,576	4,879	+303
Barge reports requested	5,786	8,164	+2,378

Table 2. *Principal Component Analysis: C1, C2 Eigenanalysis of the Correlation Matrix*

Eigenvalue	1.9990	0.0010
Proportion	1.000	0.000
Cumulative	1.000	1.000
Variable	PC1	PC2
C1	0.707	0.707
C2	0.707	-0.707

Table 3. *Principal Component Analysis: C1, C2 Eigenanalysis of the Covariance Matrix*

Eigenvalue	868214156	423172
Proportion	1.000	0.000
Cumulative	1.000	1.000
Variable	PC1	PC2
C1	0.661	0.750
C2	0.750	-0.661

The percentage of variation explained by the analysis is thus 100%.In the correlation matrix the resultant equation of variation between the two years 2007 and 2008 varies accordingly as follows:

$$Y = .707X1 + .707X2$$

Likewise in the covariance analysis result obtained from the principal component analysis, a 100% variation was obtained from the first analysis and zero percent in the second. The resultant variation equation is as follows:

$$Y = 0.661X1 + 0.750X2$$

Table 4. *Factor Analysis: C1, C2 Principal Component Factor Analysis of the Correlation Matrix Unrotated Factor Loadings and Communalities*

Variable	Factor1	Factor2	Communality
C1	1.000	0.022	1.000
C2	1.000	-0.022	1.000
Variance	1.9990	0.0010	2.0000
% Var	1.000	0.000	1.000

Table 5. *Factor Score Coefficients*

Variable	Factor1	Factor2
C1	0.500	22.475
C2	0.500	-22.475

From the factor analysis we conclude from results that the output of tanker and barge reports submitted varies effectively with a variation of:

$$Y = .5000X1 + .5000X2.$$

Thus we conclude that the reports submitted called the output yielded a satisfactory result for the periods 2007 and 2008 based on principal component analysis. Thus we now have a means through which government institutions in control of maritime operations can measure the outputs of shipping firm's compliance to instituted regulations like the ISM code.

5. Conclusion

The work analyzed safety measurement standards in the entire maritime industry and recommended new ways for assessing the outputs of safety regulations controlling safety in the industry such as the ISM code. A new way of applying new safety metrics such as the principal component analysis in assessing the outputs of the industry was demonstrated. The view of this work is that the culture of continuous improvement be adopted by all concerned as required by the ISM code while measurement metrics like the one prescribed in this work be used to ascertain the output efficiency of the industry from time to time.Modelling of safety culture in the tanker sector of the maritime industry from this work shows a way of analyzing the impact of inspections carried out in a given period.From the work ascertained the effect of these inspections carried out on both tanker and barges during the period of review.

References

[1] Lappalainen,J Transforming maritime safety culture,Publications from the centre for maritime studies,A 46/2008,Turku. (2008)

[2] Lappalainen,J and Salmi K Safety culture and maritime personnel's safety attitude interview report, Publications from the centre for maritime studies,University of Turku A 48/2009,(2009)

[3] Praised,R. From compliance culture toward safety culture: World Maritime University Malmo, Sweden. (2009)

[4]   Sagen ,A The ISM code in practice, Oslo, Norway: Tano Aschelong,AIT OHa,As Norway. (1999)

[5]   Havold,J. From safety culture to safety orientation, developing   a tool to measure safety in shipping; NorwegianUniversity of Science and Technology, NTNU(2007)

[6]   Kristiansen,S. Maritime Transportation-safety management and risk Analysis, Elsevier, Amsterdam,(2005).

[7]   Mejia, M. Performance criteria for the international safety management (ISM) code: Proceedings of the 2nd General Assembly of IAMU. International Association Maritime Universities 2/5 October 2001-kobe japan. (2001)

[8]   OCIMF, "Annual Report", Oil Companies International Marine Forum: London. (2009). Science, 198 Aschehoug, Oslo.

29

Statistical characterization of extreme hydrologic parameters for the peripheral river system of Dhaka city

Sarfaraz Alam[*], Muhammad Sabbir Mostafa Khan

Department of Water Resources Engineering, Bangladesh University of Engineering and Technology, Dhaka 1000, Bangladesh

Email address:

sarfaraz@wre.buet.ac.bd (S. Alam), mostafakhan@wre.buet.ac.bd (M. S. M. Khan)

Abstract: Selection of appropriate probability distribution function is one of the most important steps of frequency analysis. Due to the existence of large number of distributions, hydrologists follow different methods to select the best one. In this paper, annual maximum, minimum water level and discharge of five peripheral rivers, namely Buriganga, Turag, Tongi, Balu and Lakhya around Dhaka city have been analyzed to compute the basic statistics and fit them with sixty two probability density functions (PDF). Three goodness-of-fit (GoF) statistics, namely Chi-square, Kolmogorov–Smirnov and Anderson Darling were used to rank each of the distribution. Furthermore, ranks obtained from three GoF were used to compute overall rank of all distributions for each hydrologic parameter. The study reveals that, four different distributions were found best fit for four extreme cases. Dagum (4P) and Chi-square (2P) fit best for annual maximum and minimum water level respectively, whereas Cauchy and Johnson SB were found for annual maximum and minimum discharge respectively. Moreover, ranks of frequently used distributions, namely General Extreme Value (GEV), Log-Pearson III (LP3), Log-normal (LN) and Gumbels were compared with the best fit distributions and did not give satisfactory results. The method used in this study would be helpful for flood frequency analysis of other rivers of Bangladesh. This may also be used for evaluation of best fit distribution of river system for other countries as well.

Keywords: Probability Distribution, Rank, Water Level, Discharge, Dhaka City

1. Introduction

Design of different types of hydraulic structures and flood plain zoning, economic evaluation of flood protection projects, etc. require information on flood magnitudes and their frequencies (Rakesh 2005). In addition low flow frequency analysis is also required for assessing water quality, water availability, navigability etc. Moreover climate change associated with global warming got potentiality to change the probability of these events. In particular, highly populated cities surrounded by rivers are vulnerable to flooding and pollution. Urban development and industrialization demand good estimation of both water level and discharge. Thus, it is necessary to explore the water level and discharge extremes around major cities, especially which are in highly populated regions dominated by socio-economic development and prone to pollution and flooding.

Reliable flood frequency distribution selection is one of the major problems faced by the hydrologists. This issue is very much important as different distributions may produce significantly different estimates for the same return period (Coulson 1991). In order to identify the appropriate probability distribution function a wide range of researches have been conducted previously (e.g., Cunanne 1973; Stedinger 1980; Stedinger et al.1992; Vogel RM 1993; Markiewicz et al. 2006; Mitosek et al. 2006; Laio F et al. 2009; Haddad K. 2010, Rahman A S. et al. 2013). Some of the commonly used distributions for annual maximum flood series include Extreme Value Type 1 (EV1), General Extreme Value (GEV), Extreme Value Type 2 (EV2), Two component Extreme Value, Normal, Log Normal (LN), Pearson Type 3 (P3), Log Pearson Type 3 (LP3), Gamma, Exponential, Weibull, Generalised Pareto and Wakeby (Cunnane 1989; Bobee et al.1993). Furthermore, there is no theoretical basis for the use of a single distribution over others, especially when analyzing a relatively short annual flood series for predicting the magnitude of extreme events (Gumbel 1958). Goodness of fit test is often used to select appropriate probability distribution for frequency analysis. Discrepancies may also arise from the use of different parameter estimation methods, such as method of moments, maximum likelihood and probability-weighted-moments (Betül Saf 2009). Different goodness-of- fit statistics, such as Chi-square, Kolmogorov–Smirnov and Anderson Darling may also show different ranks for each type of distribution.

A number of studies have been conducted on the rivers of Bangladesh using frequently used distribution functions. Ferdows M (2005) compared three frequently used distributions, namely Log Normal (Two parameters, LN2 and three parameters, LN3); Extreme value Type-l (EVl) or Grumbel and Log-person type-3 (LP3), using annual peak flow of three major rivers (Meghna, Brahmaputra, Ganges-5 stations). In Bangladesh, four distributions are mainly used for at-sit frequency analysis of annual maximum discharge. Gumbel and LN distribution are used by Bangladesh Water Development Board and few departments/ firms respectively, whereas LP3 and GEV were used in National Water Plan and Flood Action Plan respectively (Karim MA and Chowdhury JA. 1995). Karim MA and Chowdhury JA. (1995) suggested the use of GEV for flood frequency analysis of the rivers of Bangladesh. Bari MF. et al.(2002) found that LP3 is more suitable for low flow frequency analysis for the rivers in North-West Bangladesh. So, almost all the researches on low and high flow were conducted using frequently used distributions.

In this study an attempt has been made to identify the best distributions for annual maximum, minimum water level and discharge. The study area is located at one of the most densely populated region in the world. Dhaka, the capital city of Bangladesh is surrounded by a number of rivers, namely Buriganga, Turag, Balu, Lakhya and Tongi. Annual maximum, minimum water level and discharge data of these rivers were used to compute the basic statistics and fit them with sixty two distributions. Three goodness-of-fit (GoF) statistics, namely Chi-square, Kolmogorov–Smirnov and Anderson Darling were used to rank each distribution. Furthermore, median of the ranks obtained from three GoF for each distribution were used to compute overall rank of

all PDFs for each extreme hydrologic parameter, likewise Table 5 shows the ranks for annual maximum water level . Top three distributions for each station and each type of dataset were finally computed. Ranks of frequently used distributions, namely GEV, LP3, LN and Gumbels are also shown to assess their performance for different hydrologic extreme cases.

2. Study Region and Data

2.1. Study Region

Dhaka city is highly populated and influenced by the rapid socio-economic growth. The area of the city is 360 km2 with an estimated growth rate of 4.2% per annum that labeled the city as a mega city (Haigh, 2004; Karn and Harada, 2001; Rahman S. and Hossain, F., 2007). Elevation above MSL is 4 m. Buriganga, Turag, Balu, Tongi and Lakhya are the peripheral river system around Dhaka city as shown in Fig 1. The area is greatly affected by the monsoon rainfall leading to flood and water logging. Considerable attention has been paid to understand the occurrence of flood of different return period.

2.2. Data Collection

Yearly maximum, minimum water level for 6 stations and discharge for 3 stations were collected from Bangladesh Water Development Board (BWDB). Location of the gauging stations can be referred to Fig 1. Detailed information of the dataset is given in Table 1. Other than one station on Turag river all the stations are subjected to tidal flow.

3. Methodology

In this research sixty one probability distribution functions (PDF) were used to fit the observed water level and discharge dataset. The list of PDF's used are given in Table 2.

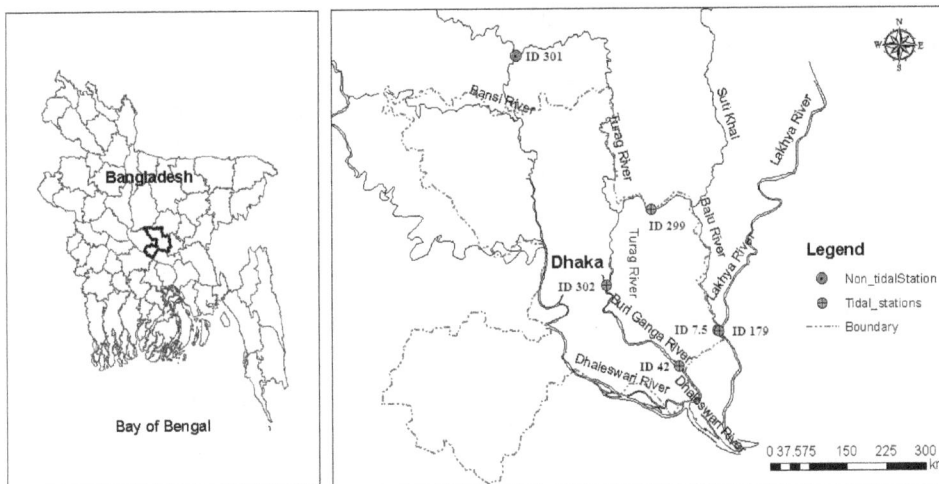

Figure 1. (a) Bangladesh map (b) Peripheral river system of Dhaka city (right).

Table 1. Dataset of water level and discharge station location

Station id	Station	Longitude	Latitude	River	Type	Time interval
301	Kaloikor	90.210	24.082	Turag	Non Tidal WL & Q	1949-2009
42	Dhaka (Mill Barak)	90.445	23.677	Buriganga	Tidal WL	1909-2009
302	Mirpur	90.338	23.783	Turag	Tidal WL	1954-2009
299	Tongi Khal	90.404	23.882	Tongi	Tidal WL	1960-2009
179	Demra	90.505	23.723	Lakhya	Tidal WL & Q	1952-2009
7.5	Demra	90.502	23.723	Balu	Tidal WL & Q	1962-2009

Table 2. Probability density functions used for fitting all hydrological dataset.

Beta	Burr	Burr (4P)	Cauchy	Chi-Squared	Chi-Squared (2P)	Dagum	Dagum (4P)
Erlang	Erlang (3P),	Error	Error Function	Exponential	Exponential (2P),	Fatigue Life	Fatigue Life (3P)
Frechet	Frechet (3P)	Gamma	Gamma (3P)	Gen. Extreme Value	Gen. Gamma	Gen. Gamma (4P)	Gen. Pareto
Gumbel Max	Gumbel Min	Hypersecant	Inv. Gaussian	Inv. Gaussian (3P)	Johnson SB	Kumaraswamy	Laplace
Lognormal	Lognormal (3P)	Nakagami	Normal	Pareto	Pareto 2	Pearson 5	Pearson 5 (3P)
Pearson 6	Pearson6 (4P)	Pert	Power Function	Rayleigh	Reciprocal	Reciprocal	Rice
Student's t	Triangular	Uniform	Weibull	Weibull(3P)	Rayleigh (2P)	Levy	Levy(2P)
Log-Gamma	Log-Logistic	Log-Logistic(3P)	Log-Pearson 3	Log-Normal	Johnson SU		

For each station annual maximum, minimum water level and discharge were plotted using all the distributions above (few were found not applicable in some cases). Goodness of fit was performed using Chi-Square, Kolmogorov-Smirnov and Anderson Darling. Ranks according to these three goodness of fit showed a great variation. Median of the ranks obtained from goodness of fit for each PDF was used to rank all the PDFs. A sample example is given in Table 3 showing ranks of maximum water level of Turag river at station 301. PDF's such as General Extreme Value (GEV), Log Pearson Type III, Log Normal and Gumbels which are mostly used for extreme value analysis are shown in Table 3 to highlight their ranking. All the distributions were fitted and ranked using the tool Easyfit. Three goodness of fit method gave separate ranking for each PDF. Median of the ranks obtained from three GoF for each distribution was used to compute the overall rank of all the PDFs. This procedure was followed for maximum, minimum water level and discharge for each station. Finally all these ranks were used to obtain the best fit distributions for each extreme hydrologic parameter. A sample example for station 301 (Turag) is shown in Table 9 (Appendix) which illustrates the best fit distributions obtained following the above procedure. Best fit distributions found from different GoF statistic and median of the ranks for station 301 are shown in Fig 2.

4. Results and Discussion

4.1. Basic Statistics

In order to understand the statistical properties of annual maximum and minimum water level and discharge of the peripheral rivers of Dhaka city, basic statistics have been computed (Table 3-4). We have described descriptive statistics such as max, min, range, min, variance, standard

deviation, coefficient of variation, standard error, skewness and excess kurtosis. Considering annual maximum water level (Table 3), station 301 (Turag) gave both highest and lowest WL values of 10.48 and 2.27 mPWD respectively with highest mean of 7.7 mPWD. Maximum and minimum range of 8.21 and 2.55 mPWD were found at station 301 and 179 (Lakhya) respectively. Standard deviation and coefficient of variance were also highest at station 301 with values 1.78 and 0.23 respectively, whereas minimum values were 0.5 and 0.09 respectively at station 179. Other than station 179 all the stations showed negative skewness. In context of excess kutosis station 299(Tongi) gave highest value of 4.01 and the lowest value 0.69 was found at station 179. Similarly considering annual minimum water level, station 301(Turag) is having highest range 2.42 mPWD (2.91-0.49) and standard deviation 0.56. Whereas station 301 (Lakhya) was found to have lowest range 0.94 mPWD and st. deviation 0.16. Highest and lowest coefficient of variation were found 0.34 and 0.18 at station 301 and 7.5 (Balu) respectively. Stations other than station 301 and 7.5 possess positive skewness. With regard to excess kurtosis station 301 has a negative value of -0.76, whereas maximum value of 3.64 is associated at station 302 (Turag).

Moreover considering annual maximum discharge of all three stations (Table 4), station 179 gives the highest and mean discharge 2610 m3/s and 1947.3 m3/s respectively with minimum standard deviation, coefficient of variance and skewness of 378.44, 0.19 and -1.35 respectively. Maximum standard deviation and coefficient of variation were found 462.94 and 0.91 at station 301 and 7.5 respectively. Excess kurtosis of 13.53 at station 7.5 was quite higher than the minimum value 2.58 found at station 301. Similarly considering annual minimum discharge of all the stations, station 179 gives highest mean and standard deviation of 647.24 m3/s and 316.94 with negative skewness and kutosis of -0.02 and -0.81 respectively.

Whereas lowest mean and standard deviation were computed at station 301, though it possess highest coefficient of variation of 1.04 and excess kurtosis 14.77. Considering all discharge dataset it was found that discharge variation is highest at station 301 (2200-0.42 m3/s), standard deviation also varied greatly (462.94 to 3.26) at this station. Station 179 has a tendency to show negative skewness which is opposite to others. So, considering all the annual water level dataset highest variation (10.48-0.49) and standard deviation (1.78, 0.56) are associated with station 301 which is non tidal in behavior. Most of the stations showed negative skewness while annual maximum WL considered, whereas the scenario was vice versa considering annual minimum WL. Other than one case (Annual minimum WL, station 301 = -0.76) all the stations showed positive excess kurtosis.

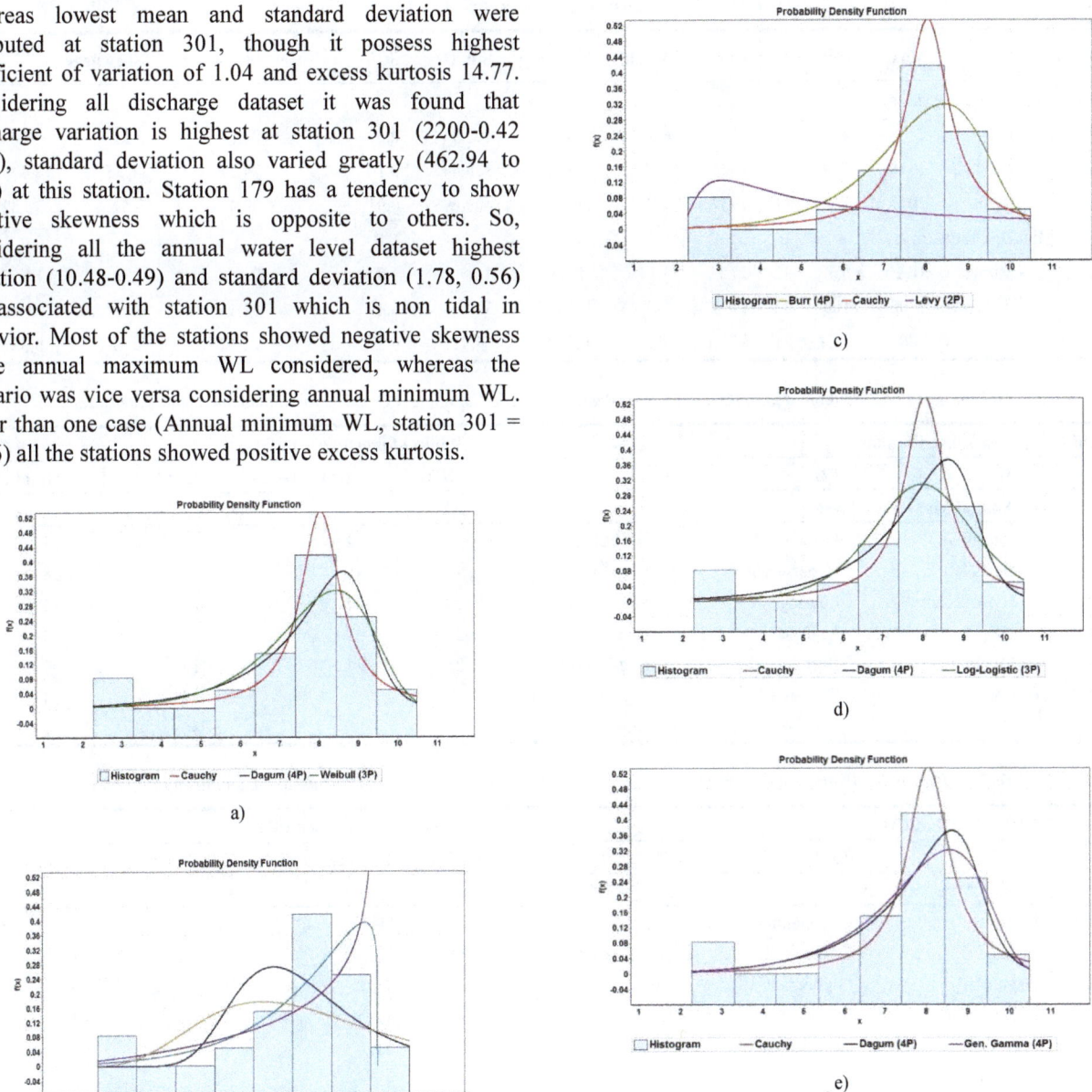

a)

b)

c)

d)

e)

Figure 2. *Probability density function plotting of station 301 for annual maximum WL a) Best 3 based on the median of ranks obtained from Goodness-of-fits b) Frequently used distributions c) Best 3 according to Kolmogorov Smirnov d) Best 3 according to Anderson Darling e) Best 3 according to Chi-Squared*

Table 3. *Descriptive statistics of annual water level (mPWD) for individual stations*

Station Id	River	Max	Min	N	Range	Mean	Variance	Std. Dev	Coeff. of variation	Std. error	Skewness	Excess Kartosis
					Annual Max. Water Level (mPWD)							
301	Turag	10.48	2.27	60	8.21	7.7	3.18	1.78	0.23	0.23	-1.8	3.38
42	Buriganga	7.58	3.2	94	4.38	5.77	0.38	0.62	0.11	0.06	-0.37	3.12
302	Turag	8.35	2.71	54	5.64	6.18	0.9	0.95	0.15	0.13	-0.63	2.57
299	Tongi	7.84	2.71	49	5.13	5.99	0.88	0.94	0.16	0.13	-1.34	4.01
179	Lakhya	7.11	4.56	48	2.55	5.84	0.25	0.50	0.09	0.07	0.39	0.69
7.5	Balu	7.09	2.57	46	4.52	5.82	0.47	0.69	0.12	0.10	-2.19	10.47
					Annual Min. Water Level (mPWD)							
301	Turag	2.91	0.49	60	2.42	1.66	0.31	0.56	0.34	0.07	-0.17	-0.76
42	Buriganga	1.23	0.24	95	0.99	0.66	0.03	0.16	0.25	0.02	0.48	1.82
302	Turag	1.71	0.12	56	1.59	0.80	0.06	0.24	0.29	0.03	0.64	3.64
299	Tongi	1.52	0.53	50	0.99	0.94	0.04	0.20	0.21	0.03	0.86	1.15
179	Lakhya	1.42	0.48	49	0.94	0.84	0.03	0.16	0.20	0.02	0.98	2.83
7.5	Balu	1.4	0.34	48	1.06	0.93	0.03	0.17	0.18	0.02	-0.29	2.96

Table 4. Descriptive statistics of discharge(m³/s) for individual stations

Station Id	River	Max	Min	N	Range	Mean	Variance	Std. Dev	Coeff. of variation	Std. error	Skewness	Excess Kartosis
Annual Max. Discharge (m³/s)												
301	Turag	2200	4	32	2196	686.11	214320	462.94	0.67	81.84	0.91	2.58
179	Lakhya	2610	774	26	1935.8	1947.3	143210	378.44	0.19	74.22	-1.35	4.3
7.5	Balu	2077.3	140	21	1937.3	451.47	167850	409.69	0.91	89.4	3.52	13.53
Annual Min. Discharge (m³/s)												
301	Turag	18.2	0.42	32	17.78	3.14	10.64	3.26	1.04	0.58	3.38	14.77
179	Lakhya	1300	106	26	1194	647.24	100450	316.94	0.49	62.16	-0.02	-0.81
7.5	Balu	288.52	16.1	21	272.42	73.33	4214.9	64.92	0.89	14.17	2.13	5.38

Table 5. Best fit distributions and ranks of GEV, Log Pearson III, Log Normal and Gumbels for annual maximum water level.

St. id	Best 3 distributions			Ranks of mostly used PDF			
	#1	#2	#3	GEV	Log Pearson III	Log Normal	Gumbels
	Annual Max. water level						
301	Cauchy	Dagum (4P)	Weibull (3P)	25	31	41	34
42	Burr (4P)	Hypersecant	Dagum	11	46	29	28
302	Hypersecant	Log-logistic	Laplace	28	47	29	35
299	Log logistic (3P)	Laplace	Dagum(4P)	27	50	35	30
179	Dagum	Burr	Log-Logistic	25	11	14	44
7.5	Cauchy	Dagum(4P)	Log logistic (3P)	29	52	34	28
All	Dagum (4P)	Log-Logistic (3P)	Burr (4P)	24	30	46	31

Table 6. Best fit distributions and ranks of GEV, Log Pearson III, Log Normal and Gumbels for annual minimum water level.

St. id	Best 3 distributions			Ranks of mostly used PDF			
	#1	#2	#3	GEV	Log Pearson III	Log Normal	Gumbels
	Annual Min. water level						
301	Dagum(4P)	Gen. Gamma	Error	10	29	16	28
42	Laplace	Erlang	Hypersecant	14	33	34	41
302	Dagum(4P)	Log-logistic	Burr	8	7	14	34
299	Chi-squared	Burr (4P)	Chi-squared(2P)	17	11	19	43
179	Cauchy	Chi-squared(2P)	Chi-squared	24	21	17	44
7.5	Chi-squared(2P)	Log logistic (3P)	Logistics	13	48	34	32
All	Chi-Squared (2P)	Burr (4P)	Chi-Squared	10	28	16	42

Table 7. Best fit distributions and ranks of GEV, Log Pearson III, Log Normal and Gumbels for annual maximum discharge.

St. id	Best 3 distributions			Ranks of mostly used PDF			
	#1	#2	#3	GEV	Log Pearson III	Log Normal	Gumbels
	Max. Discharge						
301	Error	Hypersecant	Laplace	10	27	45	16
179	Cauchy	Log-logistic(3P)	Laplace	20	40	23	29
7.5	Cauchy	Burr	Log-logistic(3P)	19	31	12	19
All	Cauchy	Laplace	Log-Logistic (3P)	17	36	25	18

Table 8. Best fit distributions and ranks of GEV, Log Pearson III, Log Normal and Gumbels for annual minimum discharge.

St. id	Best 3 distributions			Ranks of mostly used PDF			
	#1	#2	#3	GEV	Log Pearson III	Log Normal	Gumbels
	Min. Discharge						
301	Fatigue Life	Logistic	Log Normal (3P)	8	7	32	47
179	Dagum	Error	Johnson AB	4	10	40	19
7.5	Gen. Pareto	Johnson AB	Fatigue Life (3P)	4	12	2	52
All	Johnson SB	Gen. Extreme Value	Inv. Gaussian (3P)	2	8	35	50

4.2. Ranking of Probability Distribution Function

All the PDF were ranked for both water level and discharge at each station. Finally all those ranks were used to determine the ranking of PDF for similar type of dataset, such as max. wl, min. wl, max. Q and min. Q. Table 5-8 describes best three distributions for each type of dataset, ranks of mostly used distributions are also added in those tables.

Best fit distributions for annual maximum water level are listed in Table 5. It reveals great combination of various types of distributions. Considering all dataset Dagum(4P), Log-logistic(3P) and Burr (4P) were found the best three. On the other hand ranks of frequently used distributions GEV, LP3, LN and Gumbels were 24, 30, 46 and 31 respectively. Similarly, ranks of distributions for minimum WL are listed in Table 6. Frequently used distributions fell behind in this case too. The ranks obtained are 10, 28, 16 and 42 respectively. Whereas the best fit distributions are Chi-squared (2P), Burr (4P) and Chi-Squared.

Similar to water level, discharge dataset were also fitted to different distributions (Table 7-8). In case of annual maximum discharge various distributions were found best fit for different stations. Frequently used distributions didn't show good performance in this case. Ranks of frequently used distributions GEV, LP3, LN and Gumbels were 17, 36, 25 and 18 respectively. Best fit distributions were Cauchy, Laplace and Log-logistic (3P). Whereas, considering annual minimum discharge GEV (Generalized Extreme Value) gave a good result in combined ranking. But separately each station gave different best fit distributions. Best fit distributions for annual minimum water level were Johnson SB, GEV and Inv. Gaussian (3P).

5. Conclusion

In this study annual maximum, minimum water level and discharge of the peripheral river system of Dhaka city have been analyzed to identify the best fit distribution among sixty two distributions. Three goodness-of-fit statistics (GoF), namely Chi-square, Kolmogorov-Smirnov and Anderson Darling were used to rank all distributions median of the ranks obtained from three GoF for each distribution were used to compute overall rank of all PDFs for each extreme hydrologic parameter. The study reveals that, four different distributions were found best fit for four extreme hydrologic cases. Dagum (4P) and Chi-square (2P) fit best for annual maximum and minimum water level respectively, whereas Cauchy and Johnson SB were found for annual maximum and minimum discharge respectively. The ranks of frequently used distributions GEV, LP3, LN and Gumbels were not satisfactory for almost all the hydrologic parameters.

Recommendation

The method used in this study would be helpful for flood frequency analysis of other rivers of Bangladesh. This may also be used for evaluation of best fit distribution for other countries as well.

Acknowledgements

Authors are gratefully acknowledging the cooperation rendered by Bangladesh Water Development Board (BWDB) for providing the necessary data.

Appendix

Table 9. Ranking of PDF for maximum water level of Turag river at Kaloikor station (st. id 301)

Distribution	Kolmogorov Smirnov Ranking of PDF for maximum water level of Turag river at Kaloikor station (st. id 301)		Anderson Darling		Chi-Squared		Median	
	Statistic	Rank	Statistic	Rank	Statistic	Rank	Value	Rank
Beta	0.15	10	1.69	7	11.00	15	10	9
Burr	0.16	12	2.78	13	11.09	16	13	13
Burr (4P)	0.11	5	1.20	6	4.12	3	5	4
Cauchy	0.09	1	0.61	1	0.79	1	1	1
Chi-Squared	0.42	51	13.42	47	85.33	53	51	53
Chi-Squared (2P)	0.37	50	10.98	43	70.84	51	50	51
Dagum	0.16	11	1.87	9	13.21	19	11	10
Dagum (4P)	0.10	3	0.97	2	10.52	14	3	2
Erlang	0.29	42	8.43	37	23.58	37	37	36
Erlang (3P)	0.22	28	4.39	21	16.77	31	28	27
Error	0.19	18	2.73	12	8.36	10	12	12
Error Function	0.92	59	554.77	60	932.97	54	59	59
Exponential	0.47	53	17.66	51	22.18	34	51	53
Exponential (2P)	0.43	52	14.76	49	8.86	12	49	49

Distribution	Kolmogorov Smirnov Ranking of PDF for maximum water level of Turag river at Kaloikor station (st. id 301)		Anderson Darling		Chi-Squared		Median	
	Statistic	Rank	Statistic	Rank	Statistic	Rank	Value	Rank
Fatigue Life	0.31	44	8.52	38	37.72	43	43	42
Fatigue Life (3P)	0.21	25	3.96	19	16.46	26	25	22
Frechet	0.37	49	12.13	45	67.57	50	49	49
Frechet (3P)	0.28	40	7.57	34	25.91	38	38	38
Gamma	0.23	30	6.18	29	13.27	20	29	28
Gamma (3P)	0.21	26	4.40	22	17.30	32	26	24
Gen. Extreme Value	0.13	8	12.53	46	N/A		27	25
Gen. Gamma	0.26	38	6.63	31	28.76	40	38	38
Gen. Gamma (4P)	0.12	7	1.12	3	6.06	7	7	7
Gen. Pareto	0.18	16	33.68	58	N/A		37	36
Gumbel Max	0.26	36	11.84	44	16.73	30	36	34
Gumbel Min	0.14	9	1.71	8	8.17	8	8	8
Hypersecant	0.18	15	2.91	14	11.65	17	15	15
Inv. Gaussian	0.20	23	7.23	33	15.92	22	23	19
Inv. Gaussian (3P)	0.20	22	3.88	17	16.45	25	22	18
Johnson SB	0.18	14	19.27	53	N/A		33.5	33
Kumaraswamy	0.12	6	1.20	5	4.12	5	5	4
Laplace	0.19	19	2.73	11	8.36	11	11	10
Levy	0.58	57	22.39	55	36.62	42	55	57
Levy (2P)	0.49	56	16.29	50	1.48	2	50	51
Log-Gamma	0.31	45	9.64	40	38.01	44	44	45
Log-Logistic	0.30	43	8.20	36	47.62	47	43	42
Log-Logistic (3P)	0.10	2	1.97	10	5.54	6	6	6
Log-Pearson 3	0.18	17	14.32	48	N/A		32.5	31
Logistic	0.19	20	3.19	16	16.54	27	20	16
Lognormal	0.29	41	7.86	35	35.34	41	41	41
Lognormal (3P)	0.21	27	4.25	20	16.71	28	27	25
Nakagami	0.17	13	5.07	26	10.46	13	13	13
Normal	0.21	24	3.94	18	16.45	24	24	20
Pareto	0.49	55	19.40	54	8.25	9	54	56
Pareto 2	0.48	54	18.33	52	22.45	35	52	55
Pearson 5	0.32	46	9.51	39	50.62	48	46	46
Pearson 5 (3P)	0.24	33	5.02	25	14.06	21	25	22
Pearson 6	0.27	39	7.15	32	41.17	45	39	40
Pearson 6 (4P)	0.25	34	5.25	28	20.05	33	33	32
Pert	0.19	21	3.01	15	16.25	23	21	17
Power Function	0.22	29	4.74	24	12.12	18	24	20
Rayleigh	0.32	47	9.95	42	78.45	52	47	48
Rayleigh (2P)	0.35	48	9.64	41	47.13	46	46	46
Reciprocal	0.58	58	32.26	57	27.50	39	57	58
Rice	0.25	35	5.15	27	16.73	29	29	28
Student's t	0.92	60	208.24	59	1199.10	55	59	59
Triangular	0.26	37	4.69	23	22.68	36	36	34
Uniform	0.24	31	22.66	56	N/A		43.5	44
Weibull	0.24	32	6.53	30	55.19	49	32	30
Weibull (3P)	0.11	4	1.19	4	4.12	4	4	3
Johnson SU	No fit	-	-	-	-	-	-	-

References

[1] Bari MF and Sadek S., Regionalization of low-flow frequency estimates for rivers in northwest Bangladesh, FRIEND 2002—Regional Hydrology: Bridging the Gap between Research and Practice (Proceedings of the fourth International FRIEND Conference held at Cape Town. South Africa. March 2002). IAHS Publ. no. 274.

[2] Betül Saf (2009), Regional Flood Frequency Analysis Using L-Moments for the West Mediterranean Region of Turkey , Water Resour Manage 23:531–551 DOI 10.1007/s11269-008-9287-z

[3] Bobee B, Cavidas G, Ashkar F, Bernier J, Rasmussen P (1993) Towards a systematic approach to comparing distributions used in flood frequency analysis. J Hydrol 142:121–136

[4] Coulson CH (1991) Manual of operational hydrology in B.C., 2nd edn. B.C. Water Management Division, Hydrology Section, Ministry of Environment, Lands and Parks, BC, Canada

[5] Cunanne C (1973) A particular comparison of annual maxima and partial duration series methods of flood frequency prediction. J Hydrol 18:257–271

[6] Cunnane C (1989) Statistical distributions for flood frequency analysis. WMO No. 718, WMP, Geneva

[7] Ferdows M and Hossain M (2005), Flood Frequency Analysis at Different Rivers in Bangladesh: A Comparison Study on Probability Distribution Functions, Thammasat Int. J. Sc. Tech., Vol. 10, No. 3, 53-62

[8] Haddad K. and Rahman A. (2010). Selection of the best fit flood frequency distribution and parameter estimation procedure: a case study for Tasmania in Australia, Stoch Environ Res Risk Assess (2011) 25, DOI 10.1007/s00477-010-0412-1, 415–428

[9] Haigh M.J. (2004). Sustainable management of headwater resources: the Nairobi 'headwater' declaration (2002) and beyond. Asian Journal of Water, Environment and Pollution, Vol. 1, No. 1-2, 17–28.

[10] Karn S.K. and Harada, H. (2001). Surface water pollution in three urban territories of Nepal, India, and Bangladesh. Environmental Management, Vol. 28, No. 4, 483–496

[11] Karim MA and Chowdhury JA. (1995) A comparison of four distributions used in flood frequency analysis in Bangladesh, Hydrological Sciences Journal, 40:1, 55-66, DOI: 10.1080/02626669509491390

[12] Laio F, Di Baldassarre G, Montanari A (2009) Model selection techniques for the frequency analysis of hydrological extremes. Water Resour Res 45:W07416. doi:10.1029/2007/WR006666

[13] Markiewicz I, Strupczewski WG, Kochanek K, Singh V (2006) Discussion of Non-stationary pooled flood frequency analysis. J Hydrol 276:210–223

[14] Mitosek HT, Strupczewski WG, Singh VP (2006) Three procedures for selection of annual flood peak distribution. J Hydrol 323: 57–73

[15] Rahman AS., Rahman A., Zaman MA., Haddad K., Ahsan A., Imteaz M. (2013), A study on selection of probability distributions for at-site flood frequency analysis in Australia, Nat Hazards (2013) 69:1803–1813 DOI 10.1007/s11069-013-0775-y

[16] Rahman S. and Hossain, F., (2007). Spatial Assessment of Water Quality in Peripheral Rivers of Dhaka City for Optimal Relocation of Water Intake Point. Water Resources Management, Vol. 1, No. 22, 377-391.

[17] Rakesh Kumar and Chandranath Chatterjee,(2005), Regional Flood Frequency Analysis Using L-Moments for North Brahmaputra Region of India, J. Hydrol. Eng, ASCE, Vol. 10, No. 1, 1-7, DOI:10.1061/(ASCE)1084-0699(2005)10:1(1)

[18] Stedinger JR (1980) Fitting lognormal distributions to hydrologic data. Water Resour Res 16(3):481–490

[19] Stedinger JR, Vogel RM, Foufoula-Georgiou E (1992) Frequency analysis of extreme events. In: Maidment DR (ed) Handbook of hydrology. McGraw-Hill, New York

[20] Vogel RM, McMahon TA, Chiew FHS (1993) Flood flow frequency model selection in Australia. J Hydrol 146(421):449

30

Monthly predicted flow values of the Sanaga River in Cameroon using neural networks applied to GLDAS, MERRA and GPCP data

SIDDI Tengeleng[1, 2, 4], NZEUKOU Armand[1, *], KAPTUE Armel[5], TCHAKOUTIO SANDJON Alain[1, 3], SIMO Théophile[6], Djiongo Cedrigue[6]

[1]Laboratoire d'Ingénierie des Systèmes Industriels et de l'Environnement (LISIE), Institut Universitaire de Technologie Fotso Victor, Université de Dschang, P.O Box 134 Bandjoun, Cameroun

[2]Higher Institute of the Sahel, University of Maroua, P.O. Box 46 Maroua, Cameroon

[3]Laboratory for Environmental Modeling and Atmospheric Physics (LAMEPA) Department of Physics University of Yaounde 1 P.O Box 812 Yaounde, Cameroon

[4]Laboratory of Mechanics and Modeling of Physical Systems (L2MSP) - University of Dschang, Cameroon

[5]Geographic Information Science Center of Excellence, South Dakota State University, Brookling, SD 57007, USA

[6]Laboratory of Automatic and Applied Informatics (LAIA), IUT-FV, University of Dschang

Email address:

siddit2000@yahoo.fr (SIDDI T.), armand.nzeukou@gmail.com (NZEUKOU A.), armel.kaptue@sdstae.edu (KAPTUE A.), stchakoutio@yahoo.com (TCHAKOUTIO S. A.), sitheo1@yahoo.fr (SIMO T.) cedrigueboris.djiongokenfack@gmail.com (Djiongo C.)

Abstract: The aim of our study is to predict the discharge rate of the river Sanaga using neural network techniques. Our investigations have taken place in the Sanaga watershed area in Cameroon. The measurement station is situated in the locality of Edea-Song-Mbengue (04°04'15"N, 10°27'50"E) where we have obtained monthly values of the river Sanaga discharge rates that have been measured in situ from January 1989 to December 2004. We have trained neural networks (NN), each with data of parameters such as the surface albedo, the total cloud fraction, the evaporation, the outgoing longwave radiation, the air temperature, the specific humidity, the surface runoff and the precipitation height. The precipitation values have been obtained from GPCP (Global Precipitation Climatology Project) and those of the other parameters from the data assimilation systems GLDAS (Global Land Data Assimilation System) and MERRA (Modern Era-Retrospective analysis for Research and Application). As desired outputs of the NN during the learning process, we have used the measured river runoff values. After introducing temporal delays of 01 and 02 months in the learning-process, we could observe the presence of the memory effect of the parameters used on the temporal evolution of the river discharge rate. After analysis of the performance's criteria of the NN with the help of the calculated Root Means Square Errors (RMSE) and determination coefficients between predicted values and in situ observed ones, we have perceived that the NN which takes into account the two-month delay can predict the river discharge rate with a strong correlation.

Keywords: River Runoff, GLDAS, GPCP, MERRA, Neural Network, Sanaga Watershed area

1. Introduction

In sub-Saharan Africa in general and in Cameroon, in particular, electrical energy is mostly produced by hydroelectric dams. The mastering of the temporal evolution of the discharge rate of the rivers concerned allows better adapting the generators used. The river discharge rate is a hydrological parameter that can be nowadays evaluated by modern measurement instruments [1-3]. However, the acquirement, the maintenance and the operability of these logistics lead to significant costs that constitute a handicap in evaluating this parameter in sub-Saharan Africa. Collecting direct measurement of stream flow (discharge) on a continuous basis is challenging, especially during large flood events [4]. A common practice is to convert records of water-stages into discharges by using a pre-established stage-discharge relationship. Such relationships are often referred to as a rating curve. Unfortunately, the stage-discharge relationship is not always a simple unique relationship. The rating-curve has to be actualized more times in accordance with the number of the possible runoff values. Furthermore, some factors are able to modify temporarily or definitely the water flow and consequently the rating-curve [5]. Thus, it is important to regularly measure the river runoff in order to determine the relationship stage-discharge and to control its variation [6]. It is also important to emphasize that the forecast of river discharge in a given region allows better managing the problems relied to water such as floods and droughts [7].

The runoff values of a river can be estimated after using some remote sensed information [8-9]. These values could be influenced by some atmospheric parameters and soil properties like the surface albedo, the total cloud fraction, the evaporation, the outgoing long wave radiation, the air temperature, the specific humidity, the surface runoff and the precipitation height.

The estimation of atmospherical and hydrological parameters with the help of neural networks has been in the past and is still now an interesting subject of scientific research. Scientists like Deming et al. [10] have used them to validate atmospherical profiles of temperature; Moreau et al. [11] have developed a neural network named « Gated Expert » (GE) in order to retrieve liquid water content of the atmosphere over the ocean, using radiometric data; Lek et al. [12] have used the neural model to establish the relationship river runoff-rain; Tesch and Randeu [13] and Laurence et al. [14] have used them to predict the river discharge rate for short-term.

The aim of this work is to use the neural network techniques to predict the discharge rate of the river Sanaga in Cameroon, precisely at the measurement point Song-Mbengue, through assimilated data issued from data bases MERRA, GLDAS and GPCP. In the following sections, we will briefly give the functioning principle of the neural networks and present thereafter the study zone which is the watershed of Sanaga basin, the data used, the methodology required by the developed neural networks and will then end by presenting the obtained results.

2. Brief Description of Functioning Principle of Neural Networks

The neural network (NN) is formed by simple computation elements, the neurons that are connected to others by weights (synapses). During the learning phase, it learns the statistical relationship between input- and output data and determines the weights that have to be attributed to each connection between the neurons in order to approximate with accuracy this relationship [15].

In opposition to traditional methods, the neural network does need neither a detailed comprehension of the physics

behind the problem that has to be solved nor a step by step conception of a detailed model describing this physical phenomenon. Only a general comprehension of the observation and measure technique is required in order to determine the most appropriate input parameters that statistically permit a large coverage of possible cases. Once trained, the algorithms possess a higher execution speed as the physical methods that need fastidious computation steps for each inversion [16]. The neural network method has been developed and implemented by many authors and can be considered nowadays as a powerful tool for estimation in several scientific research domains [17-18].

The structure of a neural network can be defined as an oriented graph whose nodes represent the neurons and the bows the different connexions between them. The nodes are gathered in input, hidden, and output layers (see figure 1).

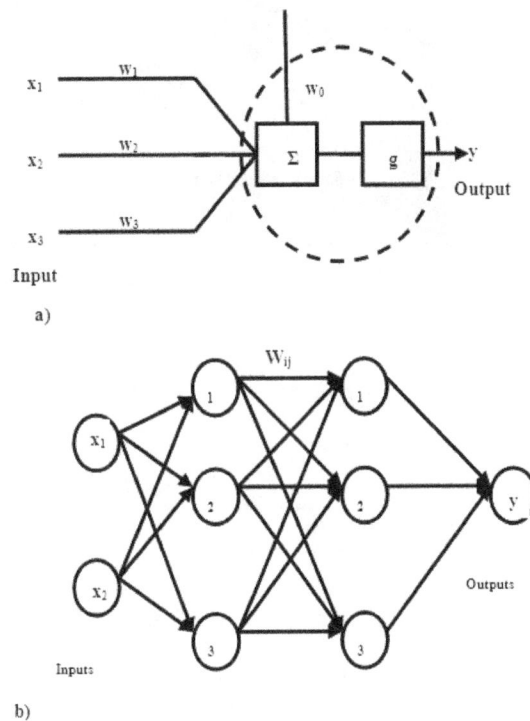

Figure 1. *a) a neuron model; b) an example of a neural network with one input layer (02 inputs), two hidden layers (03 neurons in each) and one output layer (01 neuron)*

In a given neural network, the neuron receives and adds n inputs x_j, each possessing a connection weight w_i. To the result a bias w_0 is added. The final obtained result is applied to a transfer function (or activation function) g in order to obtain the output y. If X is the input vector, the output can be calculated as follows:

$$g : \mathrm{IR}^n \to \mathrm{IR}$$

$$y(X) = g\left(\sum_{i=1}^{n} w_i \cdot x_i + w_0\right) = g\left(\sum_{i=0}^{n} w_i \cdot x_i\right) = g\left(W^T \cdot \tilde{X}\right) \quad (1)$$

Whereby

$$x_0 = 1, \ W = (w_0, w_1, \cdots, w_n)^T \ \text{and} \ \tilde{X} = (1, x_1, \cdots, x_n)^T \quad (2)$$

3. Study Zone and Used Data

3.1. Study Zone

Cameroon is situated in central Africa, between the 1st and the 13th degree north latitude, and between the 8th and the 17th east longitude (see figure 2a). The country is characterized by a dense hydrographical network. During the whole year, there is water in the Sanaga, but the discharge rate of this river varies from one year to another. As study zone, we have focused on the watershed area of Sanaga that is approximately delimited by the geographical coordinates (3,51°N, 10.10°E) and (6,33°N, 14,42°E) (see figure 2b).

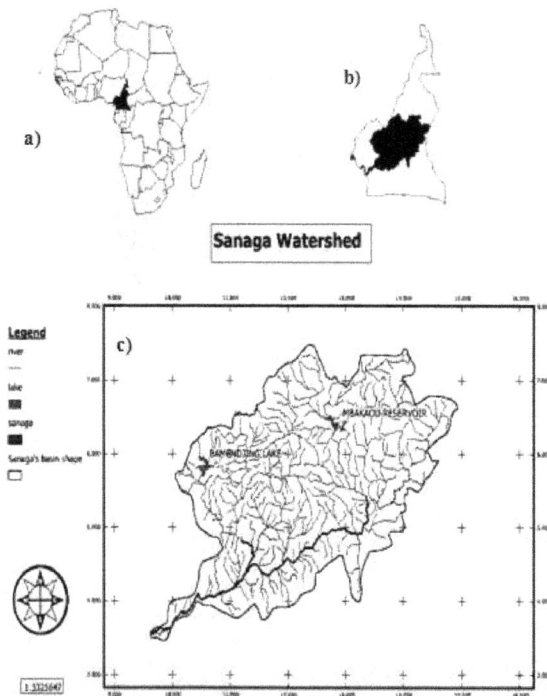

Figure 2. Sanaga Watershed area; a) localization of Cameroon in Africa; b) localization of watershed area of Sanaga in Cameroun ((3,51°N, 10.10°E); (6,33°N, 14,42°E)); c) Illustration of the watershed area of Sanaga (measure point in Song-Mbengue : 04°04'15"N, 10°27'50"E).

3.1. Used Data

At disposal, we have monthly values of the discharge rate of Sanaga (m³/s) obtained in situ in the measurement station of Song-Mbengue, between January 1989 and December 2004. We have also used the data assimilation systems MERRA (Modern Era-Retrospective analysis for Research and Application), GPCP (Global Precipitation Climatology Project) and GLDAS (Global Land Data Assimilation System) [19-22].

From MERRA, we have obtained the values of the surface albedo, total cloud fraction, evaporation, outgoing longwave radiation, air temperature, specific humidity and surface runoff. From GPCP are issued the monthly accumulated rain; and from the GLDAS system, we have obtained the surface runoff.

The general characteristics of the monthly parameter means are represented in the figure 3. They concern the standardized monthly mean values from January 1989 to December 2004. Table 1 presents the statistical values of the mean, standard deviation and variation coefficients.

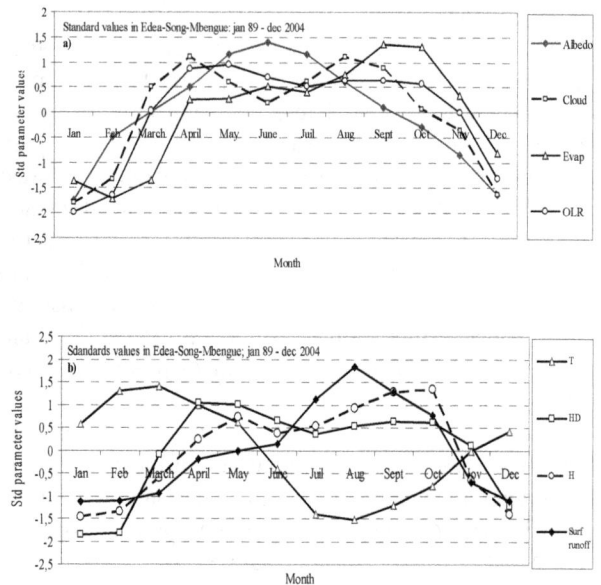

Figure 3. Evolution of standardized monthly parameter values in the watershed area of the river Sanaga: a) albedo; cloud fraction; evaporation and OLR (Outgoing Long Radiation); b) temperature; humidity; accumulated precipitation and surface runoff.

Table 1. Statistical parameter values: mean, standard deviation and variation coefficient concerning the period from January 1989 to December 2004 in the region of the Sanaga watershed area (Cameroon)

Statistical parameters	Albedo [fraction]	CLDTOT [fraction]	EVAP [kg/m²/s]	OLR [W/m²]
μ (mean)	0.123	0.826	4.60E-05	385.02
σ (std dev)	0.0058	0.0572	4.00E-06	13.57
CV (var coef)	0.047	0.0692	0.087	0.035
Statistical parameters	T [K]	HD [kg/kg]	H [mm]	Surf runoff [kg/m²/s]
μ (mean)	289.30	0.01519	140.993	6.14E-07
σ (std dev)	1.162	0.001641	88.1979	5.16E-07
CV (var coef)	0.00402	0.10808	0.6255	0.840

4. Methodology

Training of a neural network is an iterative process that ends if the NN has sufficiently learnt the relationship existing between the inputs, and the outputs presented to him. Many stop-criteria exist, but the main important ones are the iteration number, and the error calculated between the desired values and those evaluated by the NN.

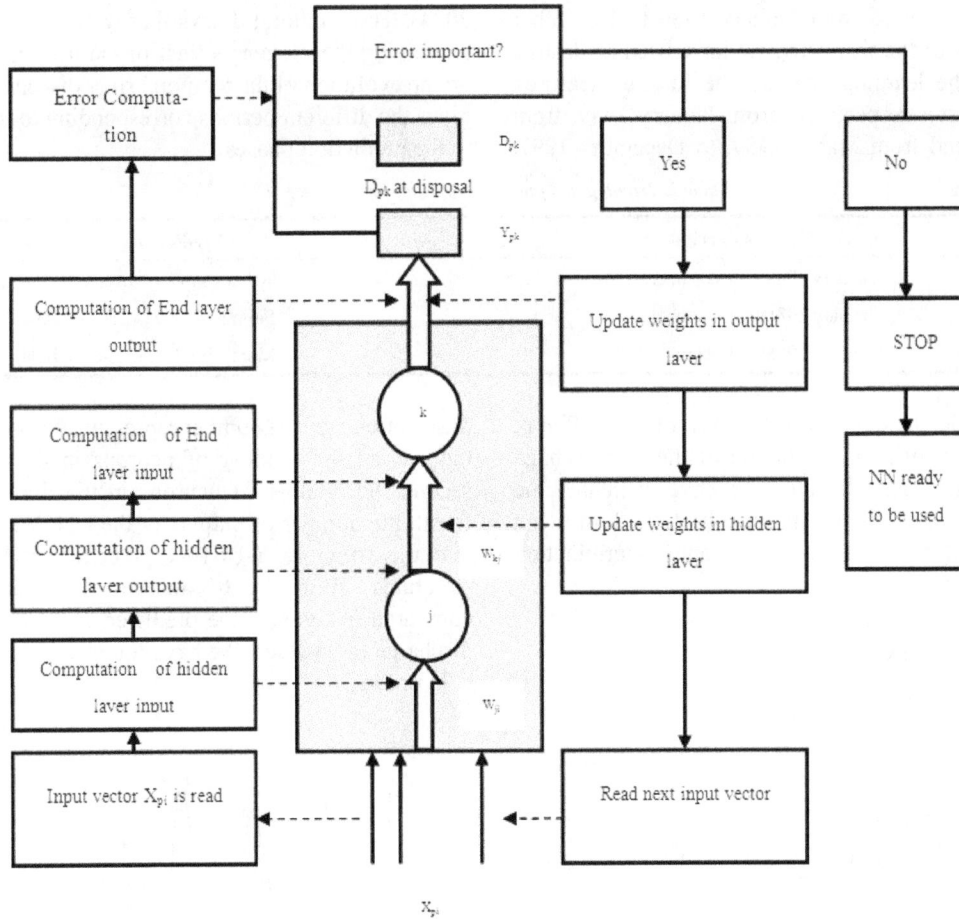

Figure 4. Learning process of a neural network

For a convenient prediction of the outputs corresponding to given inputs of a trained NN, it is necessary, during the learning process, that all possibilities of inputs / outputs combinations exist. In the figure 4, we have presented a diagram describing the learning process of a NN which can be condensed in 05 main points:

1. From the input matrix, X_{pi} is an input vector x_{pi} read (sample);
2. Input- and output values of existing hidden layers are calculated;
3. Input- and output values of the last layer (output layer) are computed;
4. The error between desired available value and NN output is evaluated;
5. If the calculated error is not significant, the learning process is stopped, and the NN can be considered as trained and used as a prediction tool for other outputs corresponding to inputs that are, up to now, unknown by the NN. If the evaluated error is important, the next

input vector (sample) is read from the available input matrix, and a new iteration begins.

For the learning of the different parameters used, we have employed the neural network toolbox available in MATLAB [23-24].

Sometimes, in accordance with different stages of the parameter values used, it is necessary to work with normalized input values situated between -1 and +1. For this purpose, we have used the normalized NN input values of parameters (equation 3):

$$X_{norm} = 2 \cdot \frac{X - X_{min}}{X_{max} - X_{min}} - 1 \qquad (3)$$

Whereby X, X_{min} and X_{max} represent respectively the real value of the parameter, its minimal value and its maximal value.

As inputs parameters of the NN we have: the albedo (%), the cloud coverage (%) (CLDTOT), the evaporation (EVAP) (kg/m^2/s), the outgoing long radiation (OLR) (W/m^2), the air temperature T (°K); the specific humidity (HD) (kg/kg), the accumulated rain (H) (mm) and the surface runoff (SURF RUNOFF) (kg/m^2/s). We have obtained the values from the NASA website and through the data assimilation systems MERRA, GPCP and GLDAS. All these NN input values correspond to the period from January 1989 to December 1996. As outputs of the three NN, we have used as desired values, during the learning process, the in situ measured river discharge rates respectively from January 1989, from February 1989 and from March 1989, to December 1997.

With the first NN, we have predicted the river discharge rates as from January 1997 to November 2004, without any time delay accordingly to the training period. The second and the third NN predict these values respectively with a temporal delay of 01 (as from February 1997 to October 2004) and 02 months (as from March 1997 to September 2004). The temporal delays that we have introduced allow perceiving the memory effect of the used input parameters on the evolution of the temporal river discharge rate. Table 2 gives the different periods corresponding to the training and to the prediction phases.

Table 2. *Training and prediction periods for the three NN*

NN	Training period	Prediction period
1	January 19889 – December 1996	January 1997 – November 2004
2	January 19889 – December 1996	February 1997 – October 2004
3	January 19889 – December 1996	March 1997 – September 2004

For better understanding the efficiency of the different neural networks during the prediction of the river Sanaga discharge rate in Song-Mbengue, we have evaluated the performance criteria through computation of the Root Mean Square Error (RMSE) (equation 4) and determination coefficient (r) (equation 5):

$$RMSE = \sqrt{\frac{\sum_{i=1}^{N}\left(x_i^n - x_i^s\right)^2}{N-1}} \qquad (4)$$

$$r = \frac{\sum_{i=1}^{N}\left(x_i^n - \overline{x}^n\right)\cdot\left(x_i^s - \overline{x}^s\right)}{\sqrt{\sum_{i=1}^{N}\left(x_i^n - \overline{x}^n\right)^2}\cdot\sqrt{\sum_{i=1}^{N}\left(x_i^s - \overline{x}^s\right)^2}} \qquad (5)$$

Whereby x_i^n and x_i^s are respectively the NN simulated values, the in situ measured ones and N the total number of samples.

5. Results and Discussions

We have trained for each temporal delay three NN with 02 hidden layers (8-10-10-1; 8-10-20-1; 8-20-10-1). In this designation the first number represents the number of inputs, the second the number of neurons in the first layer, the second the number of neurons in the third layer and the fourth the number of outputs of the NN. We have observed that the structure 8-10-10-1 is better trained as the other structures. With the 02 months' time delay, we have compared in figure 5 the predicted and the measured river discharge rate values. We have found a strong correlation of 0.95 orders.

5.1. Training Phase of the Neural Networks

We have trained for each temporal delay three NN with 02 hidden layers (8-10-10-1; 8-10-20-1; 8-20-10-1). In this

designation the first number represents the number of inputs, the second the number of neurons in the first layer, the second the number of neurons in the third layer and the fourth the number of outputs of the NN. We have observed that the structure 8-10-10-1 is better trained than other structures. With the 02 months' time delay, we have compared in figure 5 the predicted and the measured river discharge rate values. We have found a strong correlation of 0.95 orders.

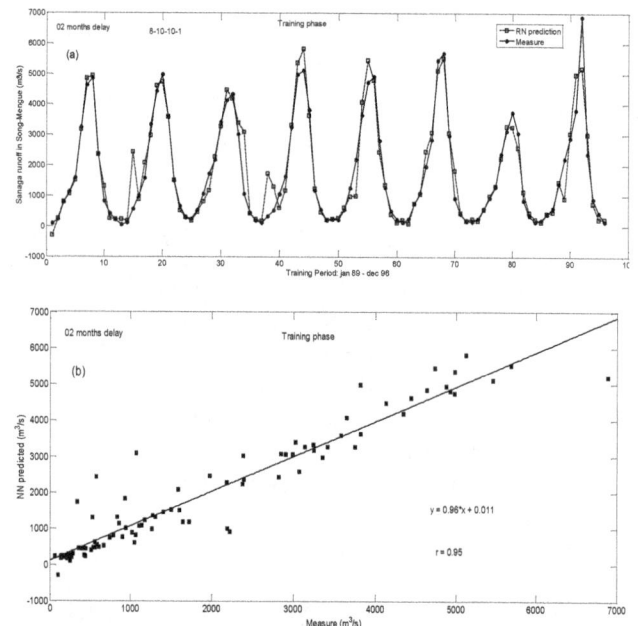

Figure 5. *Training of the neural network 8-10-10-1 with a time delay of 02 months (8 input parameters, 10 neurons in the 1st hidden, 10 neurons in the 2nd hidden layer (see the legend)*

5.2. Prediction Phase of the Neural Networks

After execution of the input values through the trained NN we can observe that the 02 months delay leads equally to better predictions. Thus, we obtain for the period from March 1997 to September 2004 the predicted river discharge values of Sanaga at the measurement point Song-Mbengue that we compare in figure 6 to the values observed in situ.

We obtain equally the existence of a strong correlation in order of 0.89 between predicted and measured values.

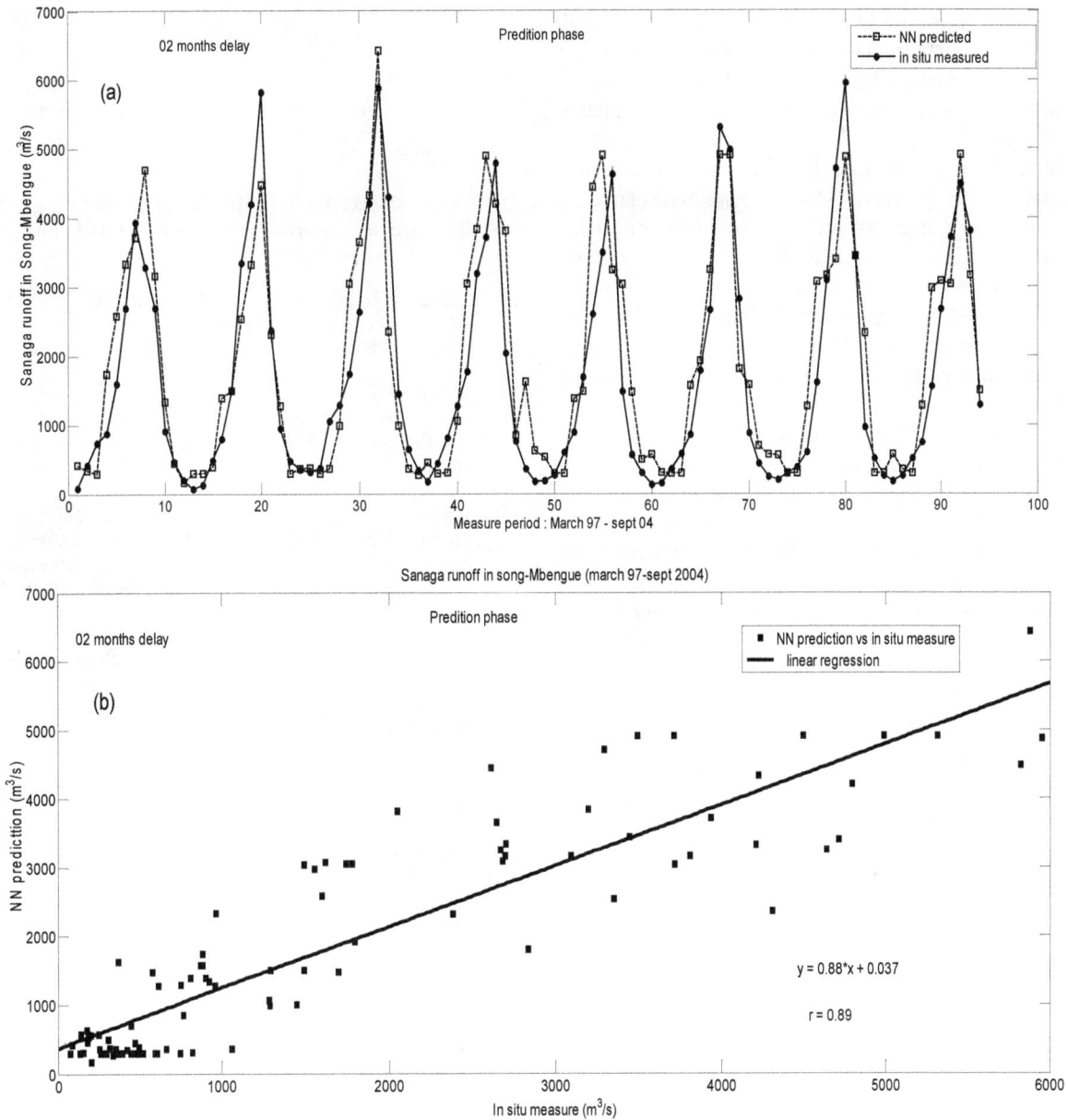

Figure 6. *a) Prediction with the 8-10-10-1 NN (squares) and measures (points) of the Sanaga discharge rate at Song-Mbengue in Cameroon during the period from March 97 to September 2004; b) Linear regression between measured and predicted values.*

The table 3 presents the different values of the performance criteria RMSE and r obtained

Table 3. *Root Mean Square Errors (RMSE) and Determination Coefficients (r) between the discharge rate values, calculated by the NN and measured in situ: training phase from January 1989 to December 1996; prediction phase from January 1997 to December 2004; use of the configuration 8-10-10-1 for all the three temporal delays (0 month, 01 months and 02 months delay)*

Delay time (months)	Training phase			Prediction phase	
	RMSE (m³/s)	r	Iteration number needed	RMSE (m³/s)	r
0	439.0	0.91	12	1056.41	0.80
1	434.0	0.92	17	882.00	0.83
2	430.9	0.95	18	735.50	0.89

6. Conclusion

In our study, we have investigated the prediction of the discharge rate of the river Sanaga in Cameroon through using assimilated data from GLDAS, MERRA and GPCP. A neural network technique has been used after the training of the models with values observed in the measurement station Song-Mbengue.

While developing the NN, several architectures have been tested, and the configuration (8-10-10-1) has been identified as the best one, independently on the temporal delay introduced. Furthermore, after analyzing the performance criteria (RMSE and r between predicted values and in situ measured ones), we have perceived that the NN which takes into account the temporal delay of 02 months produces better prediction of the river discharge rate with strong correlation.

Considering that the measures of river discharge rate values need enormous material and human resources, the use of NN models based on data from assimilation systems like GLDAS, MERRA and GPCP appears as an adequate, efficient and rapid mean with low cost that permits to predict the discharge rate values of rivers of Sanaga watershed area in Cameroon with a good spatio-temporal regularity.

Acknowledgments

We thank NASA through its service data (Goddard Earth Sciences Data and Information Services Center (GES DISC)).

Nomenclature

CLDTOT: Cloud coverage
CV: variation coefficient
EOS: Earth Observing System
EVAP: Evaporation
GES DISC: Goddard Earth Sciences Data and Information Services Center
GLDAS: Global Land Data Assimilation System
GOES: Geostationary Operational Environmental Satellite
GPCP: Global Precipitation Climatology Project
HD: Humidity
HDISC: Hydrology Data and Information System Center
μ: Mean value
MERRA: Modern Era-Retrospective analysis for Research and Application
NASA: National Atmospheric and Space Administration
NN: Neural Network
OLR: Outgoing Long Radiation
RMSE: Root Mean Square Error
σ: Standard deviation
Surf runoff: Surface runoff
T: Air temperature

References

[1] Chen, Y., C. "Flood discharge rate measurement of a mountain river - Nanshih River in Taiwan". Hydrol. Earth Syst. Sci., 2013, 17, 1951–1962.

[2] Negrel, P., Kosuth, P. and Bercher, N. "Estimating river discharge from earth observation measurements of river surface hydraulic variables". Hydrol. Earth Syst. Sci., 2011, 15, 2049–2058.

[3] EOS. "Space-Based Measurement of River Runoff Transactions", American Geophysical, Union. Eos, Vol. 86, No. 19, 10 May 2005.

[4] Emad, H. Habib and Ehab A. Meselhe. "Stage-discharge for Low-Gradient Tidal Streams Using Data-driven Models". Journal of Hydraulic Engineering, ASCE/May 2006.

[5] Lillian, Oygarden. "Erosion and surface runoff in small agricultural catchments". IAHS Publ. 1996, no. 236, Global and Regional Perspectives Centre for Soil and Environmental Research

[6] Watanabe, Fumio, Kobayashi, Yukimitsu, Suzuki, Shinji, Hotta, Tomoki and Takahashi, Satoru. "Estimating the Volume of Surface Runoff from in Situ Measured Soil Sorptivity". Journal of Arid Land Studies, 2012, 22 (1), 95-98.

[7] Pekarova, P., Pavol, Miklanek and Jan Pekar. "Spatial and temporal runoff oscillation analysis of the main rivers of the world during the 19th – 20th centuries". Journal of Hydrology, 2003, 274, 62-79.

[8] Bjerkliea David, M., Delwyn, M., Laurence, C. Smith, Lawrence Dingman. "Estimating discharge rate in rivers using remotely sensed hydraulic information". Journal of Hydrology, 2005, 309, 191–209.

[9] Tarpanelli, Angelica, Barbetta, Silvia, Brocca, Luca and Moramarco, Tommaso. "Discharge Estimation by Using Altimetry Data and Simplified Flood Routing Modeling". Remote Sens., 2013, 5, 4145-4162; doi:10.3390/rs50941 45

[10] Deming J., Chaohua D. and Weison L. "Neural networks approach to high vertical resolution atmospheric temperature profile retrieval from space borne high spectral resolution infrared sounder measurements". Proc. SPIE 6064, Image Processing: Algorithms and Systems, Neural Networks, and Machine Learning, 2006, 60641L; doi:10.1117/12.649743.

[11] Moreau, E., Mallet, C., Thiria, S., Mabbou, B., Badran, F. and Klapisz, C. "Atmospheric Liquid Water Retrieval Using a Gated Experts Neural Network". Journal of Atmospheric and Oceanic Technology, 2002, 19, 457-467.

[12] Lek, S., Dimopoulos, I., Ghachtoul, Y., El. « Rainfall-runoff modelling using artificial network". Revue des sciences de l'eau, rev. Sci. Eau, 1996, 3, 319-331.

[13] Tesch, R. and Randeu, W., L. "A neural network model for short term river flow prediction". Nat. Hazards Earth Syst. Sci., 2006, 6, 629-635.

[14] Laurence, C., Smith and Tamlin, M., Pavelsky. "Estimation of river discharge rate, propagation speed, and hydraulic geometry from space: Lena River, Siberia". Water Ressources Research, 2006, Vol. 44, doi:10.1029/2007 WR006133.

[15] Shrivastava, G., Karmakar, S., Guhathakurta, P. and Manoj Kumar Kowar, M.K.). "Application of Artificial Neural Networks in Weather Forecasting: A Comprehensive Literature Review". International Journal of Computer Applications, 2012, 51, (18), 0975-8887.

[16] Del Frate, F. and Giovanni S. "Neural Networks for the retrieval of water vapour and liquid water from radiometric data". Radio Science, 1998, 33, (5), 1373-1386.

[17] Widrow, B., Lehr, M.A. "30 years of adaptive neural networks: Perceptron, Madaline, and Backpropagation". Proceedings of the IEEE, 1990, (9), 78.

[18] Kou-Lin Shu, Xiaogang Gao, Sorooshi Sorooshia and Hoshin V. Gupta. "Precipitation Estimation from Remotely Sensed Information Using Artificial Neural Networks". Journal of applied meteorology, 1996, 36, 1176-1189.

[19] Huffman, G. J., Bolvin, D. T. "Version 1.2 GPCP One-Degree Daily Precipitation Data Set Documentation. Mesoscale", Atmospheric Processes Laboratory, NASA Goddard Space Flight Center, 2013.

[20] Rodell, M., Houser, P.R., Jambor, U., Gottschalck, J., Mitchell, K., Meng, C.J., Arsenault, K., Cosgrove, B., Radakovich, J., Bosilovich, M., Entin, J.K., Walker, J.P., Lohmann, D. and Toll, D. "The Global Land Data Assimilation System". Bulletin of American Meteorological Society, 2004, 381-394, doi: 10.1175/BAMS-85-3-381.

[21] Hongliang, F., Hrubiak, P. L., Hiroko, K., Rodell, M., Teng, W. L. and Vollmer, B.E. "Global Land Data Assimilation System (GLDAS) products from NASA, Hydrology Data and Information Services Center (HDISC)". ASPRS Annual Conference, Portland Oregon, April 28 - May 2, 2008.

[22] Rui, H., Teng W. L., Vollmer B.E., Mocko, D.M., Beaudoing, H.K. and Rodell, M. "NASA Giovanni Portals for NLDAS/GLDAS Online Visualization, Analysis and Intercomparison". American Geophysical Union, Fall Meeting, 2011.

[23] Hagan, M., Demuth, H., and Beate, M. "Neural Network design. Boston": Pws publication, 1996.

[24] Demuth, H., Beate, M., and Hogan, M. "Neural Network toolbox user's guide". The Mathworks. Inc., Natrick, USA, 2009.

31

Water quality status of recreational spots in Chittagong City

Md. Lokman Hossain[1, *], Sultana Kamrun Nahar Nahida[2], Md. Iqbal Hossain[2]

[1]Department of Global Change Ecology, Faculty of Biology, Chemistry and Geosciences, University of Bayreuth, D-95440 Bayreuth, Germany
[2]Institute of Forestry and Environmental Sciences, University of Chittagong, Chittagong 4331, Bangladesh

Email address:

lokmanbbd@gmail.com (M. L. Hossain), sknnahida@gmail.com (S. K. N. Nahida), iqbal.sohag@gmail.com (M. I. Hossain)

Abstract: The study was conducted to learn about water quality status at potential recreational spots in Chittagong City. Water samples were collected from six recreational spots. Derived parameters namely; Conductivity, Turbidity, Color, Odor, pH, Dissolve Oxygen (DO), Biological Oxygen Demand (BOD), Chemical Oxygen Demand (COD), Sulfate and Nitrate of the collected water samples were compared with the standard value of the respective parameters. The study revealed that water conductivity of Karnafully Shishu Park (1400 μs/cm) and Jatisongho Park (1430 μs/cm) could satisfy the standard level (1500μs/cm) and the other spots exceeded the limit. Water collected from Foy's lake was a bit turbid (5.07 ntu) but it did not significantly exceed the expected limit (5 ntu). The study also revealed that water of Foy's Lake was darker (55 Hazen) compared to the standard (20 Hazen). In case of odor, all the spots' water was unobjectionable except Chittagong Shishu Park and Jatisongho Park. BOD and DO content were satisfactory for all the spots. But COD was higher at Chittagong Shishu Park, Karnafully Shishu Park and Zia Smriti Complex. Nitrate concentration was also found satisfactory in water of all spots except Chittagong Shishu Park (19.33 ppm). All the parameters were under the desired levels for recreation with few exceptions. Although very small, to extend the recreational facilities for the urban people, these unwanted level of different physical and chemical parameters should be mended on an urgent basis.

Keywords: Biological Oxygen Demand, Chemical Parameter, Dissolved Oxygen, Water Quality

1. Introduction

The earth is covered by approximately 71% of water but only a total of 3% is fresh water, of which only 0.003% of freshwater is accessible to humans [1]. Around 1.2 billion people or almost one-fifth of the world's population lives in areas of physical water scarcity [2]. Water is an indispensable and one of the most precious of natural resources on this planet [3]. Water is inevitable to plant growth [4,5] and sustaining animal life [6] as well as it is the predominant inorganic constituents of living matter, forming in general nearly three quarters of the weight of the living beings [7-9]. For every living system the presence or provision of water is consequently of cardinal importance. Water is abundant in the liquid form and large quantities exist as gas (vapor) and in the form of ice and snow [10]. Water resources are one of the most critical and a valuable component of the resources of nation [11]. With marked rise in population, the demand of pure water is gradually increasing. Still now, nearly two thirds of the population does not have reasonable access to safe and ample water supply [12]. Water pollution is a phenomenon that is characterized by the deterioration of the quality of land water or seawater as a result of human activities. Human activities related with water pollution such as human mining, agriculture, stockbreeding, fish farming, forestry, urban activities, manufacturing industry, construction works and various territory industries [10]. Water pollution is a global problem, affecting both the industrialized and the developing nations. The water pollution problem in the rich and poor nations, however, is quite different in many respects [10]. The primary quality of water is vital concern for mankind and other living organisms, since it is directly related with human habitation and welfare. The quality of domestic water is most important in this respect. But hazards of water cause water born diseases epidemic still looms largely on the horizons of developing countries like Bangladesh [13]. The quality of water is one of the vital concerns for mankind since it is directly linked with human welfare [14].

Water quality concerns have often been neglected because good quality water supplies have been plentiful and readily available. The situation is now changing in many areas of the world including Bangladesh [4,15-17]. Water is used as a chemical, solvent, source of energy, coolant, transportation medium, a recreational base, habitat for fish and wildlife, a silvicutural and agricultural raw material and agent for waste removal [18]. Safe wholesome water is that free from pathogenic agents, harmful chemical substances, pleasant to taste usable to domestic and recreational purposes [19]. Recreational waters refer to those natural waters used not only for primary contact activities, such as swimming, windsurfing, and waterskiing, but also for secondary contact activities, such as boating and fishing [20]. Good quality recreational water is an essential part of the natural ecosystem. Recreational exposure to surface waters during periods of increased pathogen concentration may lead to a significantly higher risk of illness [21]. Maintaining safe recreational waters requires a concentrated effort from all of its stakeholders. From government at all levels, to local businesses and industry, to community members, all have a role to play in helping to keep recreational waters safe [20]. It is necessary to monitor water qualities of recreational water whenever these are used for recreational purpose. Considering the importance of recreation in urban life the present study was designed to measure the water quality of recreational spots.

1.1. Background Information of the Study Area

Chittagong is the gateway and the second largest city of Bangladesh [22] with a substantial, self-sustaining economic base [23]. Chittagong lies between latitude 22°14′ N and 22°24′ N longitude 91°46′ E and 91°53′ E [22] and stands on the bank of the river Karnafully. It is situated at the extreme southeastern part of the country. It comprises about 209.66 sq km [22]. It is upgraded as Municipal Corporation and finally as City Corporation in 1990 [24]. Urbanization and population growth rate is high in Chittagong city [25]. Chittagong is considered the crowning glory of Bangladesh for its some picturesque sites and elements of recreation. That is, Chittagong is replete with elements of recreation. The land, the rivers and the forests: these celestial beauties are the main elements of recreation in Chittagong. But the most unfortunate thing is that there are no scientific and systematically managed recreation spots. Chittagong is among those famous four places in the world where hills, rivers and seas are within the reach of just 10 km [26]. People flock in these places in a great number to get their holiday's comfort. However, very few works were done about recreational facilities in Chittagong. Thus, the study was undertaken to find out the recreational water quality in major recreational spots in Chittagong.

1.1.1. Foy's Lake

Foy's Lake is one of the prominent ornamental in Chittagong. It is considered as one of the most lucrative recreation spots of the city because of its extraordinary scenic beauty. It is situated 3 km north west of Chittagong City at Zakir Hossain Road. It is a naturally attractive landscape and stream area with hill ranges and gentle slope approximately 66 under above sea level running north-south director [27]. Foy's lake, a magnificent and natural scenic area with fascinating landscape green coverage of hills and blue water has become the most important criteria for recreation. The total area of the lake is 336.61 acres and others 3.55 acres. The resort surrounding the lake was first established in 1925.

1.1.2. Chittagong Shisu Park

Chittagong Shisu Park is the first man made well-designed recreational children park. It was inaugurated on November 28, 1994 at Kajir Dewri adjacent to Circuit House & Zia Smriti Museum. Total area of the park is 3 acres. Though the park was designed especially for children a number of young and aged people usually visit the park for passing their leisure period. One fourth of the park is remained as green cover.

1.1.3. Karnafuly Shisu Park

Karnafuly Shisu Park is the potential children park in the Chittagong City. It was inaugurated on 16 December, 2000. Total area of the park is 9 acres. Although there is a limitation of space, a number of facilities are provided for recreation of children. Though the park was designed especially for children a number of young and aged people usually visit the park for passing their leisure period.

1.1.4. Zia Smriti Complex

Zia Smriti Complex is the prospective place of interest for particular in recreation. The complex is situated at the opposite site of the Baddarhat bus terminal, Chittagong. The total area of this complex is 23 acres.

1.1.5. Jatisongho Park

Jatisongho Park is the impending park for all age groups of Chittagong City. It is situated at Panchlaish residential area in the City. Total area of the park is 2 acres. Although there is a limitation of space, a number of facilities are provided for recreation of city dwellers. Young and aged people of the city usually visit the park for passing their leisure period.

1.1.6. Vatiary Lake

Vatiary Lake is one of the most attractive lakes in Chittagong. It is the great source of green space. Psychologists, sociologist and mass people agree on the view that the quality of urban life depends largely on the amount and quality of green spaces within it or close to it. This green space, when it exists, is usually occupied with lake, which, apart from this aesthetic value, are increasingly considered to play a vital role in the protection of the urban environment [28].

2. Materials and Methods

2.1. Chemicals and Equipment Used

The chemicals used for parameter test include KI and $MnSO_4$, concentrated H_2SO_4, stretch as indicator, $Na_2S_2O_3$ (0.005), Alkali-iodide-Azide, buffer solution. The equipments which were used for parameter test include pH meter, pipette, burette, conical flask, cylinder, beaker, filter paper, evaporating disk, desiccators, electrical balance, BOD bottle, COD incubator, hot plate etc.

2.2. Field Collection

Nine water samples were collected from every recreational spot (total 90 samples from 10 recreational spots) as grab sampling method during March to May 2008. For proper assessment of water quality parameters, the samples were taken carefully. It was tried to make the samples representative in nature.

2.3. Analysis of Samples

The water samples were tested in the laboratory of the Institute of Forestry and Environmental Sciences, University of Chittagong and laboratory of Mohara Water Treatment Plant under Chittagong WASA, Chittagong, Bangladesh. Ten water parameters namely; pH, Conductivity, DO, BOD, COD, Sulphate, Nitrate, Color, Turbidity and Odor were tested to assess the suitability of water for recreational use. The obtained value was compared with the standard value of different parameters for recreational purpose. The water quality parameters have been analyzed following the standard analytical methods.

3. Result and Discussion

3.1. Physical Parameters

3.1.1. Conductivity

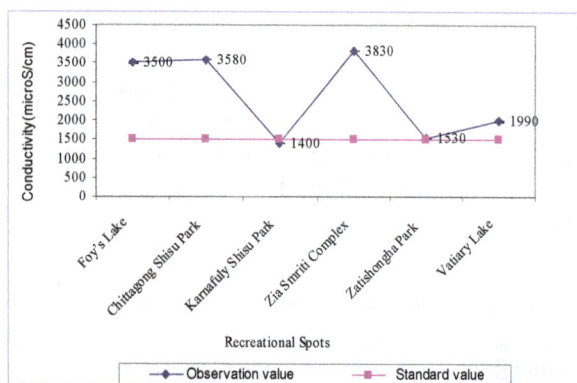

Figure 1. *Conductivity of Water at Different Recreational Spots Compared with the Standard Value.*

The study revealed that most of the recreational spot's water contains higher conductivity compared to standard value except Karnafully Shisu Park. The highest (3840

$\mu s/cm$) and lowest (1400 $\mu s/cm$) water conductivity were in Zia Smriti Complex and Karnafully Shisu Park respectively. The concentration of water conductivity of Foy's Lake, Chittagong Shisu Park and Zia Smriti Complex was about two and half times greater than that of standard value, which is very objectionable from the recreational point of view. The water conductivity of other two recreational spots namely; Vatiary Lake (1530 $\mu s/cm$) and Jatisongho Park (1990 $\mu s/cm$) also exceeded the permissible limit (Fig. 1). Therefore, for the safe use water as recreational purposes it should be maintained the standard value of conductivity in water.

3.1.2. Turbidity

Among the six recreational spots, turbidity in water in five spots found lower than the standard value (5 ntu). The highest (5.07 ntu) and lowest (0.20 ntu) turbidity were calculated in the water of Foy's Lake and Zia Smriti Complex respectively. Turbidity of Chittagong Shisu Park, Karnafully Shisu Park, Jatisongho Park and Vatiary Lake were 2.57 ntu, 1.93 ntu, 0.97 ntu and 0.87 ntu respectively (Fig. 2). Foy's lake water should be treated as it exceeded the standard value.

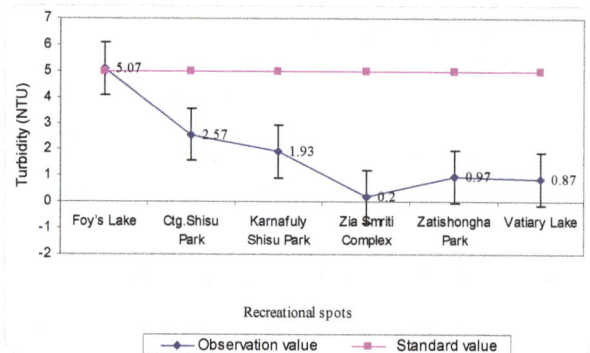

Figure 2. *Turbidity of Water at Different Recreational Spots Compared with the Standard Value.*

3.1.3. Color

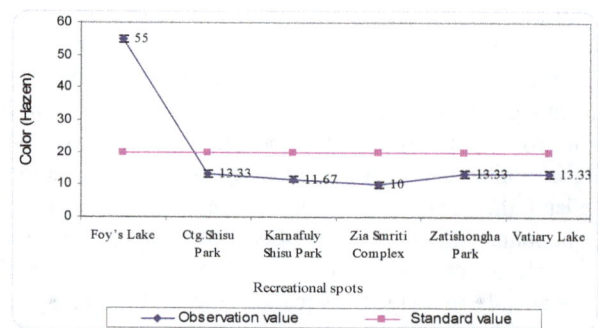

Figure 3. *Color of Water at Different Recreational Spots Compared with the Standard Value.*

The standard value of color of water is 20 Hazen [20]. The observation value of water color of six different recreational spots was not varied significantly at 1% level of significance. The highest (55 Hazen) value was found in the water of

Foy's Lake whereas; lowest (10 Hazen) was in Zia Smriti Complex (Fig. 3). The observation value was same (13.33 Hazen) in three recreational spots namely, Chittagong Shisu Park, Jatisongho Park and Vatiary Lake (Fig. 3). Therefore, from the findings it can be recommended that water of the Foy's lake is not suitable for the recreational use.

3.1.4. Odor

Odor is one of the important parameter of water, should be considered for the recreational purposes. From the recreational point of view water must be odorless. The study revealed that water of Chittagong Shisu Park and Jatisongho Park generates odor, resulting the objectionable state. Odor is arisen due to the presence of organic matters, mineral matters and microbial residues. Odor is mainly generates due to prolonged stagnant condition of water and massive alteration of water chemical parameters. To ensure the odor free water, the first and foremost task is that water should free from all sorts of organic and minerals. Since two recreational spots provides odorous breathtaking so, it is an obvious task for the concerned authority to remove odor from water for safe use of water for recreational purposes.

Table 1. *Odor of Water at Different Recreational Spots Compared with the Standard Value.*

Spots	Observation value	Standard value
Foy's Lake	Unobjectionable	
Chittagong Shisu Park	Objectionable	
Karnafully Shisu Park	Unobjectionable	Unobjectionable
Zia Smriti Complex	Unobjectionable	
Jatisongho Park	Objectionable	
Vatiary Lake	Unobjectionable	

3.2. Chemical Parameters

3.2.1. pH

The standard value of pH in the recreational water lies between 6.5 and 8.5 [20]. The highest (5.9) pH value was in the water of Vatiary Lake while, lowest (5.3) was in Jatisongho Park and Chittagong Shisu Park. pH in water of three recreational spots namely, Foys Lake, Karnafully Shisu Park and Zia Smriti Complex was given the same value (5.5) (Fig. 4). Therefore, from the observation it can be concluded that water of all of the recreational spots is acceptable for recreational purposes in respect of pH. The water of Jatisongho Park and Chittagong Shisu Park contains lower pH because water of those spots contains higher organic wastes in comparison with other spots and hence these waste matters, through the decomposition, releases acidic substances to water, accelerate the lowering the value pH.

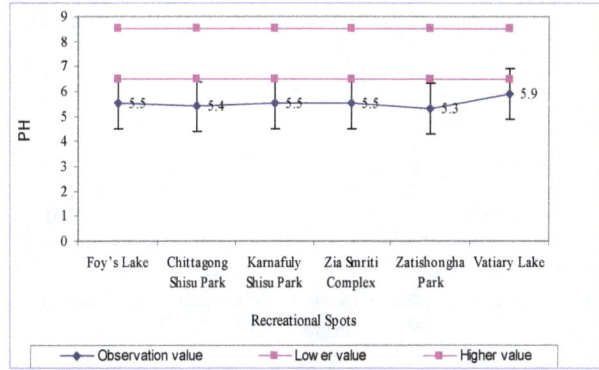

Figure 4. *pH of Water at Different Recreational Spots Compared with the Standard Value.*

3.2.2. Dissolve Oxygen (DO)

The higher the value of DO in water than standard value means greater is the acceptability of that water. That means higher the value of DO content of water, higher will be the transparency of water. The study revealed that the observed values of DO in water of different recreational sites were higher than standard value (4 mg/L) specified for recreational usage [29]. The observation values varied between 10.17 mg/L in Foy's Lake and 11.12 mg/L in Jatisongho Park (Fig. 5). The DO content in water of recreational spots was not varied at 1% level of significance.

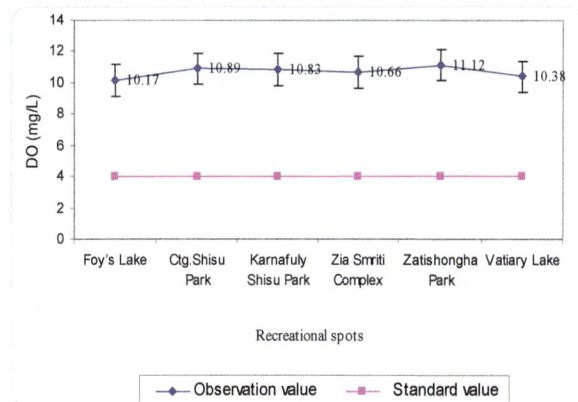

Figure 5. *DO in Water at Different Recreational Spots Compared with the Standard Value.*

3.2.3. Biological Oxygen Demand (BOD)

The mean BOD value of the study areas varies in between 2.2 to 3.15 mg/ L, but all the observation were within the standard value (8 mg/L). The highest (3.15 mg/L) BOD was found in the water of Jatisongho Park while, the lowest (2.20 mg/L) was in Foy's Lake. The BOD of water of Chittagong Shisu Park, Karnafully Shisu Park, Zia Smriti Complex and Vatiary Lake were 2.90 mg/L, 3.07 mg/L, 2.53 mg/L and 2.34 mg/L respectively (Fig. 6). The water of all the recreational spots in the study area is safe as recreational water in respect of BOD.

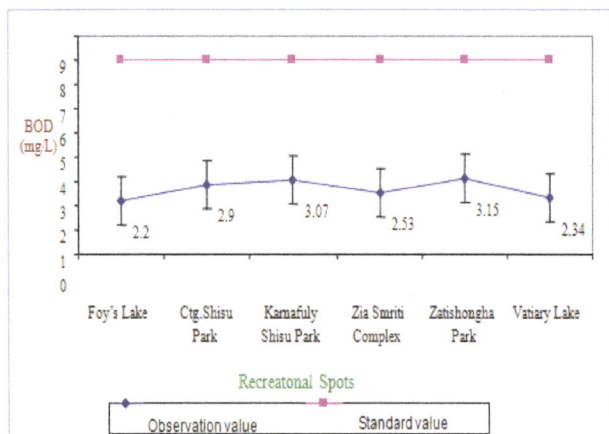

Figure 6. *BOD of water at Different Recreational Spots Compared with the Standard Value.*

3.2.4. Chemical Oxygen Demand (COD)

Water of three recreational spots maintained the acceptable level (200 mg/L), in contrary; other three spots exceeded the acceptable level. The highest (467.60 mg/L) value of COD was found in the water of Zia Smriti Complex followed by Karnafully Shisu Park (334 mg/L) and Chittagong Shisu Park (334 mg/L) (Fig. 7). However, COD of other three spots namely; Foy's Lake (66.8 mg/L), Vatiary Lake (66.8 mg/L) and Jatisongho Park (133.6 mg/L) was lower than standard limit.

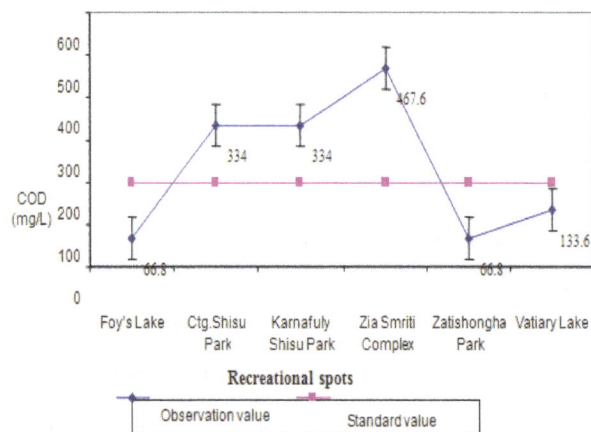

Figure 7. *COD of Water at Different Recreational Spots Compared with the Standard Value.*

3.2.5. Sulfate

The observation value of sulfate in the water is lower than that of the standard value. The lowest (195 mg/L) value was found in Foy's Lake water. The highest (246.67 mg/L) value was found in the water of Chittagong Shisu Park followed by Jatisongho Park (231.67 mg/L), Karnafully Shisu Park (223.33 mg/L), Zia Smriti Complex (213.33 mg/L) and Vatiary Lake (205 mg/L) respectively (Fig. 8). Considering sulfate standard in recreational water the study revealed that the recreational spots possess lower sulfate value compared to standard value (400 mg/L) and hence, is acceptable.

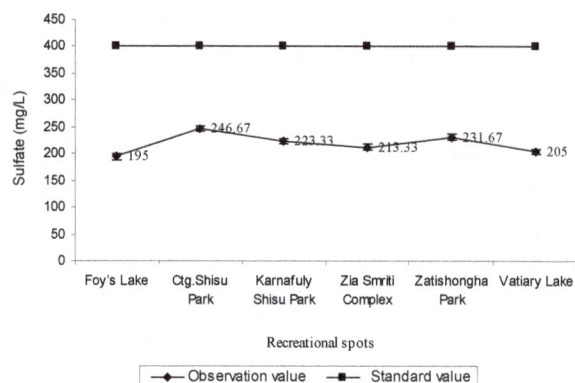

Figure 8. *Sulfate of Water at Different Recreational Spots Compared with the Standard Value.*

3.2.6. Nitrate

The highest (19.33 mg/L) concentration of nitrate was observed in the water of Chittagong Shisu Park, which is almost twice than standard value (10 mg/L). Fortunately, the presence of nitrate in the water of Karnafully Shisu Park, Zia Smriti Complex, Jatisongho Park and Vatiary Lake was zero (Fig. 9). Moreover, in Foy's lake water only 3 mg/L nitrate was present which is also lower than standard limit. Higher amount of nitrate was found due to the presence of increasing rate of organic waste, inorganic compounds from agricultural fields and mineral salts such as NH_4NO_3, $NaNO_3$, KNO_3 and $CaNO_3$ from salt made products. These salts are separated as ions and hence, increased amount of NO_3 was found in the water of Chittagong Shisu Park. Therefore organic wastes, inorganic compounds and mineral salts should be kept in control to safe use of water for recreational purposes. On the other hand, water of other five recreational spots is congenial for recreational purpose.

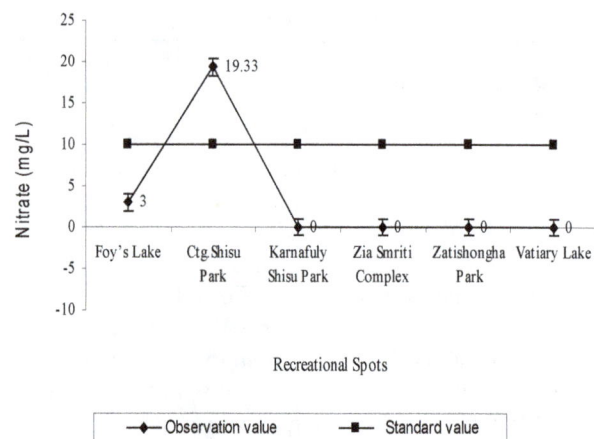

Figure 9. *Nitrate of Water at Different Recreational Spots Compared with the Standard Value.*

4. Conclusion and Recommendation

4.1. Conclusion

Water is the elixir of life. It can be used various purposes depending on unique standard for each uses. Results of different experiments on recreational water in different parts of the world exhibited that recreational water is safe when it maintains the standard value but when standard is exceeded it generates significant number of water borne diseases. In our study it was found that none of six spots maintained the standard of every parameter. If these sorts of scenario are continued the ultimately people of Chittagong will be affected by water borne diseases. In our experiment, the water quality parameters pH, DO, BOD, COD, Sulphate, Nitrate, Conductivity, Color, Odor, Turbidity were determined for analyzing whether all of the spots maintain the recreational water standard or not. Unfortunately, the experiment depicted that most of the recreational spots exceeded the standard value of conductivity except Karnafully Shisu Park. In case of Turbidity except Foy's Lake, other five recreational spots contained lower value than the standard value, which indicates that, the suitability of water for recreational purposes. However, water of five recreational spots except Foy's Lake, are suitable for recreational uses in respect of color. Two recreational spots provided odorous breathtaking. Moreover, in respect of pH, water of all of the recreational spots is congenial for recreational purposes. From the observation value of BOD it can be concluded that water of all the recreational spots is congenial as recreational water. COD was higher at Chittagong Shishu Park, Karnafully Shishu Park and Zia Smriti Complex. Nitrate value was also satisfactory except Chittagong Shisu Park. Biological parameters couldn't determine in the experiment. It is also mentioned here that only these parameters cannot make water suitable for recreation as biological parameters as well other physical and chemical parameters are important factors for recreational purposes. So for the safety of public health, these unwanted levels of different physical and chemical parameters should be maintained on an urgent basis.

4.2. Recommendation

It may recommend that necessary steps and care should be taken by the concerned authority immediately to ensure the maintenance of the quality of recreational water in Chittanong. Otherwise, hazards to human may arise in the respective areas.
- People should be aware of the standard of recreational water for their own interest.
- More researches on recreational water resource protection management have to be initiated.
- Care should be taken so that any waste not to be discharged into recreational water nearby.
- Lead agency to be set up to coordinate water related organizations and for recreational water quality control, management and planning.
- Restoration and protection of potential sites should be carried out.
- Blocked or threatened water reservoir should be urgently cleaned and managed through appropriate technology.

References

[1] Bhandari. B.B., 2003. "What is happening to our Freshwater Resources? Institute for Global Strategies. Environmental Education project". Tokyo.

[2] United Nations Food and Agriculture Organization (UNFAO), 2007. "Coping With Global Water Scarcity". Rome, FAO.

[3] Agrawal, K.C. 1993. Environmental Biology. New Delhi, India. 250p.

[4] Shamsad, S.Z.K.M., Islam, M.S., Hassan, M.Q., 1999. Ground water quality and hydrochemistry of Kushtia district, Bangladesh. *J. Asiat. Soc. Bangladesh, Sci.* 25(1):1-11.

[5] Hatfield J.L., Karlen D.L., 1994. Sustainable Agricultural Systems. Lewis Publishers. Boca Raton, Florida. USA, pp. 21-46.

[6] Doanhue, R.L., Miller, R.W., Shickluna, J.C., 1999. Soils: An Introduction to Soils and Plant Growth. 5th ed., Prentice-Hall of India (pvt.) Ltd. New Delhi, pp. 450-465.

[7] Manahan, S.E., Environmental Chemistry. CRL Press Inc. Boca Raton, USA-1994, pp.179-200

[8] Peavy, H.S., Rowe, D.R., Tchobanoglous, G., 1985. Environmental Engineering. McGraw Hill, New York, pp. 14-56.

[9] Gupta, P. K., 2000. Methods in Environmental Analysis: water, Soil and Air. Agrobios (India), Jodhpur, pp. 5-76.

[10] Trivedi, P.R., Raj, G., 1992. Encyclopedia of Environmenatal Sciences. Vol. 8. Anmol Publication (pvt) Ltd, New Delhi.311p.

[11] Siddiqui, M.H., 1992. Water Resource Development in Bangladesh, Technology and Environment. 8:(4):41-51.

[12] Ahmed, A.1999. Safe Water Supply Environmental Problems and Stratgies. Environmental Development and Management, Ed. by Pandy.

[13] Solaiman, M., 2005. Green Star Publication, Chittagong, Bangladesh, 51p.

[14] Hossain, Md.L., Islam, K.S., 2013. Assessment of Water Quality in Chandpur District of Bangladesh. *Journal of Environmental Treatment Techniques*, 1(2):91-100.

[15] Shamsad, S.Z.K.M., Islam, K.Z., Mahmud, M.S., Hakim, A., 2014. Surface Water Quality of Gorai River of Bangladesh. *Journal of Water Resources and Ocean Science.* 3(1)1:10-16. doi: 10.11648/j.wros.20140301.13.

[16] Rahman, A.A., Huq, S., Conway, G.R., 2000. Environmental Aspect of Surface Water system of Bangladesh. The University Press Limited, Dhaka, pp. 7-265.

[17] Islam, M.S., Hasan, M.Q., Shamsad, S.Z.K.M., 1998. Quality of irrigation water in the Kushtia District of Bangladesh. *J. Biol. Sci.* 7(2):129-138.

[18] De, A.K.. 2000. Environmenatal Chemistry. New Age International Publishers, India. 242-244p.

[19] Hussain, A., 2006. Pond Water Quality of Sandwip Island. B. Sc (Hons.). Project paper, Institute of Forestry and Environmental Sciences, University of Chittagong, Bangladesh.1-2p.

[20] Canadian guidelines for recreational water quality. Part-1. (http://www.hc-sc.gc.ca/Last Retrived on 01 July, 2008).

[21] Sunger, N., Teske, S.S., Nappier, S., Haas, C.N. 2012 Recreational use assessment of water-based activities, using time-lapse construction cameras. *Journal of Exposure Science and Environmental Epidemiology,* 22:(281–290), doi:10.1038/jes.2012.4

[22] Islam, S., 2003. Banglapedia: National Encyclopedia of Bangladesh. Asiatic Society of Bangladesh, Nimtali, Dhaka, Bangladesh. PP 515.

[23] Hossain, Md.L., Das, S.R., Hossain, M.K., 2014. Impact of Landfill Leachate on Surface and Ground Water Quality. *Journal of Environmental Science and Technology.* (Accepted). http://scialert.net/onlinefirst.php?issn=1994-7887. doi:10.3923/jest.2014

[24] Chittagong Development Authority (CDA), 1992. Preparation of structural plan, master plan and detailed area plan for Chittagong (UNDP/UNHS project No. BGD/88/052), Chittagong, Bangladesh.

[25] Salam, M.A., Hossain, Md. L., Das, S.R., Wahab, R., Hossain, M.K., 2012. Generation and assessing the composition of household solid waste in commercial capital city of Bangladesh. *International Journal of Environmental Science, Management and Engineering Research,* 1(4):160-171.

[26] Bhuiyan. M.H., 2006. Outdoor Recreational Potentialities in Chittagong City.B.Sc. Project Paper submitted to the Institute of Forestry and Environmental Sciences, University of Chittagong, Bangladesh.29-44p.

[27] Khan, M.O., 1979. Cutting of hills: Lack of law or lack of law enforcement? The Daily star, The 15 June, Dhaka, Bangladesh.

[28] Olembo, R.J. and P.de. Rham. 1987. Urban forestry in two different worlds. Unasylva,39(155):2635 (http://www.fao.org/DOCREP/ARTICLE/WFC/XII/0347-B 5.HTM.

[29] WWF National surface water classification criteria 2007. Compiled by freshwater and toxics programme, WWF-Pakistan. http://www.cmc.sandia.gov/cmc-papers/sand2000-080 9.pdf

Permissions

All chapters in this book were first published in WROS, by Science Publishing Group; hereby published with permission under the Creative Commons Attribution License or equivalent. Every chapter published in this book has been scrutinized by our experts. Their significance has been extensively debated. The topics covered herein carry significant findings which will fuel the growth of the discipline. They may even be implemented as practical applications or may be referred to as a beginning point for another development.

The contributors of this book come from diverse backgrounds, making this book a truly international effort. This book will bring forth new frontiers with its revolutionizing research information and detailed analysis of the nascent developments around the world.

We would like to thank all the contributing authors for lending their expertise to make the book truly unique. They have played a crucial role in the development of this book. Without their invaluable contributions this book wouldn't have been possible. They have made vital efforts to compile up to date information on the varied aspects of this subject to make this book a valuable addition to the collection of many professionals and students.

This book was conceptualized with the vision of imparting up-to-date information and advanced data in this field. To ensure the same, a matchless editorial board was set up. Every individual on the board went through rigorous rounds of assessment to prove their worth. After which they invested a large part of their time researching and compiling the most relevant data for our readers.

The editorial board has been involved in producing this book since its inception. They have spent rigorous hours researching and exploring the diverse topics which have resulted in the successful publishing of this book. They have passed on their knowledge of decades through this book. To expedite this challenging task, the publisher supported the team at every step. A small team of assistant editors was also appointed to further simplify the editing procedure and attain best results for the readers.

Apart from the editorial board, the designing team has also invested a significant amount of their time in understanding the subject and creating the most relevant covers. They scrutinized every image to scout for the most suitable representation of the subject and create an appropriate cover for the book.

The publishing team has been an ardent support to the editorial, designing and production team. Their endless efforts to recruit the best for this project, has resulted in the accomplishment of this book. They are a veteran in the field of academics and their pool of knowledge is as vast as their experience in printing. Their expertise and guidance has proved useful at every step. Their uncompromising quality standards have made this book an exceptional effort. Their encouragement from time to time has been an inspiration for everyone.

The publisher and the editorial board hope that this book will prove to be a valuable piece of knowledge for researchers, students, practitioners and scholars across the globe.

List of Contributors

K. G. Mandal, J. Padhi, A. Kumar, D. K. Sahoo, P. Majhi, S. Ghosh, R. K. Mohanty, M. Roychaudhuri
Directorate of Water Management (ICAR), Bhubaneswar, Odisha, India.

Md. Abdullah Asad, Mohammad Ahmeduzzaman, Shantanu Kar and Md. Ashrafuzzaman Khan
Dept. of Civil Engineering, Stamford University Bangladesh, Dhaka 1217, Bangladesh.

Md. Nobinur Rahman
Dept. of Civil Engineering, Rajshahi University of Engineering & Technology, Rajshahi 6204, Bangladesh.

Samiul Islam
Office Engineer, BETS Consulting Services Ltd., Dhaka, Bangladesh.

Djamal Boudieb, Kamal Mohammedi, Abdelkader Bouziane and Youcef Smaili
MESOteam, LEMI, M. Bougara University Boumerdes, Algeria.

Rajib Kamal and M. A. Matin
Department of Water Resources Engineering, Bangladesh University of Engineering and Technology, Dhaka, Bangladesh.

Sharmina Nasreen
Bangladesh Water Development Board, Dhaka, Bangladesh.

Ala H. Amiri, Ruwaya R. Alkendi and Yasser T. Ahmed
Faculty of Science, Department of Biology, United Arab Emirates University.

Mohamed Alaoui
Ministry Of Energy, Mines, Water and Environment, Department Of Water; Morocco.

Eng. Jamal Y. Al-Dadah
M.Sc. in Agriculture & Environmental Science, Head of Planning Department, Palestinian water Authority, Gaza Strip.

E. Zerrouk
GSGES, Kyoto University, Japan.

S. Senthil Kumar and Mohamed Jaabir
PG and Research Department of Biotechnology, National College (Autonomous), Tiruchirappalli, Tamil Nadu, India.

Pervin Yanikkaya Aydemir
Masters Degree Program, Yeditepe University, Anthropology Department, Istanbul-Turkey.

Muhammad Muktar Namadi, Mohammed Yau and Faruruwa Mohammed Dahiru
Chemistry Department, Nigerian Defence Academy, Kaduna, Nigeria.

Manu Haruna Isa
RightLinks Integrated Services Limited Kaduna, Nigeria.

Katsina Sani Mamman
Desertification Control Dept. Federal Ministry of Environment Abuja, Nigeria.

CHEBIHI Lakhdar and KHODJET KESBA Omar
Laboratory of Mobilization and valorization of the water resources Ecole Nationale Superieure de l Hydraulique, RN n 29, BP 31, Soumaa, Blida, Algeria.

Sang-Il Lee and Jae Young Seo
Department of Civil and Environmental Engineering, Dongguk University, Seoul, South Korea.

Sang Ki Lee
Department of Civil Engineering, University of Idaho, Boise, ID, USA.

David O. Omole
Department of Civil Engineering College of Science and Technology, Covenant University, Canaanland, Km 10 Idiroko Road, Ota, Nigeria.

Sarfaraz Alam and M. Abdul Matin
Department of Water Resources Engineering, Bangladesh University of Engineering and Technology, Dhaka, Bangladesh.

Beaven Utete and Rutendo Maria Kunhe
Chinhoyi University of Technology Department of Wildlife and Safari Management P. Bag, Chinhoyi.

V. Sivasankar
Department of Chemistry, Thiagarajar College of Engineering (Autonomous), Madurai , Tamil Nadu, India.

M. Kameswari
Department of Mathematics, Thiagarajar College of Engineering (Autonomous), Madurai, Tamil Nadu, India.

T. A. M. Msagati
Department of Applied Chemistry, University of Johannesburg, Doornfontein Campus, P. O. Box 17011, Johannesburg, South Africa.

M. Venkatapathy
Department of Chemistry, A.A Government Arts College, Musiri, Tamil Nadu, India.

M. Senthil Kumar
Department of Civil Engineering, Sethu Institute of Technology, Virudhunagar, Tamil Nadu, India.

Beaven Utete and Edmore Happison Chikova
Chinhoyi University of Technology, Department of Wildlife and Safari Management P. Bag, Chinhoyi.

Shams Al Din Al Hajjaji
American Univeristy in Cairo/ Law Department, Egyptian Public Prosecution Bureau, Egypt.

Khalid Qahman
Ministry of Environmental Affairs, Palestine.

Eman Abu-afash Mokhamer
School of Public Health, Alquds University, Palestine.

Festus Panduleni Nashima and Martin Hipondoka
University of Namibia, Private Bag, Windhoek, Namibia.

Inekela Iiyambo
Rossing Uranium of Namibia, Private Bag, Swakopmund, Namibia.

Johannes Hambia
University of Namibia, Ogongo Campus, Private Bag, Oshakati, Namibia.

Kadhim Naief Kadhim and Abbas Yasir Hussein
College of Engineering, University of Babylon-Iraq.

Chinedum Onyemechi
Dept. of Maritime Management Technology, Federal University of Technology, Owerri, Nigeria.

Lazarus Okoroji and Declan Dike
Dept of Transport Management Technology, Federal University of Technology, Owerri, Nigeria.

Sarfaraz Alam and Muhammad Sabbir Mostafa Khan
Department of Water Resources Engineering, Bangladesh University of Engineering and Technology, Dhaka, Bangladesh.

SIDDI Tengeleng
Laboratoire d Ingenierie des Systemes Industriels et de l'Environnement (LISIE), Institut Universitaire de Technologie Fotso Victor, Universite de Dschang, Bandjoun, Cameroun

Higher Institute of the Sahel, University of Maroua, P.O. Box 46 Maroua, Cameroon

Laboratory of Mechanics and Modeling of Physical Systems (L2MSP) - University of Dschang, Cameroon.

NZEUKOU Armand
Laboratoire d Ingenierie des Systemes Industriels et de l'Environnement (LISIE), Institut Universitaire de Technologie Fotso Victor, Universite de Dschang, Bandjoun, Cameroun.

KAPTUE Armel
Geographic Information Science Center of Excellence, South Dakota State University, Brooklin, USA.

TCHAKOUTIO SANDJON Alain
Laboratoire d Ingenierie des Systemes Industriels et de l'Environnement (LISIE), Institut Universitaire de Technologie Fotso Victor, Universite de Dschang, Bandjoun, Cameroun

Laboratory for Environmental Modeling and Atmospheric Physics (LAMEPA) Department of Physics University of Yaounde, Cameroon.

SIMO Theophile and Djiongo Cedrigue
Laboratory of Automatic and Applied Informatics (LAIA), IUT-FV, University of Dschang.

Md. Lokman Hossain
Department of Global Change Ecology, Faculty of Biology, Chemistry and Geosciences, University of Bayreuth, D-95440 Bayreuth, Germany.

Sultana Kamrun Nahar Nahida and Md. Iqbal Hossain
Institute of Forestry and Environmental Sciences, University of Chittagong, Chittagong, Bangladesh.

Index

www.ingramcontent.com/pod-product-compliance
Lightning Source LLC
Chambersburg PA
CBHW080531200326

41458CB00012B/4397